The Problem of Catalan

Yuri F. Bilu • Yann Bugeaud • Maurice Mignotte

The Problem of Catalan

 Springer

Yuri F. Bilu
Institute of Mathematics of Bordeaux
University of Bordeaux and CNRS
Talence, France

Yann Bugeaud
IRMA, Mathematical Institute
University of Strasbourg and CNRS
Strasbourg, France

Maurice Mignotte
IRMA, Mathematical Institute
University of Strasbourg and CNRS
Strasbourg, France

ISBN 978-3-319-36255-7   ISBN 978-3-319-10094-4 (eBook)
DOI 10.1007/978-3-319-10094-4
Springer Cham Heidelberg New York Dordrecht London

Mathematics Subject Classification (2010): 11-02, 11D41,11D61, 11J86, 11M06, 11R04, 11R18, 11R27, 11R29, 11R32, 11R33.

Printed on acid-free paper

Springer is part of Springer Science+Business Media (www.springer.com)

Preface

In 1842 the Belgian mathematician Eugène Charles Catalan asked whether 8 and 9 are the only consecutive pure powers of nonzero integers. One hundred and sixty years after, the question was answered affirmatively by the Swiss mathematician of Romanian origin Preda Mihăilescu. In other words, $3^2 - 2^3 = 1$ is the only solution of the equation $x^p - y^q = 1$ in integers x, y, p, q with $xy \neq 0$ and $p, q \geq 2$.

Since 2002, the different steps of the proof have been presented by various authors; see, for instance, the expository articles [9, 10, 82]. Complete proofs appeared in monographs by Schoof [124] and Cohen [25, 26].

In this book we give a complete and (almost) self-contained exposition of Mihăilescu's work, which must be understandable by a curious university student not necessarily specializing in number theory. We assume very modest background: a standard university course of algebra, including basic Galois theory, and working knowledge of basic algebraic number theory such as ideal decomposition, units (including the Dirichlet unit theorem), ideal classes, and finiteness of the class group. From the ramification theory we use only one basic fact: a prime number is ramified in a number field if and only if it divides the discriminant. All necessary facts from algebraic number theory are gathered (without proofs) in Appendix A.

We do not assume any knowledge about cyclotomic fields; everything needed is defined or proved in the book.

With our minimalistic approach, some omissions were inevitable. For instance, an experienced reader can notice that many arguments in this book have an obvious non-Archimedean flavor. Nevertheless, we resisted the temptation of broader use of the language of (non-Archimedean) valuations. Our main motivation was that a matured reader will easily reveal the non-Archimedean context wherever it is hidden, but abusing the non-Archimedean language may create problems for a less knowledgeable reader.

Another example is restricting to commutative groups and rings in Appendices C and D. Of course, certain results from these appendices (like the theorem of Maschke) extend to noncommutative case as well. We, however, assume commutativity, because this makes the arguments technically simpler and is sufficient for our purposes.

One more notable omission is Runge's method. We are aware, of course, that certain proofs, especially in Chaps. 3, 8, and 9, use Runge-style arguments. We feel, however, that Runge's method can be correctly explained only by using the language of algebraic curves, which is foreign in this book.

Chapters 4, 5, 7, 10, and 12 are dedicated to the general theory of cyclotomic fields. Catalan's problem is treated in Chaps. 2, 3, 6, 8, 9, and 11. Chapter 13 is quite isolated and independent of the others. In it we give a concise introduction to Baker's method and prove the Theorem of Tijdeman, which was the top achievement in Catalan's problem before Mihăilescu's work.

The book has six appendices. As we already mentioned, Appendix A is a very brief account of basic algebraic number theory. Other appendices treat miscellaneous topics in algebra and number theory used in the book. While Appendix A contains almost no proofs, in the other appendices, we prove everything we need.

Our notation is mainly standard. We denote by $|S|$ the cardinality of a finite set S. We use $\lfloor x \rfloor$ and $\lceil x \rceil$ to denote the lower and the upper integral part of $x \in \mathbb{R}$, respectively:

$$\lfloor x \rfloor = \max\{n \in \mathbb{Z} : n \leq x\}, \qquad \lceil x \rceil = \min\{n \in \mathbb{Z} : n \geq x\}.$$

Unless the contrary is stated explicitly, letters p and q denote distinct odd prime numbers; in particular, q never denotes a power of p.

On page vii we display the logical dependence of chapters and appendices. Dashed lines indicate weak dependence.

Logical Dependence of Chapters and Appendices

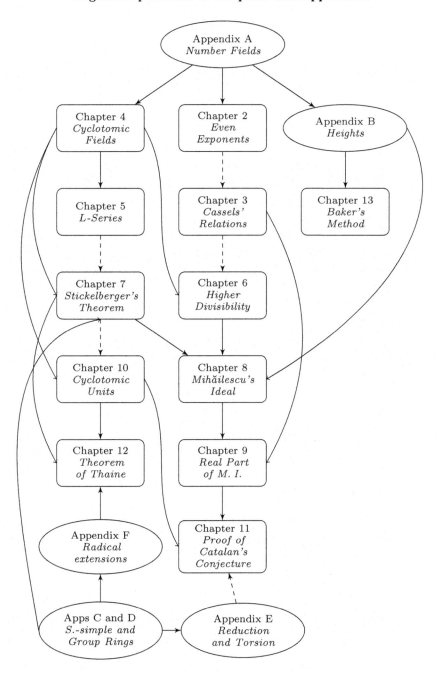

Acknowledgments

During the preparation of this book we enjoyed valuable discussions with many colleagues, including Henri Cohen, Jean Fresnel, Jean-François Jaulent, Radan Kučera, Hendrik Lenstra Jr., and Jacques Martinet, to mention only a few of them. Boris Bartolome helped us to compose the index and detected many inaccuracies in the text.

Yuri Bilu thanks Elina Wojciechowska for the support and inspiration.

Talence, France   Yuri F. Bilu
Strasbourg, France   Yann Bugeaud
Strasbourg, France   Maurice Mignotte
June 2014

Contents

Chapter 1
A Historical Account

Here we give a very brief overview of the history of Catalan's problem. For additional historical data and further references the reader may consult the corresponding sections of [117, 124].

1.1 Catalan's *Note Extraite*

The Belgian mathematician Eugène Charles Catalan (1814–1894) holds a position of "répétiteur" at the École Polytechnique when he published in 1842 the following two lines in the section *Théorèmes et problèmes* on p. 520 of the first volume of the journal *Nouvelles Annales de Mathématiques*[1] (Fig. 1.1):

> *Théorème.* Deux nombres entiers consécutifs, autres que 8 et 9, ne peuvent être des puissances exactes. (Catalan).

The theorems listed on pp. 519–521 of this journal were stated without proofs, and it is rather doubtful that Catalan had one. Indeed, 2 years later, Catalan wrote the now famous letter [21] to August Leopold Crelle, published in his journal in 1844. Here is the text of this letter, appeared as a *note extraite* just above the list of misprints detected in the previous issues of Crelle's journal (Fig. 1.2):

> **Note**
> extraite d'une lettre adressée à l'éditeur par Mr. *E. Catalan*,
> Répétiteur à l'école polytechnique de Paris.
> Je vous prie, Monsieur, de vouloir bien énoncer, dans votre recueil, le théorème suivant, que je crois vrai, bien que je n'aie pas encore réussi à le démontrer complètement: d'autres seront peut-être plus heureux:

---

[1]The journal was founded in the same year by the École Polytechnique teachers Gerono and Prouhet and was mainly addressed to the young candidates of the École Polytechnique and École Normale Supérieure.

© Springer International Publishing Switzerland 2014
Y.F. Bilu et al., *The Problem of Catalan*, DOI 10.1007/978-3-319-10094-4_1

— **520** —

48. *Théorème.* **Deux nombres entiers consécutifs, autres que 8 et 9, ne peuvent être des puissances exactes. (Catalan.)**

Fig. 1.1 Page 520 of the *Nouvelles Annales de Mathématiques* **1** (1842)

192

13.
Note
extraite d'une lettre adressée à l'éditeur par Mr. **E. Catalan**, Répétiteur à l'école polytechnique de Paris.

„**J**e vous prie, Monsieur, de vouloir bien énoncer, dans votre recueil, le „théorème suivant, que je crois vrai, bien que je n'aie pas encore réussi à „le démontrer complètement: d'autres seront peut-être plus heureux:
    „Deux nombres entiers consécutifs, autres que 8 et 9, ne peuvent être „des puissances exactes; autrement dit: l'équation $x^m - y^n = 1$, dans „laquelle les inconnues sont entières et positives, n'admèt qu'une seule „solution."

Fig. 1.2 Catalan's *note extraite* [21]

Deux nombres entiers consécutifs, autres que 8 et 9, ne peuvent être des puissances exactes ; autrement dit : l'équation $x^m - y^n = 1$, dans laquelle les inconnues sont entières et positives, n'admèt qu'une seule solution.

Most likely, these are the first two occurrences of the general Diophantine equation $x^m - y^n = 1$ in the mathematical literature. (Of course, some particular instances of this equation occurred well before; see below.)

Lionnet [73], who was probably unaware about Catalan's "note extraite," posed the same question in 1868, again in the *Nouvelles Annales de Mathématiques*.

Dickson mentioned the question in the second volume of his *History of the Theory of Numbers* [28], attributing it to Catalan: *E. Catalan expressed his belief that $x^m - y^n = 1$ holds only for $3^2 - 2^3 = 1$* [28, Chap. XXVI, p. 738].

However, it was not until the middle of the twentieth century (that is, more than 100 years after Catalan's *note extraite*) that referring to this question as the *problem* (or the *conjecture*) *of Catalan* became common. In the meantime, that is, roughly, from 1870 until 1950, one finds almost no mention of Catalan in the dozen of research papers treating the equation $x^m - y^n = 1$ or its particular instances.

In his book *Diophantine analysis*, published in 1915, Carmichael [15, p. 116, Problem 69] suggested to study the equations $x^m = y^n + c$, and in particular the case $c = 1$, but did not mention Catalan. Neither did Obláth in his series of articles [108–110], treating the case of even m.

Gloden [43, 44] used the expression *problème de Catalan* in 1952 and 1953 in a short note and in a survey on the history of this problem. But LeVeque [70] and Cassels [16], who worked on the equation $a^x - b^y = 1$ in the same years, were, most probably, unaware of Catalan's letter to Crelle. In particular, in the first section of [16], Cassels conjectured that the equation $a^x - b^y = 1$ (in four variables) could have only finitely many (nontrivial) solutions.

However, already in Chap. 4 of his book [71], published in 1956, LeVeque discussed Catalan's conjecture, and in 1960 Cassels [18] not only explicitly mentions Catalan but also gives, in Sect. 1.2, a history of the conjecture and quotes most of the articles treating (particular instances of) the Catalan equation and published before 1960.

It seems that, since 1960, the name of Catalan has been commonly attributed to the equation $x^m - y^n = 1$ and the corresponding problem.

1.2 Particular Cases

After this short discussion, we go on with a brief survey on the main contributions towards the resolution of Catalan's equation. We do not always follow the chronological order, but rather put together various results according to their types.

Actually, the story began long before Catalan enunciated his conjecture. Indeed, at least three nontrivial results solving particular cases of $x^m - y^n = 1$ have been established well before 1842.

The first of these is nearly seven centuries old: Levi ben Gershon (1288–1344), a medieval Jewish philosopher and astronomer, answering a question of the French composer Philippe de Vitry, proved that $3^m - 2^n \neq 1$ when $m \geq 3$. The proof is pretty simple: clearly $n \geq 2$, which implies that $3^m \equiv 1 \bmod 4$, whence m is even. Writing $m = 2k$ we obtain $(3^k - 1)(3^k + 1) = 2^n$, which implies that both $3^k - 1$ and $3^k + 1$ are powers of 2. But the only powers of 2 which differ by 2 are 2 and 4. Hence $k = 1$, contradicting the assumption $m \geq 3$.

Next, the amateur French mathematician Bernard Frénicle de Bessy (1605–1675) proved in 1657 that $x^2 - y^n = 1$ has no solution with $n \geq 2$ and y an odd prime number, thus answering a question asked by Pierre de Fermat in his *Deuxième Défi aux Mathématiciens*.

The third result is a remarkable theorem of Euler [31], who proved in 1738 that the equation $x^2 - y^3 = 1$ has no nontrivial solutions other than $3^2 - 2^3 = 1$. See more on this in Sect. 2.5.

In 1850, just 6 years after the publication of Catalan's *note extraite*, the French mathematician Lebesgue[2] [67] showed that the equation $x^m - y^2 = 1$ has no solution. Mention that Lebesgue quoted the "Théorème" from the 1842 volume of the *Nouvelles Annales de Mathématiques* [20] and not the *note extraite* from Crelle's journal [21].

---

[2]Victor Amédée Lebesgue (1791–1875), not the much more famous Henri Léon Lebesgue.

It then turned out that $x^2 - y^n = 1$ is much more difficult to solve. There have been partial results by the Norwegian mathematicians Nagell[3] [98, 103] and Selberg[4] [126] (who solved $x^4 - y^n = 1$), the Hungarian mathematician Obláth [108–110], and the Finnish mathematicians Inkeri[5] and Hyyrö [52]. Finally, the Chinese mathematician Ko Chao [56] showed in 1965 that this equation has no solutions with $n \geq 5$ (the case $n = 3$ already being done by Euler). See Sect. 2.3.3 for more history and bibliography on this equation.

Combined with Nagell's result [99] from 1921 asserting that $x^m - y^n = 1$ has no solutions with mn divisible by 3 other than $3^2 - 2^3 = 1$, Ko Chao's result implies that in any further solution of Catalan's equation $x^m - y^n = 1$ both the exponents m and n are greater than or equal to 5. Consequently, by 1965 the smallest unsolved Catalan's equation was $x^5 - y^7 = 1$. As we shall see in the sequel, it remained unsolved until 1990.

The theorems of Euler, Lebesgue, and Ko Chao are proved in Chap. 2 of this book.

A number of partial results were obtained with conditions imposed upon the variables x and/or y rather than the exponents m and n. Gerono[6] [39] proved in 1857 that $x^m - y^n = 1$ has no further solution with x and y prime numbers. Later [40, 41], he got the same conclusion under the weaker assumption that either x or y is a prime number. As Dickson indicates in his *History of the Theory of Numbers*, this result was rediscovered many times.

The equations $x^y - y^x = 1$ and $(x + 1)^y - x^{y+1} = 1$ were solved in 1876 by Moret-Blanc [97] and Meyl [83], respectively. (It is likely that the former equation was solved by Catalan himself, but we found no publication confirming this hypothesis.) Obláth [110] proved that there are no further solutions when all the prime divisors either of x or of y are of certain special type. Hampel [45] proved that there are no further solutions when $|x - y| = 1$. Alternative proofs and generalizations of his result were suggested by Schinzel [121], Rotkiewicz [120], and Obláth [111].

In 1952, LeVeque [70] showed that for any pair (x, y) of integers with $x \geq 2$, $y \geq 2$, and $xy > 6$, there is at most one pair (m, n) of positive integers with $x^m - y^n = 1$ and that this pair can be found explicitly if it exists. This is related to the Conjecture of Pillai discussed in Chap. 13 of the present book.

---

[3]Trygve Nagell (sometimes spelled as Nagel, 1895–1988).

[4]Sigmund Selberg (1910–1994), the elder brother of Atle Selberg.

[5]Kustaa Inkeri (1908–1997).

[6]Recall that he was one of the founders of the *Nouvelles Annales de Mathématiques*.

1.3 Cassels' Relations

The modern history of Catalan's equation began with the 1960 work of Cassels [18]. To explain Cassels' discovery, notice first of all that we may assume the exponents to be distinct prime numbers and write the equation as $x^p - y^q = 1$. Now rewriting it as

$$(x - 1)\frac{x^p - 1}{x - 1} = y^q, \qquad (1.1)$$

it is easy to see (cf. Lemma 2.8) that the factors on the left may have either 1 or p as their greatest common divisor. If the gcd is 1 then both the factors must be pure qth powers. If the gcd is p then one of the factors is a qth power times p and the other is a qth power divided by p.

Cassels proved that, when $p, q \geq 3$, the first case is impossible, that is, the greatest common divisor is p. Since, for $p \geq 3$, the second factor cannot be divisible by p^2 (see again Lemma 2.8), it follows that

$$x - 1 = p^{q-1}a^q, \quad \frac{x^p - 1}{x - 1} = p\,u^q, \quad y = p\,av \qquad (1.2)$$

with some $a, u, v \in \mathbb{Z}$.

Identities (1.2), called *Cassels' relations*, are instrumental in the study of Catalan's equation: most of the subsequent contributions rely on them. It is useful to extend the range of variables x and y from positive integers, distinct from 1, to all integers, distinct from 0 and ± 1. This makes the equation symmetric: if (x, y, p, q) is a solution with odd p and q, then so is $(-y, -x, q, p)$. Hence, in addition to (1.2), we have the symmetric triple of relations; in particular, not only $p \mid y$, but also $q \mid x$.

When one of p or q is 2, the same result (that is, $p \mid y$ and $q \mid x$) was established much earlier by Nagell [100, 103] (Theorem 2.9 in this book). Nagell's result is crucial in the proof of the Theorem of Ko Chao.

One curious consequence of Nagell's and Cassels' theorems is that three consecutive positive integers cannot be powers. This was observed independently by Hyyrö [47] and Mąkowski [76], answering a question of LeVeque [71] and Sierpiński [128]. Indeed, assume that x^p, y^q, and z^r are consecutive positive perfect powers, with p, q, and r prime numbers. Then q divides x and z, and, since $p, r \geq 2$, we find that q^2 divides $z^r - x^p = 2$, a contradiction.

Cassels' theorem is proved in Chap. 3.

1.4 Analysis: Logarithmic Forms

To solve completely Catalan's equation may be much too difficult, and one may try to prove that it has only finitely many solutions, that is, to confirm the conjecture of Cassels [16]. In his famous letter to Mordell [127], Siegel indicated a proof of the following theorem: let $f(x) \in \mathbb{Z}[x]$ be a polynomial with at least three simple roots and let $n \geq 2$ be an integer; then the equation $f(x) = y^n$ has finitely many solutions in $x, y \in \mathbb{Z}$. (If $n \geq 3$ then two simple roots would suffice.) This theorem implies that for any fixed exponents $m, n \geq 2$ the equation $x^m - y^n = 1$ has finitely many solutions. Hyyrö [48] and Evertse [30] gave explicit bounds for the number of solutions; in particular, the latter showed that there is at most $(mn)^{\min\{m,n\}}$ solutions. Thus, *to prove Cassels' conjecture, it suffices to bound the exponents in Catalan's equation.*

The above results were obtained by ineffective methods, that is, by methods which do not yield explicit upper bounds for x and y. The situation changed dramatically after Alan Baker created his theory of logarithmic forms [4]. This very powerful tool provides explicit (albeit huge) upper bounds for solutions of many types of Diophantine equations. For instance, Baker [6] gave an explicit upper bound for the solutions of the equation $f(x) = y^n$; this bound was refined many times; see [13] for the most recent results and further references.

An important advantage of Baker's method is that it allows one to treat exponents as unknowns. In 1976, Schinzel and Tijdeman [122] proved the following beautiful theorem: *given a polynomial $f(x) \in \mathbb{Z}[x]$ with at least two distinct roots, any solution of $f(x) = y^n$ in integers x, y (with $y \neq 0, \pm 1$) and a positive integer n satisfies $n \leq n_0(f)$, where n_0 can be expressed explicitly in terms of the polynomial f.*

This has the following consequence for Catalan's equation: *if we fix one of the four variables x, y, m, and n in $x^m - y^n = 1$, then the other three variables can be explicitly bounded in terms of the fixed one.* The same result is true for the more general *Pillai equation* $ax^m + by^n = c$, where a, b, and c are (fixed) nonzero integers.

This result is already quite remarkable, but it is still very far even from Cassels' conjecture (telling that Catalan's equation has finitely many solutions in the *four* variables x, y, m, and n.), let alone the complete solution of Catalan's problem. Therefore it was quite a sensation when in the very same year 1976 Tijdeman [134] proved Cassels' conjecture, by a very ingenious application of Baker's method to Catalan's equation. Notice that Tijdeman's argument does not apply to more general equations; for instance, it is still unknown whether the equation $x^m - y^n = 2$ has finitely or infinitely many solutions.

Since Tijdeman uses Baker's method, it is not difficult to derive from his proof an explicit (very huge) upper bound for x, y, m, and n. This was done by Langevin [62], who showed that the greatest prime divisor of mn is less than 10^{107} and that

$$|x^m|, |y^n| < 10^{10^{10^{10^{320}}}}.$$

Thus, Tijdeman's result (together with Langevin's computation) implies that Catalan's problem is *decidable*: one can find all solutions of Catalan's equation just by verifying all numbers below Langevin's bound. Of course, in practice such a calculation is totally unrealistic, and in spite of this impressive progress Catalan's problem remained open.

Application of Baker's method to Catalan's (and Pillai's) equation is the subject of Chap. 13 of this book.

1.5   Algebra: Cyclotomic Fields

In a different direction, there was a quest for algebraic criteria on the pair of prime exponents (p, q) such that the Diophantine equation $x^p - y^q = 1$ has no nontrivial solution. The first result of this kind, obtained in 1964 by Inkeri [50], was of limited use, because of the assumption that one of the exponents p or q is congruent to 3 modulo 4. In 1990, Inkeri obtained a more general criterion [51] involving the class number h_p of the pth cyclotomic field. He proved that if p and q are distinct odd prime numbers such that the equation $x^p - y^q = 1$ has a solution in nonzero integers x and y, then either $q \mid h_p$ or $p^{q-1} \equiv 1 \bmod q^2$ (call them the *class number condition* and the *Wieferich condition*,[7] respectively).

Inkeri's criterion implies at once that $x^5 - y^7 = 1$ has no nontrivial solutions, since $h_5 = 1$ and $5^6 \equiv 43 \bmod 7^2$. A similar verification shows that $x^p - y^q = 1$ has no nontrivial solution with $5 \le p, q < 89$.

Inkeri's theorem is proved in Sect. 6.2 of this book.

Inkeri's criteria have been extended and refined by Mignotte [86] and Schwarz [125], who managed to relax the class number condition. In particular, Schwarz replaced the full class number h_p by the relative class number h_p^-. This was of great importance, because, unlike h_p, the number h_p^- is quite easy to compute. This opened the road for using electronic computations in Catalan's problem.

In a different direction, Bugeaud and Hanrot [14], inspired by some ideas from [11], managed to completely remove the Wieferich condition: they proved that a solution (x, y, p, q) of Catalan's equation with prime exponents satisfies either $q \le p$ or $q \mid h_p^-$. Notice that this implies a bound on q in terms of p; before [14] such a bound could be proved only using Baker's method.

The theorem of Bugeaud and Hanrot is proved in Sect. 8.6 of the present book.

1.6   Numerical Results

As we have already remarked, solving Catalan's problem just by enumerating all possible solutions below Langevin's bound was unrealistic. However, one can apply a cleverer strategy: verify Inkeri's (or similar) conditions for the pair (p, q) below Langevin's bound 10^{107}. Unfortunately, given the huge value of this bound, this task was also unrealistic, even with the most powerful computers.

---

[7]A similar condition occurred in the work of Wieferich [138] on the Fermat equation; see Sect. 2.3.3.

In early 1990s, spectacular improvements on estimates for logarithmic forms lead to dramatical refinement of Langevin's bound. For instance, Mignotte [84] replaced 10^{107} by $1.31 \cdot 10^{18}$, and, independently, a slightly weaker result was obtained by Glass et al. [42]. It was still computationally impossible to check all possible pairs (p, q), but it became possible to check all pairs with certain "small" values of p and with every q below the new upper bound. On this way, Glass et al. [42]   showed that, given a solution (x, y, p, q) of Catalan's equation with prime exponents, we have $\min\{p, q\} \geq 17$ (and even $\min\{p, q\} \geq 37$, if one excludes several explicitly given pairs (p, q)).

There have been many subsequent improvements of this lower bound, particularly in the series of articles by Mignotte and Roy [89–91]. After several months of very heavy computations they managed to prove [91] that

$$\min\{p, q\} \geq 10^5. \tag{1.3}$$

However, it eventually became clear that one could not overcome the gap between the lower and upper bounds just by brute force, that is, by refining the upper bound and the algebraic criteria and performing heavy computations with increasingly more powerful computers. One needed new ideas to solve Catalan's problem.

1.7   The Final Attack

Guillaume Hanrot presented the result of [14] at the **XXI** *Journées Arithmétiques* held at the *Pontificia Universitas Lateranensis* in July 1999. Preda Mihăilescu was in the audience and that was the start of the story: he had the feeling that his knowledge on cyclotomic fields could perhaps help to bring something new. The time showed that he was perfectly right!

No later than in September 1999, he announced [92] a cardinal improvement of Inkeri's criterion, in the direction "orthogonal" to that of [14]. He finally managed to remove the class number condition from the criterion, by proving that every solution (x, y, p, q) satisfies $p^{q-1} \equiv 1 \bmod q^2$. By symmetry, we must also have $q^{p-1} \equiv 1 \bmod p^2$, that is, (p, q) is a *double Wieferich pair*; see Sect. 6.5.

With this new criterion, the verification of (1.3) can be done in a few hours of computations, and with one month of computations, Mignotte and Roy   managed to prove that $\min\{p, q\} \geq 10^7$. The double Wieferich pairs, which resisted Mihăilescu's test, were disposed of using the criteria from [14, 85].

Simultaneously, upper bounds were also refined, and by 2001 it was known that $10^7 \leq \min\{p, q\} \leq 7.2 \cdot 10^{11}$ and $\max\{p, q\} \leq 7.8 \cdot 10^{16}$; see [87]. Later on, Grantham and Wheeler   made extensive computations showing that $\min\{p, q\} \geq 3.2 \cdot 10^8$; see [8]. Still, a gap between the lower and the upper bound persisted.

And then Mihăilescu came back. In December 2001 he sent to Yuri Bilu a manuscript with a complete proof of Catalan's conjecture. In April 2002, after several months of thorough verification, Bilu confirmed the validity of Mihăilescu's proof and prepared his own exposition of Mihăilescu's argument, which was sent, together with the original text of Mihăilescu, to a number of colleagues. In May 2002 Bilu gave a talk on Mihăilescu's work in the Erwin Schrödinger Institut (Vienna). This was, probably, the first public announcement of the solution of Catalan's problem.

Mihăilescu's original proof [9, 93] required an estimate (1.3) and thereby depended on the theory of logarithmic forms and some (not too heavy) electronic computations. He found subsequently an alternative approach for the part of the proof where these results were used and was thus able to solve the Catalan equation without logarithmic forms and electronic computation; see [10, 94] and Chap. 8 of the present book.

A journalist report of the events around the proof of Catalan's conjecture can be found in [116].

Chapter 2
Even Exponents

In this chapter we consider Catalan's equation $x^p - y^q = 1$ when one of the exponents p and q is even. This reduces to the case when one of p and q is equal to 2, and the other is an odd prime number.

2.1 The Equation $x^p = y^2 + 1$

Six years after Catalan's *note extraite*, the French mathematician Lebesgue [67] made the first step in the long way towards the solution of Catalan's problem. He proved that Catalan's equation $x^p - y^q = 1$ has no solutions with $q = 2$.

Theorem 2.1 (V.A. Lebesgue). *Let $p \geq 3$ be an odd number. Then the equation $x^p = y^2 + 1$ has no solutions in nonzero integers x and y.*

(One may further assume that p is prime, but this is not needed for the proof.)

Proof. If y is odd and x is even, then $y^2 + 1 \equiv 2 \bmod 4$ and $x^p \equiv 0 \bmod 8$, a contradiction. Hence y is even and x is odd. Write

$$x^p = (1 + iy)(1 - iy). \tag{2.1}$$

The greatest common divisor of $1 + iy$ and $1 - iy$ (in the ring of Gaussian integers $\mathbb{Z}[i]$) divides the sum $(1 + iy) + (1 - iy) = 2$. Since $1 + y^2$ is odd, the numbers $1 + iy$ and $1 - iy$ are coprime.

Since $\mathbb{Z}[i]$ is a unique factorization ring, every factor in (2.1) is equal to a pth power times a unit of $\mathbb{Z}[i]$; that is,

$$1 + iy = \varepsilon \alpha^p, \qquad 1 - iy = \bar{\varepsilon} \tilde{\alpha}^p,$$

© Springer International Publishing Switzerland 2014

Y.F. Bilu et al., *The Problem of Catalan*, DOI 10.1007/978-3-319-10094-4_2

where $\alpha \in \mathbb{Z}[i]$ and $\varepsilon \in \mathbb{Z}[i]^\times = \{\pm 1, \pm i\}$. Since p is odd, every unit of $\mathbb{Z}[i]$ is a pth power of another unit. Writing $\varepsilon = \varepsilon_1^p$ and $\beta = \varepsilon_1 \alpha$, we obtain

$$1 + iy = \beta^p, \qquad 1 - iy = \bar{\beta}^p.$$

Write $\beta = a + ib$, where $a, b \in \mathbb{Z}$. Since p is odd, the number $2a = \beta + \bar{\beta}$ divides $\beta^p + \bar{\beta}^p = 2$. Hence $a = \pm 1$. Further, $1 + y^2 = (a^2 + b^2)^p = (1 + b^2)^p$ is an odd number, which implies that b is even. It follows that

$$1 + iy = (a + ib)^p \equiv a^p + ipa^{p-1}b \bmod 4,$$

and, in particular, $1 \equiv a^p \bmod 4$, which rules out the possibility $a = -1$.

Thus, $\beta = 1 + ib$. Comparing the real parts in the equality

$$1 + iy = (1 + ib)^p,$$

we obtain

$$1 = \sum_{k=0}^{(p-1)/2} (-1)^k \binom{p}{2k} b^{2k},$$

which can be rewritten as

$$-\binom{p}{2} b^2 + \sum_{k=2}^{(p-1)/2} (-1)^k \binom{p}{2k} b^{2k} = 0. \tag{2.2}$$

We shall use Lemma A.1 to show that (2.2) is impossible.

For $1 \le k \le (p-1)/2$ we have

$$\binom{p}{2k} b^{2k} = \binom{p}{2} b^2 \frac{1}{k(2k-1)} \binom{p-2}{2k-2} b^{2k-2}.$$

Hence

$$\mathrm{Ord}_2\left(\binom{p}{2k} b^{2k}\right) - \mathrm{Ord}_2\left(\binom{p}{2} b^2\right) \ge (2k-2)\mathrm{Ord}_2 b - \mathrm{Ord}_2 k$$

$$\ge 2k - 2 - \log_2 k.$$

Since $2k - 2 > \log_2 k$ for $k \ge 2$, we have $\mathrm{Ord}_2\left(\binom{p}{2k} b^{2k}\right) > \mathrm{Ord}_2\left(\binom{p}{2} b^2\right)$ for $k = 2, \ldots, (p-1)/2$. Now Lemma A.1 implies that the left-hand side of (2.2) cannot vanish. The theorem is proved.   □

2.2 Units of Real Quadratic Rings

This section is auxiliary. In it we recall the structure of the unit group of the ring $\mathbb{Z}[\sqrt{D}]$, where D is a positive integer, which is not a square (we do not assume that D is square-free).

Theorem 2.2. *The multiplicative group $\mathbb{Z}[\sqrt{D}]_+^\times$ of positive units of the ring $\mathbb{Z}[\sqrt{D}]$ is infinite cyclic.*

It follows that the group $\mathbb{Z}[\sqrt{D}]^\times$ of all units of $\mathbb{Z}[\sqrt{D}]$ is ± 1 times an infinite cyclic group.

The ring $\mathbb{Z}[\sqrt{D}]$ is not, in general, the ring of integers of the quadratic field $\mathbb{Q}(\sqrt{D})$. Hence Theorem 2.2 is not a formal consequence of the *Dirichlet unit theorem*, as stated in Appendix A.2. Of course, it may be deduced from the slightly more general *Dirichlet unit theorem for maximal orders*; see [12, Sect. 2.4]. However, we sketch here a short independent proof, for the reader's convenience.

The proof uses the famous *Dirichlet approximation theorem*.

Theorem 2.3 (Dirichlet approximation theorem). *Let α be a real number and $Y > 0$. Then there exist integers x, y such that $0 < y \leq Y$ and $|y\alpha - x| < Y^{-1}$.*

Informally, this means that the rational number x/y is a "good approximation" for α.

Proof. It is well known and simple: on the quotient group $\mathbb{R}/\mathbb{Z}$ consider the images of the real numbers $m\alpha$, where m runs through the integers satisfying $0 \leq m \leq Y$. We obtain $\lfloor Y \rfloor + 1$ points on $\mathbb{R}/\mathbb{Z}$, which split $\mathbb{R}/\mathbb{Z}$ into $\lfloor Y \rfloor + 1$ disjoint intervals. Since $\lfloor Y \rfloor + 1 > Y$, at least one of these intervals is of length $< Y^{-1}$. This means that there exist integers m_1, m_2, and x such that $0 \leq m_1 < m_2 \leq Y$ and $|m_2\alpha - m_1\alpha - x| < Y^{-1}$. Putting $y = m_2 - m_1$, we complete the proof. □

The following consequence is immediate.

Corollary 2.4. *Let α be a real number. Then there exist infinitely many couples $(x, y) \in \mathbb{Z}^2$ with $y > 0$ and $|y\alpha - x| < y^{-1}$.*

We return to the proof of Theorem 2.2. We denote the group $\mathbb{Z}[\sqrt{D}]_+^\times$ of positive units by U. We denote by $\alpha \mapsto \bar{\alpha}$ the nontrivial automorphism of $\mathbb{Z}[\sqrt{D}]$ (that is, $\overline{x + y\sqrt{D}} = x - y\sqrt{D}$) and by $\mathcal{N} : \mathbb{Z}[\sqrt{D}] \to \mathbb{Z}$ the norm map: $\mathcal{N}\alpha = \alpha\bar{\alpha}$.

Proof of Theorem 2.2. First of all, we prove that $U \neq \{1\}$. By the Dirichlet approximation theorem, there exist infinitely many couples of positive integers x and y such that $\left| x - y\sqrt{D} \right| \leq y^{-1}$. For any such x and y we have $x \leq y\sqrt{D} + 1$. Hence $\alpha = x + y\sqrt{D}$ satisfies $0 < \alpha \leq 2y\sqrt{D} + 1$ and $|\bar{\alpha}| \leq y^{-1}$, which implies $|\mathcal{N}\alpha| = |\alpha\bar{\alpha}| \leq 2\sqrt{D} + 1$.

We have proved that $\mathbb{Z}[\sqrt{D}]$ contains infinitely many positive elements of norm bounded by $2\sqrt{D} + 1$. Therefore there exists a nonzero $a \in \mathbb{Z}$ such that $\mathbb{Z}[\sqrt{D}]$ contains infinitely many positive elements of norm equal to a. Since there are only finitely many residue classes $\mod a$, there exist distinct positive $\alpha, \beta \in \mathbb{Z}[\sqrt{D}]$ such that $\mathcal{N}\alpha = \mathcal{N}\beta = a$ and $\alpha \equiv \beta \mod a$. Then

$$\alpha\bar{\beta} \equiv \beta\bar{\beta} \equiv a \equiv 0 \mod a.$$

Put $\eta = \alpha/\beta$. Then $\eta \neq 1$, because α and β are distinct. On the other hand, $\eta = \alpha\bar{\beta}/a \in \mathbb{Z}[\sqrt{D}]$, and, similarly, $\eta^{-1} \in \mathbb{Z}[\sqrt{D}]$. We have found an element $\eta \in U$ distinct from 1. Hence $U \neq \{1\}$.

Next, we prove that U is an infinite cyclic group. The logarithmic map $\log : U \to \mathbb{R}$ defines an injective homomorphism of U into the additive group of real numbers. Since $U \neq \{1\}$, the image $\log U$ is a nonzero subgroup of $\mathbb{R}$.

Further, if $\eta \in U$ satisfies $\eta > 1$, then $\eta = x + y\sqrt{D}$ with $x, y > 0$, whence $\eta > \sqrt{D}$. It follows that any positive element of $\log U$ is greater than $\log \sqrt{D}$, which implies that $\log U$ is a discrete subgroup of $\mathbb{R}$.

Since any nonzero discrete subgroup of $\mathbb{R}$ is infinite cyclic, the theorem follows. $\square$

The unit $\eta > 1$, generating the group $\mathbb{Z}[\sqrt{D}]_+^{\times}$, is called the *basic* (or *fundamental*) unit of the ring $\mathbb{Z}[\sqrt{D}]$. Usually, it is not easy to find the basic unit or to decide whether a given unit is basic. In some cases this can be done using the following simple observation.

Proposition 2.5. Let $\eta = a + b\sqrt{D}$ be the basic unit of $\mathbb{Z}[\sqrt{D}]$, and let $\theta = x + y\sqrt{D}$ be any other unit. Then $b \mid y$.

Proof. We may assume that $x, y > 0$. Then $\theta = \eta^n$, where n is a positive integer. It follows that $2b\sqrt{D} = \eta - \bar{\eta}$ divides $2y\sqrt{D} = \eta^n - \bar{\eta}^n$, whence the result. $\square$

The following consequence is immediate.

Corollary 2.6. Assume that $D = a^2 \pm 1$, where a is a positive integer (satisfying $a > 1$ if $D = a^2 - 1$). Then $a + \sqrt{D}$ is a basic unit of $\mathbb{Z}[\sqrt{D}]$.

It is worth mentioning that Corollary 2.6 is the simplest particular case of the famous theorem of Størmer (1897).

Theorem 2.7 (Størmer). Let $a + b\sqrt{D}$ be a unit of $\mathbb{Z}[\sqrt{D}]$ such that $a, b > 0$ and every prime divisor of b divides D. Then it is a basic unit.

We do not prove this theorem since we do not need it. An interested reader can find the proof in Ribenboim's book [117, Sect. A.4].

To conclude, let us mention that the results of this section are often interpreted in terms of the "Pell Diophantine equation" $x^2 - Dy^2 = \pm 1$.

2.3   The Equation $x^2 - y^q = 1$ with $q \geq 5$

The equation $x^2 - y^q = 1$ with an odd (prime) exponent q is much more difficult than $x^p - y^2 = 1$. The case $q = 3$ was settled already by Euler, but, in spite of some partial results, the general case remained open until 1965, when the Chinese mathematician Ko Chao[1] proved [55, 56] that, for a prime $q \geq 5$, the equation $x^2 - y^q = 1$ has no solutions in positive integers x and y.

In 1976 Chein [23] discovered another proof, simpler than Ko Chao's and based on a totally different idea. Chein's proof is reproduced below (with some changes). Ko Chao's proof can be found in Mordell's book [96, Sect. 30].

Both the arguments of Ko Chao and Chein work for a prime $q \geq 5$ and do not extend to $q = 3$. This case is solved in Sect. 2.5 by a totally different argument.

Most of the known proofs of the theorem of Ko Chao rely on a result of Nagell about the arithmetical structure of the solutions of the equation $x^2 - y^q = 1$. We prove this theorem in Sect. 2.3.1. The theorem of Ko Chao will be proved (following Chein) in Sect. 2.3.2. In Sect. 2.3.3 we briefly describe the history of the equation.

2.3.1   Nagell's Theorem

We start with an elementary lemma, which will be used in the next chapter as well.

Lemma 2.8. *Let A and B be distinct coprime rational integers and p a prime number.*

1. *If p divides one of the numbers $(A^p - B^p)/(A - B)$ and $A - B$, then it divides the other as well.*
2. *Put $d = \gcd\left((A^p - B^p)/(A - B), A - B\right)$. Then $d \in \{1, p\}$.*
3. *If $p > 2$ and $d = p$, then $\mathrm{Ord}_p\left((A^p - B^p)/(A - B)\right) = 1$.*

Proof. All the three statements easily follow from the identity

$$\frac{A^p - B^p}{A - B} = \frac{((A - B) + B)^p - B^p}{A - B} = \sum_{k=1}^{p} \binom{p}{k}(A - B)^{k-1}B^{p-k}. \qquad (2.3)$$

Rewriting it as

$$\frac{A^p - B^p}{A - B} = \sum_{k=1}^{p-1} \binom{p}{k}(A - B)^{k-1}B^{p-k} + (A - B)^{p-1},$$

we obtain $(A^p - B^p)/(A - B) \equiv (A - B)^{p-1} \bmod p$, which proves part (1).

---

[1] Sometimes spelled as Ko Zhao.

Rewriting (2.3) as

$$\frac{A^p - B^p}{A - B} = pB^{p-1} + (A - B) \sum_{k=2}^{p} \binom{p}{k}(A - B)^{k-2} B^{p-k}, \tag{2.4}$$

we observe that d divides pB^{p-1}. Since A and B are coprime, d and B are coprime as well, and we conclude that $d \mid p$, which proves part (2).

Finally, if $d = p$ then p divides $A - B$ and does not divide B. Rewriting (2.3) as

$$\frac{A^p - B^p}{A - B} = pB^{p-1} + (A - B) \left(\sum_{k=2}^{p-1} \binom{p}{k}(A - B)^{k-2} B^{p-k} + (A - B)^{p-2} \right),$$

we obtain $(A^p - B^p)/(A - B) \equiv pB^{p-1} \bmod p^2$, which proves part (3). □

Next, we establish the following preliminary result, due to Nagell [100].

Theorem 2.9 (Nagell). *Let x, y be positive integers and q an odd prime number satisfying $x^2 - y^q = 1$. Then $2 \mid y$ and $q \mid x$.*

Proof. Write $(x - 1)(x + 1) = y^q$. The greatest common divisor of $x - 1$ and $x + 1$ divides 2. If y is odd then they are coprime; hence both are qth powers: $x - 1 = a^q$ and $x + 1 = b^q$. We obtain $b^q - a^q = 2$, which is impossible. This proves that $2 \mid y$.

Further, write

$$\frac{y^q + 1}{y + 1}(y + 1) = x^2.$$

By Lemma 2.8, the greatest common divisor of the factors in the left-hand side is either 1 or q. If x is not divisible by q then the factors are coprime, which means that each of them is a complete square. Thus, there exist positive integers a and b such that

$$y + 1 = a^2, \quad \frac{y^q + 1}{y + 1} = b^2, \quad x = ab.$$

On the other hand, equality $x^2 - y^q = 1$ means that $x + y^{(q-1)/2} \sqrt{y}$ is a unit of the ring $\mathbb{Z}[\sqrt{y}]$, and Corollary 2.6 implies that $a + \sqrt{y}$ is the basic unit of this ring. Hence there exists a positive integer n such that

$$x + y^{(q-1)/2} \sqrt{y} = (a + \sqrt{y})^n. \tag{2.5}$$

We want to show that this is impossible.

First of all, let us prove that n is even. Expanding $(a + \sqrt{y})^n$ by the binomial formula, and reducing $\bmod y$, we find

$$(a + \sqrt{y})^n \equiv a^n + na^{n-1}\sqrt{y} \bmod y$$

in the ring $\mathbb{Z}[\sqrt{y}]$. Combining this with (2.5), we obtain

$$na^{n-1} \equiv y^{(q-1)/2} \equiv 0 \bmod y.$$

Since y is even and a is odd, this implies that n is even.

Next, reducing (2.5) $\bmod a$ and using the congruences

$$x = ab \equiv 0 \bmod a, \quad y = a^2 - 1 \equiv -1 \bmod a,$$

we obtain

$$(-1)^{(q-1)/2}\sqrt{y} \equiv (-1)^{n/2} \bmod a,$$

which means that a divides one of the numbers $1 \pm \sqrt{y}$ in the ring $\mathbb{Z}[\sqrt{y}]$. Since $a > 1$, this is impossible. The theorem is proved. $\square$

We learned this argument from Hendrik Lenstra (private communication). It was also independently discovered by Nesterenko and Zudilin [106]. Nagell himself used the Theorem of Størmer (see Theorem 2.7) to show that both units $a + \sqrt{y}$ and $x + y^{(q-1)/2}\sqrt{y}$ should be basic, which is a contradiction.

2.3.2   Chein's Proof of the Theorem of Ko Chao

Now we are ready to prove the theorem of Ko Chao.

Theorem 2.10 (Ko Chao). *The equation $x^2 - y^q = 1$ has no solutions in positive integers x, y and prime $q \geq 5$.*

Proof (Chein). Theorem 2.9 implies that x is odd. Assume that $x \equiv 3 \bmod 4$. Equality $(x - 1)(x + 1) = y^q$ implies that there exist positive integers a and b such that

$$x + 1 = 2^{q-1}a^q, \quad x - 1 = 2b^q.$$

Notice that $a^q = (b^q + 1)/2^{q-2} < b^q$, which implies $a < b$.

We have

$$(b^2 + 2a)\frac{b^{2q} + (2a)^q}{b^2 + 2a} = b^{2q} + (2a)^q = \left(\frac{x-1}{2}\right)^2 + 2(x+1) = \left(\frac{x+3}{2}\right)^2.$$

$$(2.6)$$

We again invoke Theorem 2.9, this time the statement $q \mid x$. Since $q > 3$, this implies that the right-hand side of (2.6) is not divisible by q. Now Lemma 2.8 yields that the factors on the left-hand side are coprime. Hence they are complete squares.

Since $b^2 + 2a$ is a complete square, we have $2a \geq (b+1)^2 - b^2 = 2b + 1$, which contradicts the previously established inequality $a < b$.

If $x \equiv 1 \bmod 4$ then $x - 1 = 2^{q-1}a^q$ and $x + 1 = 2b^q$. We obtain

$$b^{2q} - (2a)^q = \left(\frac{x-3}{2}\right)^2,$$

and the rest of the argument is the same.   □

2.3.3 Some Historical Remarks

Here we give some historical and bibliographical remarks on the equation

$$x^2 - y^q = 1 \tag{2.7}$$

with $q \geq 5$.

Before Ko Chao, the problem attracted many mathematicians. In 1921 Nagell [100] observed that a theorem of Lebesgue [66] asserting that the equation $x^5 + y^5 = 8z^5$ has no solutions implies that $x^2 - y^5 = 1$ has no solutions.

In the same article Nagell [100] presented several conditions involving the congruence class of q modulo 16 and the arithmetic of the number field $\mathbb{Q}(\sqrt{-q})$ under which the equation has no solution. In particular, he proved that there are no solutions with $q \leq 101$, except maybe with $q = 31, 59, 73, 83$ or 89. In 1934, he [103] improved upon the latter result, by showing that there is no solution if $q \not\equiv 1 \bmod 8$. Thus, below 101 only 73 and 89 remained untreated.

In 1932, Selberg [126] solved completely the Diophantine equation $x^4 - y^q = 1$, answering a question posed by Nagell [98] in 1919.

In 1940, Obláth [108, 109] showed that (2.7) has no solution except, possibly, when

$$2^{q-1} \equiv 1 \pmod{q^2}, \qquad 3^{q-1} \equiv 1 \pmod{q^2}. \tag{2.8}$$

His starting point was the key observation that it is sufficient to solve the equation $2^{q-2}a^q - b^q = \pm 1$, as it is clear from the proof of Theorem 2.10. Then, he combined the results of Lubelski [75] on the Diophantine equation $x^q + y^q = cz^q$ with Theorem 2.9 to get (2.8).

The abovementioned work of Lubelski generalizes the famous results of Wieferich [138] and Mirimanoff [95] on the Fermat equation $x^q + y^q + z^q = 0$. Wieferich showed that if the latter equation has a solution with q not dividing xyz (the "first case" of the Fermat theorem), then $2^{q-1} \equiv 1 \bmod q^2$, and Mirimanoff showed that $3^{q-1} \equiv 1 \bmod q^2$. Wieferich-type conditions systematically occur in Catalan's problem; see Sects. 6.2 and 6.5.

Obláth [110] combined his conditions with Nagell's results to show that there is no solution with $q < 25000$.

In 1961, Inkeri and Hyyrö [52] improved upon Nagell's Theorem 2.9: they showed that existence of a nontrivial solution of (2.7) implies that $q^2 \mid x$ and $q^3 \mid (y - 1)$. Using this result, they proved that there is no nontrivial solutions with $q < 100000$.

Equation $x^2 - y^q = 1$ continued to attract researchers even after the work of Ko Chao. We have already seen Chein's contribution. In 2004 Mignotte [88] suggested yet another proof of the theorem of Ko Chao. He adapted the classical argument of Kummer to show that (2.7) has no solution when $q \geq 5$ is a regular[2] prime.

The first irregular primes congruent to 1 modulo 8 are 233 and 257. This gives a weaker result than Obláth's; however, modern sharp estimates [65] for binary logarithmic forms imply that (2.7) is impossible for $q > 200$. This gives an alternative proof of Ko Chao's theorem.

One can also apply a deep theorem of Ribet [118] on the Diophantine equation $a^p + 2^\alpha b^p + c^p = 0$ which implies that (2.7) has no nontrivial solutions. This is not the easiest way to prove Ko Chao's theorem, since Ribet's work uses the advanced machinery of Galois representations.

We are left with the equation $x^2 - y^3 = 1$. It will be solved in Sect. 2.5, after some preparation in Sect. 2.4.

2.4   The Cubic Field $\mathbb{Q}(\sqrt[3]{2})$

In this auxiliary section we determine the ring of integers $\mathcal{O}_K$ and the group of units $\mathcal{U}_K = \mathcal{O}_K^\times$ of the cubic field $K = \mathbb{Q}(\sqrt[3]{2})$. We shall use this in Sect. 2.5.

Everywhere throughout this section we use the notation

$$\pi = 1 + \sqrt[3]{2}, \quad \eta = \sqrt[3]{2} - 1.$$

Notice that η is a unit of K.

We start from a simple observation.

Proposition 2.11. *The principal ideal (π) is a prime ideal of K. It satisfies $(\pi)^3 = (3)$.*

Proof. One verifies that $3 = \pi^3 \eta$. Since η is a unit, this implies that $(3) = (\pi)^3$. Since a rational prime number cannot split in K into more than 3 primes, the ideal (π) is prime. $\square$

---

[2]An odd prime number ℓ is *regular* if it does not divide the class number of the ℓth cyclotomic field and *irregular* if it does.

In the sequel, we write Ord_π instead of $\mathrm{Ord}_{(\pi)}$.

Proposition 2.12. *The ring of integers $\mathcal{O}_K$ is $\mathbb{Z}[\sqrt[3]{2}]$.*

Proof. Denote by $\mathcal{D}$ the discriminant of K and put $d = \left[\mathcal{O}_K : \mathbb{Z}[\sqrt[3]{2}]\right]$. Then $\mathcal{D}d^2$ is the discriminant of the ring $\mathbb{Z}[\sqrt[3]{2}]$. Since the conjugates of $\sqrt[3]{2}$ are $\xi\sqrt[3]{2}$ and $\xi^2\sqrt[3]{2}$, where $\xi = (-1 + \sqrt{-3})/2$ is a primitive cubic root of unity, we have

$$\mathcal{D}d^2 = \left(\det\left[\left(\xi^k\sqrt[3]{2}\right)^\ell\right]_{0 \le k,\ell \le 2}\right)^2 = -2^2 \cdot 3^3.$$

It follows that $d \mid 6$. If d is even, then $\mathcal{D}$ is odd, which is impossible because 2 ramifies in K. Thus, d divides 3.

Since $\mathbb{Z}[\sqrt[3]{2}] = \mathbb{Z}[\pi]$, every $\alpha \in \mathcal{O}_K$ can be written as $\alpha = a_0 + a_1\pi + a_2\pi^2$, where $a_0, a_1, a_2 \in \frac{1}{3}\mathbb{Z}$. If we show that

$$\mathrm{Ord}_3(a_k) \ge 0 \qquad (k = 0, 1, 2), \tag{2.9}$$

it would follow that $\alpha \in \mathbb{Z}[\pi]$, proving the proposition.

Observe that $\mathrm{Ord}_\pi\left(a_k\pi^k\right) \equiv k \bmod 3$. It follows that the numbers

$$\mathrm{Ord}_\pi\left(a_k\pi^k\right) \qquad (k = 0, 1, 2)$$

are pairwise distinct. Lemma A.1 implies that

$$\mathrm{Ord}_\pi(\alpha) = \min_{0 \le k \le 2} \mathrm{Ord}_\pi\left(a_k\pi^k\right).$$

But $\mathrm{Ord}_\pi(\alpha) \ge 0$, because α is an algebraic integer. Hence $\mathrm{Ord}_\pi\left(a_k\pi^k\right) \ge 0$ for $k = 0, 1, 2$, which is only possible if $\mathrm{Ord}_3(a_k) \ge 0$ for all k. This proves (2.9) and the proposition. $\square$

Proposition 2.13. *The unit group $\mathcal{U}_K$ is generated by -1 and η.*

Proof. The field K has a real embedding and a pair of complex conjugate embeddings. We identify K with its real embedding and denote the complex embeddings by σ and $\bar{\sigma}$, so that

$$|\sigma(\eta)|^2 = \eta^{-1} = 1 + \sqrt[3]{2} + \sqrt[3]{4}.$$

The Dirichlet unit theorem (Appendix A.2) implies that the rank of the unit group is 1. Also, since K has a real embedding, it cannot contain roots of unity other than ± 1. Thus, $\mathcal{U}_K$ is generated by -1 and a unit θ, where we may assume that

$$0 < \theta < 1. \tag{2.10}$$

Then $\eta = \theta^m$, where $m \geq 1$. This implies the inequality

$$|\sigma(\theta)| = |\bar{\sigma}(\theta)| \leq |\sigma(\eta)| = \left(1 + \sqrt[3]{2} + \sqrt[3]{4}\right)^{1/2} < 2. \qquad (2.11)$$

Proposition 2.12 implies that $\theta = a_0 + a_1 \sqrt[3]{2} + a_2 \sqrt[3]{4}$, where $a_0, a_1, a_2 \in \mathbb{Z}$. We have

$$a_k = \frac{1}{3} \mathrm{Tr}\left(\theta\left(\sqrt[3]{2}\right)^{-k}\right) \qquad (k = 0, 1, 2),$$

where $\mathrm{Tr} : K \to \mathbb{Q}$ is the trace map. Using inequalities (2.10) and (2.11), we obtain $|a_0|, |a_1|, |a_2| \leq 5/3$. Thus, $a_0, a_1, a_2 \in \{0, \pm 1\}$. Also, $a_0 \neq 0$, since otherwise θ is not a unit.

Among the 18 remaining possibilities, only ± 1, $\pm \eta$, and $\pm \eta^{-1}$ are units. Among the latter, only η belongs to the interval $(0, 1)$. Thus, $\theta = \eta$. $\qquad\square$

Remark 2.14. The following more refined argument shows that $a_2 = 0$ (which means that only six possibilities are to be verified instead of 18). Assume that η is not a fundamental unit. Then $\eta = \theta^m$ with $m \geq 2$. This implies, instead of (2.11), the inequality

$$|\sigma(\theta)| = |\bar{\sigma}(\theta)| \leq |\sigma(\eta)|^{1/2} = \left(1 + \sqrt[3]{2} + \sqrt[3]{4}\right)^{1/4} < 1.5.$$

Hence

$$|a_2| = \frac{1}{3}\left|\mathrm{Tr}\left(\theta\left(\sqrt[3]{4}\right)^{-1}\right)\right| \leq 0.9,$$

that is, $a_2 = 0$.

2.5 The Equation $x^2 - y^3 = 1$

This equation has a long history. Already Fermat stated (as usual, without a proof) that it has no solutions in positive integers except the obvious $3^2 - 2^3 = 1$. Euler [31] was the first to prove this. This proof, quite involved, is reproduced in Ribenboim's book [117, Sect. A.2].

Theorem 2.15 (Euler). *The only solution of the equation $x^2 - y^3 = 1$ in nonzero integers x, y is $(\pm 3)^2 - 2^3 = 1$.*

Actually, Euler proved much more: *the equation $x^2 - y^3 = 1$ has no solutions in **rational** numbers x, y other than $(\pm 1, 0)$, $(0, -1)$ and $(\pm 3, 2)$. A reader familiar with the notion of elliptic curve can express this as *the elliptic curve $x^2 - y^3 = 1$ has rank 0 and torsion 6 over $\mathbb{Q}$.*

Later Euler [32, Vol. 2, Article 247] and Legendre [68, pp. 406–409] used an "infinite descent" argument to prove the following theorem.

Theorem 2.16 (Euler, Legendre). *The equation $u^3 + 1 = 2v^3$ has no solutions in integers u, v with $v \neq 0, 1$.*

Theorem 2.15 is an easy consequence of Theorem 2.16.

Proof of Theorem 2.15 (Assuming Theorem 2.16). Rewrite our equation as $(x - 1)(x + 1) = y^3$. If x is even then the factors on the left are coprime. Hence they are complete cubes, which is impossible, because two cubes cannot differ by 2.

Now assume that x is odd. Then y is even, and, replacing x by $-x$, we may assume that $x \equiv 3 \bmod 4$. Rewrite the equation as

$$\frac{x-1}{2}\frac{x+1}{4} = \left(\frac{y}{2}\right)^3.$$

The factors on the left are coprime rational integers, hence both cubes: $x - 1 = 2u^3$ and $x + 1 = 4v^3$, where u and v are nonzero integers. We obtain $u^3 + 1 = 2v^3$, whence $v = 0$ or 1 by Theorem 2.16. It follows that $x = -1$ or $x = 3$, which proves Theorem 2.15. □

Here again, both Euler and Legendre proved much more: *the equation $u^3 + 1 = 2v^3$ has no nontrivial solutions in $u, v \in \mathbb{Q}$* (and even in $u, v \in \mathbb{Q}(\sqrt{3})$). This also implies the "rational" version of Theorem 2.15 mentioned above.

In 1922, Nagell [101, Sect. 10] suggested an alternative proof of Euler's theorem; see also [102].

Here we give a different proof, due to McCallum [80]. McCallum's argument is more transparent than the proofs of Euler, Euler-Legendre, and Nagell, but it does not extend to rational solutions.

(Recently Notari [107] suggested yet another proof, which is totally elementary and, like McCallum's proof, works only for integer solutions.)

Denote by K the cubic field $\mathbb{Q}(\sqrt[3]{2})$. Recall that $\eta = \sqrt[3]{2} - 1$ generates the group of positive units of K.

Notice that integers u, v satisfy $u^3 + 1 = 2v^3$ if and only if $v\sqrt[3]{2} - u$ is a positive unit of the field K. Hence Theorem 2.16 is equivalent to the following statement.

Theorem 2.17. *The field K has no positive units of the form $a_0 + a_1\sqrt[3]{2}$ (with $a_0, a_1 \in \mathbb{Z}$) other than 1 and η.*

Theorem 2.17 is a particular case of the famous result of Delaunay [27]: *given a cube-free integer d, the ring $\mathbb{Z}[\sqrt[3]{d}]$ has at most one nontrivial positive unit of the form $a_0 + a_1\sqrt[3]{d}$; if such a unit exists, then it generates the group of positive units.* Skolem [130, pp. 114–120] gave another proof of the first part of Delaunay's theorem using his local method. See also [96, Theorems 23.5 and 24.5]. The proof of McCallum, reproduced below in Sect. 2.5.2, can be viewed as a simplified version of Skolem's local argument for $d = 2$.

For the proof, we need a preparatory statement on the $\mathfrak{p}$-adic convergence of binomial series.

2.5.1 Binomial Series

In this subsection K is an arbitrary number field, not just $\mathbb{Q}(\sqrt[3]{2})$.

If α is a complex number with $|\alpha| < 1$ then the binomial series $\sum_{k=0}^{\infty} \binom{n}{k} \alpha^k$ converges to $(1 + \alpha)^n$. We need the $\mathfrak{p}$-adic generalization of this: if $\mathrm{Ord}_{\mathfrak{p}}(\alpha) > 0$ then the series $\sum_{k=0}^{\infty} \binom{n}{k} \alpha^k$ "$\mathfrak{p}$-adically converges" to $(1 + \alpha)^n$.

Thus, let $\mathfrak{p}$ be a prime ideal of the field K, and let $\mathcal{O}_{\mathfrak{p}}$ be the local ring of $\mathfrak{p}$:

$$\mathcal{O}_{\mathfrak{p}} = \{\alpha \in K : \mathrm{Ord}_{\mathfrak{p}}(\alpha) \geq 0\}.$$

Given $\alpha, \beta \in \mathcal{O}_{\mathfrak{p}}$, we say that $\alpha \equiv \beta \bmod \mathfrak{p}^N$ if $\mathrm{Ord}_{\mathfrak{p}}(\beta - \alpha) \geq N$.

Proposition 2.18. *Let n be an integer, N a nonnegative integer, and $\mathfrak{p}$ a prime ideal of a number field K. Then for any $\alpha \in K$ with $\mathrm{Ord}_{\mathfrak{p}}(\alpha) > 0$ we have*

$$\sum_{k=0}^{N} \binom{n}{k} \alpha^k \equiv (1 + \alpha)^n \bmod \mathfrak{p}^{N+1}.$$

Proof. If $n \geq 0$ then the assertion is an obvious consequence of the binomial formula. Now assume that $n < 0$, and write $n = -m - 1$ with $m \geq 0$. It will be convenient to replace α by $-\alpha$. Since $(-1)^k \binom{-m-1}{k} = \binom{m+k}{m}$, we have to prove that

$$\sum_{k=0}^{N} \binom{m + k}{m} \alpha^k \equiv (1 - \alpha)^{-m-1} \bmod \mathfrak{p}^{N+1}. \tag{2.12}$$

Deriving m times the identity

$$(1 - t)^{-1} - \sum_{k=0}^{m+N} t^k = t^{m+N+1} (1 - t)^{-1},$$

and dividing by $m!$, we obtain the identity

$$(1 - t)^{-m-1} - \sum_{k=0}^{N} \binom{m + k}{m} t^k = t^{N+1} P(t)(1 - t)^{-m-1}, \tag{2.13}$$

where $P(t)$ is a polynomial, depending on m and N. Notice that

$$t^{N+1} P(t) = 1 - (1 - t)^{m+1} \sum_{k=0}^{N} \binom{m + k}{m} t^k \in \mathbb{Z}[t],$$

which implies that $P(t) \in \mathbb{Z}[t]$. Hence

$$\mathrm{Ord}_{\mathrm{p}}\left(\alpha^{N+1} P(\alpha)(1-\alpha)^{-m-1}\right) \geq (N+1)\mathrm{Ord}_{\mathrm{p}}(\alpha) \geq N+1,$$

and (2.12) follows upon substituting $t = \alpha$ in (2.13).   □

2.5.2   Proof of Theorem 2.17

Now we are ready to prove Theorem 2.17. We again use the notation $K = \mathbb{Q}[\sqrt[3]{2}]$, $\eta = \sqrt[3]{2} - 1$, and $\pi = \sqrt[3]{2} + 1$ from Sect. 2.4.

Every element of the field K can be uniquely written as $a_0 + a_1 \sqrt[3]{2} + a_2 \sqrt[3]{4}$, where $a_0, a_1, a_2 \in \mathbb{Q}$. This defines the $\mathbb{Q}$-linear "coefficient functions"

$$a_0, a_1, a_2 : K \to \mathbb{Q}.$$

Since $\pi^3 \eta = 3$, we have $\pi^{3m} = 3^m \eta^{-m}$ for any integer m. Hence for any positive integer k the number π^k is divisible by $3^{\lfloor k/3 \rfloor}$ in the ring $\mathbb{Z}[\sqrt[3]{2}]$. It follows that

$$\mathrm{Ord}_3\left(a_\ell\left(\pi^k\right)\right) \geq \lfloor k/3 \rfloor \qquad (\ell = 0, 1, 2), \qquad (2.14)$$

for $k \geq 0$, which will be used throughout the proof.

Let θ be a positive unit of K with $a_2(\theta) = 0$. We have to prove that $\theta = 1$ or $\theta = \eta$. Proposition 2.13 implies that $\theta = \eta^n$, where $n \in \mathbb{Z}$. We assume that $n \neq 0, 1$ and obtain a contradiction.

Fix a large positive integer N, to be specified later. Since $\eta = -2 + \pi$, we have $(-2)^{-n}\theta = (1 - \pi/2)^n$. Proposition 2.18 implies that

$$\sum_{k=0}^{N} \binom{n}{k} \left(-\frac{\pi}{2}\right)^k \equiv (-2)^{-n}\theta \bmod \pi^{N+1}. \qquad (2.15)$$

Applying the coefficient function a_2 to both sides of (2.15), and using (2.14), we obtain

$$\sum_{k=0}^{N} \binom{n}{k} \frac{a_2\left(\pi^k\right)}{(-2)^k} \equiv (-2)^{-n}a_2(\theta) \bmod 3^{\lfloor (N+1)/3 \rfloor}. \qquad (2.16)$$

Since $a_2(\theta) = 0$, the right-hand side of congruence (2.16) vanishes. Since $a_2(1) = a_2(\pi) = 0$, so do the summands on the left of (2.16), corresponding to $k = 0$ and $k = 1$. Also, for $k \geq 2$, we have

$$\binom{n}{k} = \frac{n(n-1)}{k(k-1)} \binom{n-2}{k-2}.$$

Hence, for $k \geq 2$, the kth summand on the left of (2.16) is equal to $n(n-1)A_k$, where

$$A_k = \frac{1}{k(k-1)(-2)^k} \binom{n-2}{k-2} a_2 \left(\pi^k\right).$$

Thus, (2.16) can be rewritten as

$$n(n-1)\left(A_2 + \cdots + A_N\right) \equiv 0 \bmod 3^{\lfloor (N+1)/3 \rfloor}.$$

Since n is distinct from 0 and 1, we may choose N so large that

$$\lfloor (N+1)/3 \rfloor > \mathrm{Ord}_3(n(n-1)).$$

We obtain

$$A_2 + \cdots + A_N \equiv 0 \bmod 3. \tag{2.17}$$

On the other hand, again using (2.14), we find

$$\mathrm{Ord}_3(A_k) \geq \lfloor k/3 \rfloor - \mathrm{Ord}_3(k(k-1)) \geq \lfloor k/3 \rfloor - \log_3 k. \tag{2.18}$$

As one can easily verify, the right-hand side of (2.18) is positive for $k \geq 6$. Hence $\mathrm{Ord}_3(A_k) > 0$ for $k \geq 6$. Also, $\mathrm{Ord}_3(k(k-1)) = 0$ for $k = 5$, which implies that $\mathrm{Ord}_3(A_5) > 0$.

We have proved that $\mathrm{Ord}_3(A_k) > 0$ for $k \geq 5$. Combining this with (2.17), we obtain

$$A_2 + A_3 + A_4 \equiv 0 \bmod 3.$$

Since $a_2(\pi^2) = 1$, $a_2(\pi^3) = 3$, and $a_2(\pi^4) = 6$, we have

$$A_2 + A_3 + A_4 = \frac{1}{8} - \frac{n-2}{16} + \frac{(n-2)(n-3)}{32} = \frac{n^2 - 7n + 14}{32}.$$

It follows that $n^2 - 7n + 14 \equiv 0 \bmod 3$, which is impossible for $n \in \mathbb{Z}$. The theorem is proved. $\qquad\square$

Remark 2.19. The Diophantine equation $y^2 - x^3 = k$, where k is a nonzero integer, is usually called *Mordell's equation.* Many results on it can be found in Chap. 26 of Mordell's book [96]. Modern techniques based on logarithmic forms [11] or forms in elliptic logarithms [33] allow one to solve Mordell's equation completely for small values of k, using electronic computations. For instance, this was done for $|k| \leq 10^4$ in [33].

Chapter 3
Cassels' Relations

Due to the theorems of Euler, Lebesgue, and Ko Chao, proved in the previous chapter, we may assume that the exponents of Catalan's equation are odd. Since every odd number greater than 1 has an odd prime divisor, this reduces Catalan's problem to the following assertion.

Theorem 3.1. *Equation*

$$x^p - y^q = 1 \tag{3.1}$$

has no solutions in nonzero integers x, y and odd primes p, q.

Proving this theorem is the main objective of the rest of this book.

Starting from this point, a *solution of Catalan's equation* (or, simply, a *solution*) is, by definition, a quadruple (x, y, p, q), where x, y are nonzero integers and p, q are (distinct) odd prime numbers, satisfying $x^p - y^q = 1$. Notice that we no longer assume x and y positive. This allows us to symmetrize the problem: if (x, y, p, q) is a solution, then so is $(-y, -x, q, p)$. This symmetry will be repeatedly used in the sequel.

In this chapter we make the first step towards the proof of Theorem 3.1. Following Cassels, we reduce (3.1) to several more complicated equations, which are, however, easier to deal with. We also show, following Hyyrö [48], that Cassels' relations imply lower bounds for $|x|$ (and $|y|$) in terms of p and q.

3.1 Cassels' Divisibility Theorem and Cassels' Relations

Rewrite (3.1) as

$$(x-1)\frac{x^p - 1}{x - 1} = y^q. \tag{3.2}$$

© Springer International Publishing Switzerland 2014
Y.F. Bilu et al., *The Problem of Catalan*, DOI 10.1007/978-3-319-10094-4_3

Lemma 2.8 implies that the greatest common divisor of the factors on the left is 1 or p. More precisely, we have the following statement.

Proposition 3.2. *Let (x, y, p, q) be a solution of Catalan's equation. Then*

$$\gcd\left(\frac{x^p - 1}{x - 1}, x - 1\right) = \begin{cases} p, & \text{if } p \mid y, \\ 1, & \text{otherwise.} \end{cases} \tag{3.3}$$

Proof. If p does not divide y then it does not divide any of the numbers $x - 1$ and $(x^p - 1)/(x - 1)$. Part (2) of Lemma 2.8 implies now that

$$\gcd\left(\frac{x^p - 1}{x - 1}, x - 1\right) = 1.$$

If $p \mid y$ then p divides one of the numbers $x - 1$ and $(x^p - 1)/(x - 1)$. Part (1) of Lemma 2.8 implies that it divides both, and part (2) yields that

$$\gcd\left(\frac{x^p - 1}{x - 1}, x - 1\right) = p.$$

$\square$

Cassels [18] showed that the second option in (3.3) is impossible.

Theorem 3.3 (Cassels). *Let (x, y, p, q) be a solution of Catalan's equation. Then $p \mid y$ (and $q \mid x$ by symmetry).*

We postpone the proof of this theorem until Sect. 3.3 and formulate now the most important consequence of Theorem 3.3, known as *Cassels' relations.*

Theorem 3.4 (Cassels). *Let (x, y, p, q) be a solution of Catalan's equation. Then there exist a nonzero integer a and a positive integer u such that*

$$x - 1 = p^{q-1}a^q, \quad \frac{x^p - 1}{x - 1} = pu^q, \quad y = pau. \tag{3.4}$$

Symmetrically, there exist a nonzero integer b and a positive integer v such that

$$y + 1 = q^{p-1}b^p, \quad \frac{y^q + 1}{y + 1} = qv^p, \quad x = qvb. \tag{3.5}$$

Proof (Assuming Theorem 3.3). Since $p \mid y$, Proposition 3.2 implies that

$$\gcd\left(x - 1, \frac{x^p - 1}{x - 1}\right) = p.$$

In view of (3.2), there exist rational integers a and u such that $x - 1 = p^\alpha a^q$ and $(x^p - 1)/(x - 1) = p^\beta u^q$, one of the exponents α, β being 1 and the other $q - 1$. Notice also that u is positive, because so is $(x^p - 1)/(x - 1)$, and that $p \nmid u$; otherwise we would contradict part (3) of Lemma 2.8.

Part (3) of Lemma 2.8 implies that $\beta = 1$, and thereby $\alpha = q - 1$. This proves the first two relations in (3.4). The third one follows from them at once. Relations (3.5) follow by symmetry. The theorem is proved. $\quad\square$

Another proof of Cassels' divisibility theorem was suggested by Hyyrö [49].

3.2 Binomial Power Series

In this section we establish some basic properties of the power series

$$(1 + t)^\nu = \sum_{k=0}^{\infty} \binom{\nu}{k} t^k, \tag{3.6}$$

to be used in the proof of Theorem 3.3, as well as in other parts of the book. Recall that for any real ν the coefficient $\binom{\nu}{k}$ is defined by

$$\binom{\nu}{k} = \frac{\nu(\nu - 1) \cdots (\nu - k + 1)}{k!}.$$

First, we study the arithmetic of the coefficients when the exponent ν is a rational number. We show that the only primes appearing in the denominator of $\binom{\nu}{k}$ are those dividing the denominator of ν. We also calculate the exact order of every such prime in the denominator of $\binom{\nu}{k}$.

Recall that for any prime number p and any nonnegative integer k

$$\mathrm{Ord}_p(k!) = \left\lfloor \frac{k}{p} \right\rfloor + \left\lfloor \frac{k}{p^2} \right\rfloor + \cdots < \frac{k}{p} + \frac{k}{p^2} + \cdots = \frac{k}{p - 1}. \tag{3.7}$$

Lemma 3.5. *Let ν be a rational number with denominator b.*

1. *For any nonnegative integer k, there exists a positive integer N such that $b^N \binom{\nu}{k} \in \mathbb{Z}$.*
2. *If q is a prime divisor of b, then for any nonnegative integer k*

$$\mathrm{Ord}_q \binom{\nu}{k} = -k\,\mathrm{Ord}_q b - \mathrm{Ord}_q(k!) > -k\,\mathrm{Ord}_q b - \frac{k}{q - 1}. \tag{3.8}$$

In particular, the sequence $\{\mathrm{Ord}_q \binom{v}{k}\}_{k=0,1,\dots}$ *is strictly decreasing:*

$$0 = \mathrm{Ord}_q \binom{v}{0} > \mathrm{Ord}_q \binom{v}{1} > \mathrm{Ord}_q \binom{v}{2} > \cdots . \tag{3.9}$$

Proof. Write $v = a/b$. Then

$$\binom{v}{k} = \frac{a(a-b)\cdots(a-(k-1)b)}{b^k k!}. \tag{3.10}$$

For any prime number p not dividing b we have

$$\mathrm{Ord}_p\left(a(a-b)\cdots(a-(k-1)b)\right) \geq \left\lfloor \frac{k}{p} \right\rfloor + \left\lfloor \frac{k}{p^2} \right\rfloor + \cdots = \mathrm{Ord}_p(k!),$$

via (3.7). Hence the only prime factors of the denominator of $\binom{v}{k}$ are the prime divisors of b. This proves part (1).

For part (2) observe that if q divides b then it does not divide a. Hence it does not divide the numerator in (3.10), which implies (3.8).

Finally, (3.9) is an immediate consequence of (3.8). $\square$

We also need an estimate for the remainder term of the binomial power series.

Lemma 3.6. *Let v be a real number and m a nonnegative integer. Then for any real t satisfying $|t| < 1$ we have*

$$\left| (1+t)^v - \sum_{k=0}^{m} \binom{v}{k} t^k \right| \leq \max\left\{ 1, (1+t)^{v-m-1} \right\} \left| \binom{v}{m+1} \right| |t|^{m+1}. \tag{3.11}$$

In particular, for $m = 0$, we have

$$|(1+t)^v - 1| \leq \max\left\{ 1, (1+t)^{v-1} \right\} |vt|. \tag{3.12}$$

Proof. By the Taylor formula with the Lagrange error term, we have

$$\left| (1+t)^v - \sum_{k=0}^{m} \binom{v}{k} t^k \right| \leq \sup_{0 \leq \theta \leq 1} \left| \left(\frac{d^{m+1}(1+T)^v}{dT^{m+1}} \right)_{T=\theta t} \right| \frac{|t|^{m+1}}{(m+1)!}$$

$$= \max\left\{ 1, (1+t)^{v-m-1} \right\} \left| \binom{v}{m+1} \right| |t|^{m+1},$$

as wanted. $\square$

As an immediate application to Catalan's equation, we estimate the difference $\left|y - x^{p/q}\right|$. (Notice that, since q is odd, $x^{p/q}$ is a well-defined real number.)

Lemma 3.7. *Let* (x, y, p, q) *be a solution of Catalan's equation. Then*

$$\left|y - x^{p/q}\right| \leq \frac{1.1}{q}|x|^{p/q-p}. \tag{3.13}$$

Proof. Applying (3.12) with $\nu = 1/q$ and $t = x^{-p}$, we obtain

$$y = x^{p/q}\left(1 + x^{-p}\right)^{1/q} = x^{p/q}(1 + r)$$

with $|r| \leq (1 - |x|^{-p})^{1/q-1}q^{-1}|x|^{-p}$. One verifies by inspection that $|x|^p > 16$. Hence $(1 - |x|^{-p})^{1/q-1} \leq 1.1$, and the result follows.   □

3.3   Proof of the Divisibility Theorem

In this section we prove Theorem 3.3. The argument splits into two cases: $p < q$ and $p > q$. Proofs in both cases are similar, but the latter is technically much more complicated than the former.

We start with the simpler case $p < q$. Thus, we are going to prove the following.

Proposition 3.8. *Let* (x, y, p, q) *be a solution of Catalan's equation with* $p < q$. *Then* $p \mid y$.

Proof. Assume that p does not divide y. Proposition 3.2 implies that the numbers $(x^p - 1)/(x - 1)$ and $x - 1$ are coprime. In view of (3.2), each of them is a complete qth power.

Write $x - 1 = a^q$, where a is a nonzero integer. If $|a| = 1$ then $|x| = 2$, and $|y| \leq (|x|^p + 1)^{1/q} < 2$, because $p < q$. We obtain $|y| = 1$, a contradiction. This proves that $|a| \geq 2$.

Equality $(1 + a^q)^p = y^q + 1$ suggests that y should be close to a^p. Indeed, using (3.12), we obtain

$$x^{p/q} = a^p\left(1 + a^{-q}\right)^{p/q} = a^p(1 + r),$$

where

$$|r| \leq (1 - |a|^{-q})^{p/q-1}\frac{p}{q}|a|^{-q} \leq 1.1|a|^{-q}.$$

Thus, $\left|x^{p/q} - a^p\right| \leq 1.1|a|^{p-q} < 1/3$. Further $\left|y - x^{p/q}\right| \leq 0.1$, as easily follows from (3.13). Combining the last two inequalities, we obtain $|y - a^p| < 1$. Hence $y = a^p$, because both are rational integers.

We obtain $x^p - u^{pq} = 1$, which is impossible (two pth powers cannot differ by 1). The proposition is proved.   □

We are left with the more difficult case $p > q$. First of all, we have to establish a lower bound for x. The following statement is, essentially, due to Hyyrö [48].

Proposition 3.9 (Hyyrö). *Let (x, y, p, q) be a solution of Catalan's equation with $p > q$. Then $|x| \geq q^{p-1} + q$.*

Proof. Since $q < p$, Proposition 3.8 applies to the solution $(-y, -x, q, p)$. Hence $q \mid x$, and we have the second series (3.5) of Cassels' relations. Rewriting the relation $(y^q + 1)/(y + 1) = qv^p$ as

$$\left((-y)^{q-1} - 1\right) + \left((-y)^{q-2} - 1\right) + \cdots + (-y - 1) = q\,(v^p - 1),$$

we deduce that $y + 1$ divides $q\,(v^p - 1)$. Now the relation $y + 1 = q^{p-1}b^p$ implies that $v^p \equiv 1 \bmod q^{p-2}$.

Now we again use the assumption $p > q$, this time to observe that p does not divide the order $q^{p-3}(q - 1)$ of the multiplicative group $\bmod\, q^{p-2}$. This implies that $v \equiv 1 \bmod q^{p-2}$.

Since $|y| \geq 2$ and $(y, q) \neq (2, 3)$, an easy estimate shows that

$$\frac{y^q + 1}{y + 1} > q.$$

It follows that $v > 1$. Together with the previously established congruence $v \equiv 1 \bmod q^{p-2}$, this implies $v \geq q^{p-2} + 1$. Since $x = qvb$, we have

$$|x| \geq qv \geq q^{p-1} + q,$$

as wanted.   □

Now we are ready to consider the case $p > q$.

Proposition 3.10. *Let (x, y, p, q) be a solution of Catalan's equation with $p > q$. Then $p \mid y$.*

Proof. In the proof of Proposition 3.8 we used (3.12) to approximate $x^{p/q}$ by the integer a^p. Due to (3.13), this shows that y is close to a^p. Since both are integers, they are equal, which leads to a contradiction.

In this proof, we again approximate $x^{p/q}$. However, (3.12) is no longer sufficient to get a good approximation, and we use the general inequality (3.11) to approximate $x^{p/q}$ (and thereby y) by a certain rational number A/B. We shall see that the approximation is good enough to conclude that $y = A/B$. On the other hand, we shall show that A/B is not an integer, a contradiction.

Now let us proceed. As in the proof of Proposition 3.8, we find out that $x - 1 = a^q$, where $a \in \mathbb{Z}$. The estimate $|x| \geq q^{p-1} + q$, obtained in Proposition 3.9, implies that

$$|a|^q > q^{p-1}. \tag{3.14}$$

Applying (3.11) with $v = p/q$ and $m = \lceil p/q \rceil$, we obtain

$$x^{p/q} = a^p (1 + a^{-q})^v = a^p \left(\sum_{k=0}^m \binom{v}{k} a^{-qk} + r \right),$$

where

$$|r| \le (1 - |a|^{-q})^{v-m-1} \left| \binom{v}{m+1} \right| |a|^{-q(m+1)}.$$

Using (3.14), we estimate

$$(1 - |a|^{-q})^{v-m-1} \le \left(1 - q^{-(p-1)} \right)^{-2} \le 1.1.$$

Also,

$$\left| \binom{v}{m+1} \right| = \frac{v}{m} \frac{v-1}{m-1} \cdots \frac{v-m+1}{1} \frac{|v-m|}{m+1} \le \frac{1}{m+1}.$$

Hence $|r| \le 0.5|a|^{-q(m+1)}$. It follows that

$$\left| x^{p/q} - \sum_{k=0}^m \binom{v}{k} a^{p-qk} \right| \le |a|^p |r| \le 0.5 |a|^{p-(m+1)q}. \tag{3.15}$$

Now write

$$\sum_{k=0}^m \binom{v}{k} a^{p-qk} = \frac{A}{B},$$

where A and B are coprime integers. By part (2) of Lemma 3.5, the greatest common denominator of the binomial coefficients is $q^{m + \mathrm{Ord}_q(m!)}$. Hence B divides the number $q^{m + \mathrm{Ord}_q(m!)} a^{mq-p}$. Since $\mathrm{Ord}_q(m!) < m/(q-1)$, this implies the inequality

$$|B| \le q^{m + \mathrm{Ord}_q(m!)} |a|^{mq-p} < q^{mq/(q-1)} |a|^{mq-p}. \tag{3.16}$$

On the other hand, (3.9) implies that

$$\mathrm{Ord}_q \left(\binom{v}{m} a^{p-qm} \right) < \mathrm{Ord}_q \left(\binom{v}{k} a^{p-qk} \right) \quad (k = 0, 1, \ldots, m-1).$$

Hence $\mathrm{Ord}_q(A/B) = \mathrm{Ord}_q\left(\binom{v}{m}a^{p-qm}\right) < 0$, and, in particular, A/B **is not an integer**.

Now we are going to replace $x^{p/q}$ by y. We have trivially

$$|x|^{p-p/q} \geq |x|^{10/3} > (|x|+1)^2 \geq |a|^{2q} \geq |a|^{(m+1)q-p}.$$

Hence, using (3.15) and (3.13), we obtain

$$\left|y - \frac{A}{B}\right| \leq \frac{1.1}{q}|x|^{p/q-p} + 0.5|a|^{p-(m+1)q} < |a|^{p-(m+1)q}. \tag{3.17}$$

It is easy to see that the right-hand side of (3.17) is strictly less than $|B|^{-1}$. Indeed,

$$p - 1 \geq \frac{p+q}{q-1} \geq \frac{qm}{q-1}.$$

Hence we may use (3.14) to obtain $|a|^q > q^{p-1} \geq q^{qm/(q-1)}$. Using (3.16), we find that

$$|a|^{p-(m+1)q} = |a|^{-q}|a|^{p-mq} < q^{-mq/(q-1)}|a|^{p-mq} \leq |B|^{-1}.$$

Thus, $|y - A/B| < |B|^{-1}$. Hence $y = A/B$, which is impossible since A/B is not an integer.

This contradiction proves Proposition 3.10, completing thereby the proof of Cassels' divisibility theorem. □

One may view Cassels' divisibility theorem as the extension of Nagell's Theorem 2.9 to the case of odd exponent p. The proof of Nagell's theorem, however, does not seem to extend to odd p. On the other hand, Cassels' argument does extend to $p = 2$. It might be a good exercise for the reader to work out the details of the proof of Nagell's theorem, using Cassels' method.

3.4 Hyyrö's Lower Bounds

Cassels' relations (3.4) and (3.5) imply lower bounds for the variables x and y in terms of the exponents p and q. For instance, the relation $x - 1 = p^{q-1}a^q$ implies the inequality $|x| \geq p^{q-1} - 1$.

Hyyrö [48, Hilfssatz 2] obtained two less obvious lower bounds. One of them has already been reproduced in this book as Proposition 3.9. Now we can formulate Hyyrö's result in its complete form.

Theorem 3.11. *Let (x, y, p, q) be a solution of Catalan's equation. Then*

$$|x| \geq \max \{q^{p-1} + q, \, p^{q-1}(q-1)^q + 1\} \tag{3.18}$$

(and, symmetrically, $|y| \geq \max \{p^{q-1} + p, \, q^{p-1}(p-1)^p + 1\}$).

Hyyrö's actual estimate is slightly sharper, but (3.18) is easier to prove and sufficient for our purposes.

Proof. We start with the inequality $|x| \geq p^{q-1}(q-1)^q + 1$. We shall use the following obvious fact: the four numbers x, y, a, b in Theorem 3.4 are either altogether positive (*the positive case*) or altogether negative (*the negative case*).
  Since $q \mid x$, we have

$$p^{q-1}a^q = x - 1 \equiv -1 \bmod q.$$

Since $p^{q-1} \equiv 1 \bmod q$, this implies $a^q \equiv -1 \bmod q$, which is equivalent to $a \equiv -1 \bmod q$. Similarly, $b \equiv 1 \bmod p$.
  Now, in the positive case, we have $a \geq q - 1$ and $x \geq p^{q-1}(q-1)^q + 1$. In the negative case we have either $a \leq -q - 1$, which implies that

$$|x| \geq p^{q-1}(q+1)^q - 1 > p^{q-1}(q-1)^q + 1,$$

or $a = -1$.
  It remains to show that the last option is impossible. Thus, assume that $a = -1$, which implies $1 - x = 1 + |x| = p^{q-1}$. Since we are in the negative case, we have $b \leq 1 - p$, and

$$|y| = (|x|^p + 1)^{1/q} \leq (1 + |x|)^{p/q} < p^p < 2^{p-1}(p-1)^p$$
$$< q^{p-1}|b|^p = |1 + y| < |y|,$$

a contradiction. This proves the inequality $|x| \geq p^{q-1}(q-1)^q + 1$.
  We are left with the inequality $|x| \geq q^{p-1} + p$. In the case $p > q$ it has already been established in Proposition 3.9. In the case $p < q$, we have

$$|x| \geq p^{q-1}(q-1)^q + 1 > 2(q-1)^q > 2p^q > 2q^p > q^{p-1} + q.$$

The theorem is proved.   □

  Hyyrö's lower bounds will be widely used in this book.

Chapter 4
Cyclotomic Fields

Let m be a positive integer, and let ζ_m be a primitive mth root of unity. The number field $K_m = \mathbb{Q}(\zeta_m)$ is called the *mth cyclotomic field*. We cannot give a comprehensive treatment of the theory of cyclotomic fields in this book; for this purpose, we refer to the famous monographs of Lang [60] and Washington [136]. Here we develop only a few fragments of this beautiful theory, required for Catalan's problem.

In particular, except a few sporadic points, we use only the special case $m = p$, an odd prime number, which is technically simpler than the general case.

Thus, in this chapter (except Sect. 4.7), we fix an odd prime number p, a primitive pth root of unity ζ_p, and denote by $K_p = \mathbb{Q}(\zeta_p)$ the corresponding cyclotomic field. Since p is fixed, we shall write (when this does not confuse) ζ instead of ζ_p and K instead of K_p.

4.1 Degree and Galois Group

Any conjugate of ζ over $\mathbb{Q}$ is again a primitive pth root of unity. It follows that $[K:\mathbb{Q}] \le p - 1$, which is the number of primitive pth roots.

It is not difficult to show that the degree is equal to $p - 1$. The quickest way is to observe that the cyclotomic polynomial $\Phi_p(t) = t^{p-1} + \cdots + t + 1$ is irreducible, applying the Eisenstein criterion with the prime number p to the polynomial

$$\Phi_p(t + 1) = \frac{(t + 1)^p - 1}{t}.$$

It is more instructive, however, to obtain the equality $[K:\mathbb{Q}] = p - 1$ as part of the following assertion.

© Springer International Publishing Switzerland 2014

Y.F. Bilu et al., *The Problem of Catalan*, DOI 10.1007/978-3-319-10094-4_4

Proposition 4.1. *Denote by* $\mathfrak{p}$ *the principal ideal* $(1 - \zeta)$.

1. *For any integers* k *and* ℓ *not divisible by* p, *the number* $(1 - \zeta^k)/(1 - \zeta^\ell)$ *is a unit of the field* K. *In particular, for any* $k \not\equiv 0 \mod p$, *we have the equality of ideals* $(1 - \zeta^k) = \mathfrak{p}$.
2. *The ideal* $\mathfrak{p}$ *is a prime ideal of the field* K. *It satisfies* $\mathfrak{p}^{p-1} = p$. *In particular,* p *totally ramifies in* K.
3. *We have* $[K:\mathbb{Q}] = p - 1$. *In particular, all primitive pth roots of unity are conjugate over* $\mathbb{Q}$.

Proof. Since ℓ is not divisible by p, there exists an integer m such that $\ell m \equiv k \mod p$. It follows that

$$\frac{1 - \zeta^k}{1 - \zeta^\ell} = \frac{1 - \zeta^{\ell m}}{1 - \zeta^\ell} = 1 + \zeta^\ell + \cdots + \zeta^{\ell(m-1)}$$

is an algebraic integer. By symmetry, so is $(1 - \zeta^\ell)/(1 - \zeta^k)$. This proves part (1).

Further, substituting $t = 1$ into the identity

$$t^{p-1} + \cdots + t + 1 = (t - \zeta)(t - \zeta^2) \cdots (t - \zeta^{p-1}),$$

we obtain $(p) = (1 - \zeta)(1 - \zeta^2) \cdots (1 - \zeta^{p-1}) = \mathfrak{p}^{p-1}$. Since p cannot split in K into more than $[K:\mathbb{Q}]$ prime factors, and $[K:\mathbb{Q}] \le p - 1$, this is possible only if $[K:\mathbb{Q}] = p - 1$ and $\mathfrak{p}$ is a prime ideal. This proves parts (2) and (3).   □

The following immediate consequence of part (1) will be often used.

Corollary 4.2. *Let* $\mathfrak{q}$ *be a prime ideal of* K_p *other than* $\mathfrak{p}$. *Then distinct pth roots of unity cannot be equal* $\mod \mathfrak{q}$. *(That is, if* ζ, ζ' *are pth roots of unity satisfying* $\zeta \equiv \zeta' \mod \mathfrak{q}$ *then* $\zeta = \zeta'$.)

Since K contains all the conjugates of ζ, it is a Galois extension of $\mathbb{Q}$. Denote by G its Galois group. Part (3) implies that for any integer a, not divisible by p, there exists a unique $\sigma_a \in G$ such that $\sigma_a(\zeta) = \zeta^a$. Obviously, $\sigma_{ab} = \sigma_a \circ \sigma_b$. Also, $\sigma_a = \sigma_{a'}$ if and only if $a \equiv a' \mod p$, which implies that

$$G = \{\sigma_1, \sigma_2, \ldots, \sigma_{p-1}\}.$$

Mention also that σ_{p-1} is the complex conjugation. We shall usually denote it by ι, so that

$$\iota \circ \sigma_a = \sigma_a \circ \iota = \sigma_{p-a}.$$

Sometimes, to simplify notation, we shall write $\bar{\sigma}$ instead of $\iota \circ \sigma$.

For $a \in \mathbb{Z}$ denote by $a^* \in \mathbb{Z}/p\mathbb{Z}$ the residue class of a modulo p. Since σ_a depends only on its class modulo p, we have the well-defined map

$$(\mathbb{Z}/p\,\mathbb{Z})^{\times} \to G \qquad (4.1)$$

$$a^{*} \mapsto \sigma_a.$$

The following statement is obvious.

Proposition 4.3. *The map (4.1) defines an isomorphism $(\mathbb{Z}/p\,\mathbb{Z})^{\times} \simeq G$. In particular, G is a cyclic group.*

4.2 Integral Basis and Discriminant

As in the previous section, ζ stands for a primitive pth root of unity and $K = \mathbb{Q}(\zeta)$ is the pth cyclotomic field.

Our next purpose is to determine the ring of integers $\mathcal{O}_K$ and to calculate the discriminant $\mathcal{D}_K$. First of all, notice that $\mathcal{O}_K \supseteq \mathbb{Z}[\zeta]$. Hence $\mathcal{D}_K$ divides the discriminant of $1, \zeta, \ldots, \zeta^{p-2}$, defined by

$$\mathcal{D}(1, \zeta, \ldots, \zeta^{p-2}) = \left(\det \left[\sigma_j \left(\zeta^{k-1} \right) \right]_{1 \le j, k \le p-1} \right)^2. \qquad (4.2)$$

After expanding the Vandermonde determinant in the right-hand side, we obtain ± 1 times a product of $(p-1)(p-2)$ terms of the type $\zeta^m - \zeta^s$ (with $m \not\equiv s \bmod p$). Hence we have the equality of ideals

$$(\mathcal{D}') = \mathfrak{p}^{(p-1)(p-2)} = \left(p^{p-2} \right),$$

where we denote by $\mathcal{D}'$ the right-hand side of (4.2). Since $\mathcal{D}'$ is a rational integer, we obtain $\mathcal{D}' = \pm p^{p-2}$. This has the following consequence.

Proposition 4.4. 1. *The discriminant $\mathcal{D}_K$ is ± 1 times a power of p. The only prime number ramified in K is p.*
2. *For distinct odd primes p and q, the fields K_p and K_q are disjoint (that is, $K_p \cap K_q = \mathbb{Q}$). Also, the fields K_p and $\mathbb{Q}(i)$ are disjoint.*
3. *The field K_{p^2} is a proper extension of K_p. In fact, $[K_{p^2} : K_p] = p$.*
4. *The group of roots of unity of the field K_p is generated by $-\zeta$.*

Proof. Part (1) follows from the already mentioned fact that $\mathcal{D}_K \mid \mathcal{D}'$.

To prove part (2), observe that, since p is totally ramified in K_p, it is ramified in any subfield of K_p distinct from $\mathbb{Q}$. On the other hand, p is unramified in K_q by part (1). Hence the only common subfield of K_p and K_q is $\mathbb{Q}$. Similarly, since (2) is the only prime ramified in $\mathbb{Q}(i)$, this field is disjoint with K_p.

Next, the congruence

$$(1 - \zeta_{p^2})^p \equiv 1 - \zeta_{p^2}^p \equiv 1 - \zeta_p \bmod p$$

implies that $\wp^p = \mathfrak{p}$, where $\wp$ is the principal ideal $(1 - \zeta_{p^2})$. It follows that $\mathfrak{p}$ ramifies in K_{p^2}, and the ramification index is at least p. On the other hand, it is obvious that the degree of K_{p^2} over K_p is at most p. Hence both the degree and the ramification index are equal to p. This proves part (3).

Parts (2) and (3) yield that K_p cannot contain roots of unity other than the powers of $-\zeta$. This proves part (4).   □

Here is a trivial but frequently used consequence of part (4).

Corollary 4.5. *Assume that p and q are distinct odd prime numbers. Then every root of unity from K_p is a qth power in K_p.*

Part (1) of Proposition 4.4 can be refined as follows.

Theorem 4.6. *We have $\mathcal{O}_K = \mathbb{Z}[\zeta]$. In particular, $|\mathcal{D}_K| = p^{p-2}$.*

(One can show that $\mathcal{D}_K = (-1)^{(p-1)/2} p^{p-2}$ (see [136, Proposition 2.1]), but the sign of $\mathcal{D}_K$ will not be used in the present book.)

Proof. Fix $\alpha \in \mathcal{O}_K$ and write

$$\alpha = a_0 + a_1 \lambda + \cdots + a_{p-2} \lambda^{p-2},$$

where $\lambda = 1 - \zeta$ and $a_0, \ldots, a_{p-2} \in \mathbb{Q}$. Since $\mathbb{Z}[\zeta] = \mathbb{Z}[\lambda]$, we have to prove that $a_0, \ldots, a_{p-2} \in \mathbb{Z}$.

Since the index $[\mathcal{O}_K : \mathbb{Z}[\lambda]] = [\mathcal{O}_K : \mathbb{Z}[\zeta]] = \mathcal{D}'/D_K$ is a power of p, the denominators of the rational numbers $a_0, \ldots, a_{p-2}$ are powers of p. Now notice that for $k = 0, 1, \ldots, p-2$ we have $\operatorname{Ord}_{\mathfrak{p}}(a_k \lambda^k) \equiv k \bmod (p-1)$. It follows that the $p-1$ numbers

$$\operatorname{Ord}_{\mathfrak{p}}(a_k \lambda^k) \quad (k = 0, 1, \ldots, p-2)$$

are pairwise distinct. Lemma A.1 implies that

$$\operatorname{Ord}_{\mathfrak{p}}(\alpha) = \min_{0 \le k \le p-2} \operatorname{Ord}_{\mathfrak{p}}(a_k \lambda^k).$$

Since α is an algebraic integer, we have $\operatorname{Ord}_{\mathfrak{p}}(\alpha) \ge 0$. Hence $\operatorname{Ord}_{\mathfrak{p}}(a_k \lambda^k) \ge 0$ for all k, which is only possible if $\operatorname{Ord}_p(a_k) \ge 0$ for all k.

Thus, $a_0, \ldots, a_{p-2} \in \mathbb{Z}$, as wanted. The theorem is proved.   □

Corollary 4.7. *For any $\sigma \in G$ and any $\alpha \in \mathcal{O}_K$ we have $\alpha \equiv \sigma(\alpha) \bmod \mathfrak{p}$.*

(In other words, the inertia group of $\mathfrak{p}$ over $\mathbb{Q}$ is G.)

Proof. Since $\mathcal{O}_K = \mathbb{Z}[\zeta]$, it suffices to verify the statement for $\alpha = \zeta$, in which case it is obvious. $\qquad\qquad\qquad\qquad\qquad\qquad\qquad\qquad\qquad\qquad\qquad\qquad\qquad\qquad$ $\square$

4.3 Decomposition of Primes

Let q be a prime number. We are interested in its decomposition in the pth cyclotomic field K. So far, this has been done only for $q = p$, in Proposition 4.1(2).

Proposition 4.8. *Let q be a prime number distinct from p, the case $q = 2$ being included. Then q is unramified in K. Further, let f be the order of q in the multiplicative group $(\mathbb{Z}/p\mathbb{Z})^\times$ and $s = (p-1)/f$. Then $q = \mathfrak{q}_1 \cdots \mathfrak{q}_s$, the degree of each of the prime ideals $\mathfrak{q}_i$ being f.*

In particular, q splits completely in K if and only if $q \equiv 1 \bmod p$.

Proof. As we have already seen in Proposition 4.4:1, the prime q is unramified in K. Further, let $\mathfrak{q}$ be a prime ideal above q and $\varphi = \varphi_{\mathfrak{q}} \in \mathrm{Gal}(K/\mathbb{Q})$ its Frobenius element (see Appendix A.7). Then the degree of $\mathfrak{q}$ is equal to the order of φ.

By the definition of the Frobenius element, $\varphi(\alpha) \equiv \alpha^q \bmod \mathfrak{q}$ for any $\alpha \in \mathcal{O}_K$. In particular,

$$\varphi(\zeta) \equiv \zeta^q \bmod \mathfrak{q}. \qquad\qquad (4.3)$$

However, $\varphi(\zeta)$ is also a pth root of unity. If $\varphi(\zeta) \ne \zeta^q$ then $(\varphi(\zeta) - \zeta^q) = \mathfrak{p}$, and (4.3) is impossible. Hence $\varphi(\zeta) = \zeta^q$ and $\varphi = \sigma_q$. Since $\sigma_q^m = \sigma_{q^m}$, the order of σ_q is equal to the order of q in $\mathbb{Z}/p\mathbb{Z}$, that is, to f. The proposition is proved. $\quad\square$

4.4 Units

We continue to denote by ζ a primitive pth root of unity, and we let $K = \mathbb{Q}(\zeta)$ be the pth cyclotomic field. We denote by $\mathcal{O} = \mathcal{O}_K$ its ring of integers.

The Dirichlet unit theorem implies that the group $\mathcal{U} = \mathcal{U}_K = \mathcal{O}_K^\times$ of units of K is the direct product of the torsion subgroup $\Omega = \Omega_K$ (which is generated by $-\zeta$ by Proposition 4.4(4)) and a free abelian group of rank $r = (p-3)/2$. It is quite remarkable that the latter can be chosen canonically.

Theorem 4.9. *The group $\mathcal{U}$ of units of K is the direct product of the group Ω of roots of unity and the group $\mathcal{U}_+$ of positive real units of K.*

In other words, each unit can be uniquely presented as the product of a root of unity and a positive unit.

Another equivalent statement: if we view a unit as a complex number, then its argument is a multiple of π/p.

First of all, we show that the argument cannot be equal to $\pm\pi/2$. More precisely, we prove the following.

Lemma 4.10. *If $\theta \in \mathcal{O}$ is purely imaginary, then $\theta \in \mathfrak{p}$ (and, in particular, θ is not a unit).*

Proof. We have $\bar{\theta} = -\theta$ by the assumption and $\bar{\theta} \equiv \theta \bmod \mathfrak{p}$ by Corollary 4.7. It follows that $2\theta \equiv 0 \bmod \mathfrak{p}$, which completes the proof, since $\mathfrak{p}$ is coprime with 2.
□

We shall also need the following lemma.

Lemma 4.11. *Let η be a unit of the cyclotomic field K. Then $\eta/\bar{\eta}$ is a root of unity.*

Proof. Put $\xi = \eta/\bar{\eta}$. Since the automorphisms of K commute with the complex conjugation, we have $\sigma(\xi) = \sigma(\eta)/\overline{\sigma(\eta)}$ for any $\sigma \in G$. Hence $|\sigma(\xi)| = 1$ for any $\sigma \in G$, and Kronecker's theorem (see Appendix A.1) implies that ξ is a root of unity.
□

Proof of Theorem 4.9. It suffices to show that any unit is a product of a root of unity and a real unit. Fix $\eta \in \mathcal{U}$. Lemma 4.11 implies that $\xi = \eta/\bar{\eta}$ is a root of unity. Write $\xi = \pm\zeta^k$, where we may assume that k is even (replacing it, if necessary, by $k + p$). Put $\theta = \zeta^{-k/2}\eta$. If $\xi = -\zeta^k$ then $\theta/\bar{\theta} = -1$, which is impossible by Lemma 4.10. Hence $\xi = \zeta^k$ and $\theta/\bar{\theta} = 1$, which completes the proof.
□

4.5   The Real Cyclotomic Field and the Class Group

The pth cyclotomic field $K = \mathbb{Q}(\zeta)$ has a totally real subfield

$$K^+ = K \cap \mathbb{R} = \mathbb{Q}(\zeta + \zeta^{-1}),$$

called the *pth real cyclotomic field*. Obviously,

$$[K : K^+] = 2, \qquad [K^+ : \mathbb{Q}] = \frac{p-1}{2}.$$

One expects the arithmetic of K^+ to have very much in common with that of K. Indeed, p is the only prime number ramified in K^+; moreover, since it totally ramifies in K, it totally ramifies in K^+ as well: we have $(p) = \wp^{(p-1)/2}$, where $\wp = \left((\zeta - \zeta^{-1})^2\right)$. In K, we have $\wp = \mathfrak{p}^2$.

As for the other primes, we have the following analogue of Proposition 4.8.

Proposition 4.12. *Let q be a prime number distinct from p, the case $q = 2$ being included. Let f be the order of q in the multiplicative group $(\mathbb{Z}/p\mathbb{Z})^\times$. Put*

$$f^+ = \begin{cases} f & \text{if } f \text{ is odd,} \\ f/2 & \text{if } f \text{ is even.} \end{cases}$$

Then q decomposes in K^+ into a product of $(p-1)/(2f^+)$ prime ideals, each of degree f^+.

Proof. We use the notation

$$G = \mathrm{Gal}(K/\mathbb{Q}), \qquad G^+ = \mathrm{Gal}(K^+/\mathbb{Q}).$$

Since $\mathrm{Gal}(K/K^+) = \langle \iota \rangle$ (where ι, as usual, denotes the complex conjugation), we have $G^+ = G/\langle \iota \rangle$.

Fix a prime ideal $\mathfrak{q}$ of K above q, and let $\mathfrak{q}^+$ be the prime ideal of K^+ below $\mathfrak{q}$. Further, let $\varphi \in G$ and $\varphi^+ \in G^+$ be the Frobenius elements of $\mathfrak{q}$ and of $\mathfrak{q}^+$, respectively. As we have seen in the proof of Proposition 4.8, the order of φ is f. Hence the order of φ^+ is $f/2$ if $\iota \in \langle \varphi \rangle$, and the order of φ^+ is f if $\iota \notin \langle \varphi \rangle$.

However, ι is the only element of G of order 2. It follows that $\iota \in \langle \varphi \rangle$ if and only if f is even. Thus, the order of φ^+ is f^+, whence the result. $\qquad \square$

The discriminant and the regulator of K^+ are closely related to the corresponding quantities of K.

Proposition 4.13. *The discriminant of K^+ is $\pm p^{(p-3)/2}$.*

Proof. We have $K = K^+(\zeta)$, and

$$f(x) = x^2 - (\zeta + \zeta^{-1})x + 1$$

is a minimal polynomial of ζ over K^+. It follows that the different ∂_{K/K^+} divides $f'(\zeta) = \zeta - \zeta^{-1}$. Since $(\zeta - \zeta^{-1}) = \mathfrak{p}$, there are two options: $\partial_{K/K^+} = (1)$ or $\partial_{K/K^+} = \mathfrak{p}$. But $\mathfrak{p}$ is ramified over K^+, which means that $\partial_{K/K^+} = \mathfrak{p}$.

Thus, $\partial_{K/\mathbb{Q}} = \mathfrak{p}\partial_{K^+/\mathbb{Q}}$. Taking the norm, we obtain $|\mathcal{D}_K| = p|\mathcal{D}_{K^+}|^2$. Since $|\mathcal{D}_K| = p^{p-2}$, this implies that $|\mathcal{D}_{K^+}| = p^{(p-3)/2}$. $\qquad \square$

Proposition 4.14. *The regulator of K^+ is $2^{-(p-3)/2}$ times the regulator of K.*

Proof. Let $\eta_1, \ldots, \eta_r$ (where $r = (p-3)/2$) be a system of fundamental units of K^+ and $\sigma_1, \ldots, \sigma_{r+1}$ be the embeddings of K^+. Since K^+ is totally real, its regulator is $\mathcal{R}^+ = \left| \det \left[\log |\sigma_i(\eta_j)| \right]_{1 \le i,j \le r} \right|$.

Theorem 4.9 implies that $\eta_1, \ldots, \eta_r$ is a system of fundamental units of K as well. Since K is totally imaginary, its regulator is

$$\mathcal{R} = \left| \det \left[2 \log |\sigma_i(\eta_j)| \right]_{1 \le i,j \le r} \right| = 2^r \mathcal{R}^+.$$

This proves the proposition. $\qquad \square$

Much more interesting and nontrivial is the relation between the class numbers and class groups of the two fields. Denote by $\mathcal{H}$ and $\mathcal{H}^+$ the class groups of K and K^+, respectively. Since every ideal of K^+ defines an ideal of K, we have a natural map $\mathcal{H}^+ \to \mathcal{H}$.

Theorem 4.15. *The natural map $\mathcal{H}^+ \to \mathcal{H}$ is injective. That is, if $\mathfrak{a}$ is an ideal of K^+ which becomes principal in K, then it is principal already in K^+.*

We need the following generalization of Lemma 4.10.

Lemma 4.16. *Let $\beta \in K$ be purely imaginary. Then $\mathrm{Ord}_{\mathfrak{p}}(\beta)$ is odd.*

Proof. Assume that β is a purely imaginary element of K of even p-adic order. Multiplying β by a suitable rational integer,[1] we may assume that $\beta \in \mathcal{O}$ and, in particular, $\mathrm{Ord}_{\mathfrak{p}}(\beta) \geq 0$.

Write $\mathrm{Ord}_{\mathfrak{p}}(\beta) = 2m$ with a nonnegative integer m. Since $\mathfrak{p}^2$ is generated by the real number $\gamma = (\zeta - \zeta^{-1})^2$, the algebraic number $\theta = \beta\gamma^{-m}$ is a purely imaginary element of $\mathcal{O} \setminus \mathfrak{p}$, which contradicts Lemma 4.10. $\square$

Proof of Theorem 4.15. The proof is analogous to that of Theorem 4.9. Let $\mathfrak{a}$ be an ideal of K^+ which becomes principal in K. Let α be its generator in K. Since $\mathfrak{a}$ comes from K^+, we have $\bar{\mathfrak{a}} = \mathfrak{a}$. Hence $\bar{\alpha}$ generates the same ideal. It follows that $\alpha/\bar{\alpha}$ is a unit, and, as in the proof of Theorem 4.9, we find that it is a root of unity.

Write $\alpha/\bar{\alpha} = \pm\zeta^k$ with an even k and put $\beta = \zeta^{-k/2}\alpha$. Then $\mathfrak{a} = (\beta)$. Notice that $\mathrm{Ord}_{\mathfrak{p}}\beta = \mathrm{Ord}_{\mathfrak{p}}\mathfrak{a}$ is even, because $\mathfrak{a}$ comes from K^+. Now, if $\alpha/\bar{\alpha} = -\zeta^k$ then $\beta/\bar{\beta} = -1$, which is impossible by Lemma 4.16. Hence $\alpha/\bar{\alpha} = \zeta^k$ and $\beta \in K^+$, which completes the proof. $\square$

We identify $\mathcal{H}^+$ with its image in $\mathcal{H}$ and put $\mathcal{H}^- = \mathcal{H}/\mathcal{H}^+$. One usually calls $\mathcal{H}^+$ the *real class group*, and $\mathcal{H}^-$ the *relative class group*. Their cardinalities are called the *real class number*, and the *relative class number*, and are denoted, respectively, by h^+ and h^- (or h_p^+ and h_p^-, if we want to indicate the dependence in p).

There is also the "norm homomorphism" $\mathcal{H} \to \mathcal{H}^+$, defined by $\mathfrak{a} \mapsto \mathfrak{a}\bar{\mathfrak{a}}$. One can show that this homomorphism is surjective. We do not use it in the present book.

4.6 Cyclotomic Extensions of Number Fields

We wish to extend some of the results of the previous sections to cyclotomic extensions of number fields. Thus, in this section L is a number field, and we study the cyclotomic extension $L(\zeta)$, where $\zeta = \zeta_p$ is a primitive pth root of unity.

Of course, in general one cannot have the equality $[L(\zeta):L] = p - 1$. However, this equality does hold if p is unramified in L.

---

[1]Since the p-adic order of a rational integer is divisible by $p - 1$, multiplication by a rational integer does not change the parity of the p-adic order.

Proposition 4.17. *Assume that p is unramified in L, and let $\wp$ be a prime ideal of L above p. Further, let $\mathfrak{P}$ be the ideal of $L(\zeta)$ defined by $\mathfrak{P} = (\wp, 1 - \zeta)$. Then $\mathfrak{P}^{p-1} = \wp$, the ideal $\mathfrak{P}$ is prime, and $[L(\zeta):L] = p - 1$.*

Proof. Since $(1 - \zeta)^{p-1} = (p)$, we have

$$\mathfrak{P}^{p-1} \mid \left(\wp^{p-1}, (1-\zeta)^{p-1}\right) = \left(\wp^{p-1}, p\right).$$

Since p is unramified in L, we have $\left(\wp^{p-1}, p\right) = \wp$. Thus, $\mathfrak{P}^{p-1} \mid \wp$. However, $[L(\zeta):L] \le p - 1$, and $\wp$ cannot decompose in $L(\zeta)$ into more than $[L(\zeta):L]$ prime ideals. Hence $\mathfrak{P}^{p-1} = \wp$, the ideal $\mathfrak{P}$ is prime, and $[L(\zeta):L] = p - 1$. $\qquad\square$

(The assumption "p is unramified in L" can be relaxed: it suffices to have $\wp$ unramified over $\mathbb{Q}$.)

As in Sect. 4.1, we find the Galois group of $L(\zeta)/L$.

Corollary 4.18. *Assume that p is unramified in L. Then*

$$\text{Gal}(L(\zeta)/L) = \{\sigma_1, \ldots, \sigma_{p-1}\},$$

where σ_a is defined by $\sigma_a(\zeta) = \zeta^a$.

We also show that the inertia group of $\mathfrak{P}$ is the whole Galois group (an extension of Corollary 4.7).

Proposition 4.19. *In the setup of Proposition 4.17 denote by $\mathcal{O}_{\mathfrak{P}}$ the local ring of $\mathfrak{P}$ in $L(\zeta)$. Then for any $\alpha \in \mathcal{O}_{\mathfrak{P}}$ and $\sigma \in \text{Gal}(L(\zeta)/L)$ we have $\sigma(\alpha) \equiv \alpha \bmod \mathfrak{P}$.*

Proof. Denote by $\mathcal{O}_{\wp}$ the local ring of $\wp$ in L. Arguing as in the proof of Theorem 4.6, we show that $\mathcal{O}_{\mathfrak{P}} = \mathcal{O}_{\wp}(\zeta)$. Hence it suffices to show that $\sigma(\zeta) \equiv \zeta \bmod \mathfrak{P}$, which is obvious. $\qquad\square$

Finally, we determine the prime decomposition of p in $L(\zeta)$. The following is a straightforward consequence of Proposition 4.17.

Proposition 4.20. *Assume that p is unramified in L, and write the prime decomposition of p in L as $(p) = \wp_1 \cdots \wp_s$. Then p decomposes in $L(\zeta)$ as $(p) = \mathfrak{P}_1^{p-1} \cdots \mathfrak{P}_s^{p-1}$, where $\mathfrak{P}_i = (\wp_i, 1 - \zeta)$.*

4.7 General Cyclotomic Fields

In this section we extend some of the previous results to the general cyclotomic field $K_m = \mathbb{Q}(\zeta_m)$, where m is a positive integer and ζ_m is a primitive mth root of unity. The results of this section will be used only at a few isolated points.

Contrary to the rest of the chapter, in this section, p stands for *any* prime number, including $p = 2$, not just for an odd prime.

4.7.1 Ramified Primes

To begin with, we determine the ramified primes of the mth cyclotomic field.

Theorem 4.21. *Let m be a positive integer, and let $K_m = \mathbb{Q}(\zeta_m)$ be the mth cyclotomic field. Then an odd prime number p is ramified in K_m if and only if p divides m, and 2 is ramified in K_m if and only if $4 \mid m$.*

Proof. The "if" statement is obvious. If an odd p divides m, then K_m contains K_p, where p is ramified. Hence p is ramified in K_m. Similarly, if $4 \mid m$ then K_m contains $K_4 = \mathbb{Q}(i)$, where 2 is ramified.

For the "only if" statement we need a simple lemma.

Lemma 4.22. *Let ζ and ζ' be two distinct mth roots of unity (not necessarily primitive). Then $\zeta - \zeta'$ divides m.*

Proof. It suffices to show the following: if $\zeta \neq 1$ is an mth root of unity then $1 - \zeta$ divides m. Substituting $t = 1$ into the identity

$$1 + t + \cdots + t^{m-1} = (t - \zeta_m)(t - \zeta_m^2) \cdots (t - \zeta_m^{m-1}),$$

we obtain

$$m = (1 - \zeta_m)(1 - \zeta_m^2) \cdots (1 - \zeta_m^{m-1}).$$

Since $1 - \zeta$ is one of the factors on the right, the lemma follows. □

We return to the proof of Theorem 4.21. We argue as in the beginning of Sect. 4.2. Since $\mathcal{O}_{K_m} \subset \mathbb{Z}[\zeta_m]$, the discriminant of K_m divides the discriminant of the ring $\mathbb{Z}[\zeta_m]$. The latter discriminant is the product of several terms of the form $\zeta - \zeta'$, where ζ and ζ' are distinct mth roots of unity. By Lemma 4.22, the discriminant of $\mathbb{Z}[\zeta_m]$ divides a power of m. Hence the discriminant of K_m divides a power of m as well. It follows that every prime ramified in K_m divides m. This proves the "only if" statement for odd primes.

We are left with $p = 2$. We have already proved that 2 is unramified in K_m when m is odd. But we have to show that 2 is ramified only when $4 \mid m$. Thus, assume that 2 divides m, but 4 does not. Then $m = 2n$, where n is odd. Then $\zeta_m = -\zeta_n$, and $K_m = K_n$. Since n is odd, the prime 2 is unramified in K_n. The theorem is proved. □

One can show that $\mathcal{O}_{K_m} = \mathbb{Z}[\zeta_m]$ and calculate the discriminant of K_m: see [136, Theorem 2.6 and Proposition 2.7].

4.7.2 Degree and Galois Group

Next, we determine the degree and the Galois group of K_m over $\mathbb{Q}$. Since all conjugates of ζ_m lie among the primitive mth roots of unity, we have $[K_m:\mathbb{Q}] \leq \varphi(m)$, the Euler's function of m. We shall show that in fact $[K_m:\mathbb{Q}] = \varphi(m)$.

We start with the prime power case.

Proposition 4.23. *Let p be a prime number and k a positive integer. Then the principal ideal $\mathfrak{p}_k = (1 - \zeta_{p^k})$ satisfies*

$$\mathfrak{p}_k^{p^{k-1}(p-1)} = (p). \tag{4.4}$$

Proof. We use induction in k. For $k = 1$ and odd p equality (4.4) is established in Proposition 4.1(2), and for $k = 1$ and $p = 2$ it is obvious. Now assume that $k > 1$. Since $\zeta_{p^k}^p = \zeta_{p^{k-1}}$, we have

$$\left(1 - \zeta_{p^k}\right)^p \equiv 1 - \zeta_{p^{k-1}} \bmod p,$$

which implies $\mathfrak{p}_k^p = \mathfrak{p}_{k-1}$. But $\mathfrak{p}_{k-1}^{p^{k-2}(p-1)} = (p)$ by induction. This proves the proposition. $\qquad\square$

(The case $k = 2$ has already been done before; see Proposition 4.4(3).)
Since

$$[K_{p^k}:\mathbb{Q}] \leq \varphi(p^k) = p^{k-1}(p-1)$$

and since p cannot decompose in K_{p^k} into more than $[K_{p^k}:\mathbb{Q}]$ prime ideals, Proposition 4.23 implies that $[K_{p^k}:\mathbb{Q}] = \varphi(p^k)$ and that $\mathfrak{p}_k$ is a prime ideal in K_{p^k}.

Arguing as in the proof of Proposition 4.17, we obtain even more.

Proposition 4.24. *Let L be a number field, p a prime number unramified in L, and k a positive integer. Further, let $\wp$ be a prime ideal of L above p, and let $\mathfrak{P}$ be the ideal of $L(\zeta_{p^k})$ defined by $\mathfrak{P} = (\wp, 1 - \zeta_{p^k})$. Then $\mathfrak{P}^{p^{k-1}(p-1)} = \wp$, the ideal $\mathfrak{P}$ is prime, and $[L(\zeta_{p^k}):L] = p^{k-1}(p-1)$.*

Now we are ready to establish the equality $[K_m:\mathbb{Q}] = \varphi(m)$.

Theorem 4.25. *Let m be a positive integer. Then $[K_m:\mathbb{Q}] = \varphi(m)$.*

Proof. We use induction in m. For $m = 1$ the statement is obvious. If $m > 1$ then we can write $m = np^k$, where p does not divide n and $k > 0$. Since $n < m$ we have $[K_n:\mathbb{Q}] = \varphi(n)$ by induction. By Theorem 4.21, the prime p is unramified in K_n. Applying Proposition 4.24, we find $[K_n(\zeta_{p^k}):K_n] = \varphi(p^k)$. Since $K_n(\zeta_{p^k}) = K_m$, we obtain

$$[K_m:\mathbb{Q}] = [K_m:K_n] \cdot [K_n:\mathbb{Q}] = \varphi(p^k) \cdot \varphi(n) = \varphi(m),$$

as wanted. $\qquad\square$

Thus, all primitive mth roots of unity are conjugate over $\mathbb{Q}$. It follows that for every integer a, coprime with m, there exists a unique $\sigma_a \in \mathrm{Gal}(K_m/\mathbb{Q})$ such that $\sigma_a(\zeta_m) = \zeta_m^a$. Also, the map

$$(\mathbb{Z}/m\mathbb{Z})^\times \to \mathrm{Gal}(K_m/\mathbb{Q})$$
$$a^* \mapsto \sigma_a,$$

(where a^* is the image of a in $\mathbb{Z}/m\mathbb{Z}$) is well defined and gives an isomorphism between the groups $(\mathbb{Z}/m\mathbb{Z})^\times$ and $\mathrm{Gal}(K_m/\mathbb{Q})$.

We shall also use the following statement. Its proof, which is a straightforward application of the Chinese Remainder Theorem, is left to the reader.

Proposition 4.26. *Let m and n be coprime integers. Then for any a, coprime with m, and any b, coprime with n, there exists a morphism $\sigma \in \mathrm{Gal}(K_{mn}/\mathbb{Q})$ such that $\sigma(\zeta_m) = \zeta_m^a$ and $\sigma(\zeta_n) = \zeta_n^b$.*

4.7.3   Decomposition of Primes

Proposition 4.27. *Let q be a prime number not dividing m, and let f be the order of q in the multiplicative group $(\mathbb{Z}/m\mathbb{Z})^\times$. Then q splits in K_m into $\varphi(m)/f$ prime ideals, each of degree f.*

In particular, q splits completely in K_m if and only if $q \equiv 1 \bmod m$.

Proof. Exactly as of Proposition 4.8. We leave the details to the reader.   □

4.7.4   Units, Class Groups, etc.

Most of the properties of prime cyclotomic fields established in Sects. 4.4 and 4.5 have analogues in the general case. For instance, Theorem 4.15 completely extends to the general case, with almost the same proof: see [136, Theorem 4.14]. In particular, we can decompose the class number as a product $h_m^+ h_m^-$ of the real and the relative part.

On the other hand, the extension of Theorem 4.9 is not straightforward: see [136, Theorem 4.12 and Corollary 4.13]. We refer to Chaps. 3 and 4 of Washington's book [136] for more details on the general cyclotomic fields.

Chapter 5
Dirichlet L-Series and Class Number Formulas

In this chapter we use analytic tools (the Dirichlet L-series and Dedekind ζ-function) to obtain one of the most beautiful results of the nineteenth-century number theory: explicit formulas for the real and the relative class numbers of a cyclotomic field.

It must be pointed out that one does not need the full strength of these results for the solution of Catalan's problem. However, the class number formulas naturally come out in our context, and there is no reason to leave them out once all the necessary machinery is here.

We develop only the very minimum of the ζ- and L-functions theory; in particular, we do not use complex variables.

As in the previous chapter, we study in detail only the prime case. In the final section we indicate the changes to be made to extend the results to the composite case.

Since we use characters of finite abelian groups, the reader is advised to look through Appendix D.2 before studying this chapter.

There is one source of confusion in this chapter. The same letter ζ is used to denote the ζ-function and the primitive pth root of unity. Fortunately, ζ-functions appear only in Sects. 5.1 and 5.2, while ζ as a root of unity occurs starting from Sect. 5.3.

In this chapter p is an odd prime number.

5.1 Dirichlet Characters and L-Series

Let χ^* be a $\mathbb{C}$-character of the multiplicative group $(\mathbb{Z}/p\mathbb{Z})^\times$, that is, a homomorphism $\chi^* : (\mathbb{Z}/p\mathbb{Z})^\times \to \mathbb{C}^\times$. Recall that χ^* is called *trivial* if $\chi^*(x) = 1$ for all $x \in (\mathbb{Z}/p\mathbb{Z})^\times$ and nontrivial otherwise. We associate to χ^* a complex function on $\mathbb{Z}$, denoted by χ and defined as follows. If $a \in \mathbb{Z}$ is not divisible by p then we put $\chi(a) := \chi^*(a^*)$, where a^* is the image of a in $\mathbb{Z}/p\mathbb{Z}$. If a is a multiple of p

© Springer International Publishing Switzerland 2014
Y.F. Bilu et al., *The Problem of Catalan*, DOI 10.1007/978-3-319-10094-4_5

then it is common to put $\chi(a) = 0$. However, following Washington [136], we put, for a divisible by p,

$$\chi(a) = \begin{cases} 0 & \text{if } \chi^* \text{ is nontrivial,} \\ 1 & \text{if } \chi^* \text{ is trivial.} \end{cases} \tag{5.1}$$

The function χ, defined this way, is called a *Dirichlet character* mod p . The *trivial* Dirichlet character is the one corresponding to the trivial character of $(\mathbb{Z}/p\mathbb{Z})^\times$. With convention (5.1) the trivial character is identically 1 on $\mathbb{Z}$. It will be denoted by 1 (when this does not confuse).

A good example of a nontrivial Dirichlet character is the Legendre symbol $\left(\frac{x}{p}\right)$. For every p there is exactly $p - 1$ Dirichlet characters mod p, one trivial and $p - 2$ nontrivial characters.

It follows immediately from the definition that a Dirichlet character is totally multiplicative: for any $x, y \in \mathbb{Z}$ we have $\chi(xy) = \chi(x)\chi(y)$.

Now we give the basic definition of this chapter. The *L-series*, associated to a Dirichlet character χ, is

$$L(s, \chi) = \sum_{n=1}^{\infty} \chi(n) n^{-s}. \tag{5.2}$$

It converges absolutely for $s > 1$. For the trivial character the sum of the associated L-series is nothing but Riemann's ζ-function:

$$L(s, 1) = \zeta(s).$$

Multiplicativity of Dirichlet characters implies for the L-series Euler product expansions similar to that for the Riemann ζ-function.

Theorem 5.1. *Let χ be a Dirichlet character. Then for any $s > 1$ we have*

$$L(s, \chi) = \prod_{q}(1 - \chi(q)q^{-s})^{-1},$$

where the product extends to all prime numbers q.

Proof. For a positive integer N we put

$$S_N = \sum_{n=1}^{N} \chi(n) n^{-s}, \qquad T_N = \prod_{q \leq N}(1 - \chi(q)q^{-s})^{-1},$$

and denote by M_N the set of all positive integers composed from primes not exceeding N. Since χ is multiplicative, we have

$$T_N = \prod_{q \le N} \sum_{k=1}^{\infty} \chi(q^k) q^{-ks} = \sum_{n \in M_N} \chi(n) n^{-s},$$

the multiplication being justified since the convergence is absolute. Since $M_N \supseteq \{1, \ldots, N\}$, this implies the inequality

$$|T_N - S_N| \le \sum_{n=N+1}^{\infty} n^{-s}. \tag{5.3}$$

Since the series $\sum_{n=1}^{\infty} n^{-s}$ converges for $s > 1$, the right-hand side of (5.3) tends to 0 as N tends to infinity. Hence $T_N - S_N \to 0$ as $N \to \infty$. But $S_N \to L(s, \chi)$, whence the result. □

It is important that for a nontrivial character, the corresponding *L*-series converges also for $0 < s \le 1$.

Theorem 5.2. *Let χ be a nontrivial Dirichlet character. Then the L-series (5.2) converges for $s > 0$ to an infinitely differentiable function (called the L-function).*

Of course, a much stronger statement is true: the *L*-series converges for all complex s with $\operatorname{Re} s > 0$ to a function, holomorphic on the right half-plane, which extends to a function holomorphic on the entire complex plane $\mathbb{C}$ (see any manual of analytic number theory). However, Theorem 5.2 is more than sufficient for our purposes: what we need is the mere fact that the *L*-function is defined and continuous at 1.

Proof of Theorem 5.2. It suffices to prove that for every $k \ge 0$ the series $\sum_{n=1}^{\infty} \chi(n)(-\log n)^k n^{-s}$ converges to a continuous function on $(0, +\infty)$. This is a standard application of the *Abel summation formula*: if $(a_n)_{n \ge 1}$, $(b_n)_{n \ge 1}$ are two sequences and $A_n = a_1 + \cdots + a_n$ then for $m > n > 1$

$$\sum_{j=n}^{m} a_j b_j = -A_{n-1} b_n + A_m b_m + \sum_{j=n}^{m-1} A_j (b_j - b_{j+1}).$$

In particular, if for $n \ge n_0$ the terms b_n are nonnegative and nonincreasing, and $|A_n| \le A$ for all n, then for $m > n > n_0$

$$\left| \sum_{j=n}^{m} a_j b_j \right| \le 2 A b_n.$$

We apply this with $a_n = \chi(n)$ and $b_n = b_n(s) = (\log n)^k n^{-s}$. Since χ is a nontrivial character, we have $\chi(1) + \cdots + \chi(p) = 0$. Since χ is p-periodic, we obtain $|A_n| \le p$ for all n.

Further, fix $s_0 > 0$. Then there exists n_0 such that for every $s \geq s_0$ we have $b_{n'}(s) \leq b_n(s)$ when $n' \geq n \geq n_0$. It follows that for $s \geq s_0$ and $m > n > n_0$

$$\left| \sum_{j=n}^{m} a_j b_j(s) \right| \leq 2pn^{-s}(\log n)^k \leq 2pn^{-s_0}(\log n)^k.$$

By the Cauchy criterion, the series

$$\sum_{n=1}^{\infty} a_n b_n(s) = \sum_{n=1}^{\infty} \chi(n)(\log n)^k n^{-s}$$

converges uniformly on the interval $[s_0, +\infty)$, whence the result.   □

5.2 Dedekind ζ-Function of the Cyclotomic Field

The significance of L-functions for the cyclotomic theory stems from the fact that their product is the Dedekind ζ-function[1] of the cyclotomic field.

Theorem 5.3. *Let K be the pth cyclotomic field. Then for $s > 1$ we have*

$$\zeta_K(s) = \prod_{\chi} L(s, \chi), \tag{5.4}$$

where the product is over all the Dirichlet characters mod p.

The proof of this theorem relies on a simple but useful lemma.

Lemma 5.4. *Let G be a finite abelian group and $\hat{G}$ its dual group. Further, let g be an element of G of order m. Then we have the polynomial identity*

$$\prod_{\chi \in \hat{G}} (1 - \chi(g)T) = (1 - T^m)^{|G|/m}. \tag{5.5}$$

Proof. Since g is of order m, every $\chi(g)$ is an mth root of unity, and all mth roots of unity occur as $\chi(g)$ equally often, that is, $|G|/m$ times. It follows that the left-hand side of (5.5) is $\prod_{k=0}^{m-1} (1 - \xi^k T)^{|G|/m}$, where ξ is a primitive mth root of unity. Since

---

[1] The reader may consult Appendix A.9 for the definition and basic properties of the Dedekind ζ-function of a number field.

$$\prod_{k=0}^{m-1}\left(1 - \xi^k T\right) = 1 - T^m,$$

the lemma follows.   □

Proof of Theorem 5.3. We compare the Euler product expansions for both parts of (5.4). Rewrite the Euler product (A.7), grouping together the prime ideals above the same prime:

$$\zeta_K(s) = \prod_q \prod_{\mathfrak{q}|q}\left(1 - \mathcal{N}(\mathfrak{q})^{-s}\right)^{-1} \qquad (s > 1).$$

Also,

$$\prod_\chi L(s, \chi) = \prod_q \prod_\chi (1 - \chi(q)q^{-s})^{-1}.$$

Hence it suffices to prove that for any prime number q

$$\prod_{\mathfrak{q}|q}(1 - \mathcal{N}(\mathfrak{q})^{-s}) = \prod_\chi (1 - \chi(q)q^{-s}). \qquad (5.6)$$

The case $q = p$ is trivial: both sides of (5.6) are equal to $1 - p^{-s}$. Assume now that $q \neq p$, and denote by f the order of q in the multiplicative group $(\mathbb{Z}/p\mathbb{Z})^\times$. By Proposition 4.8 there exist exactly $(p-1)/f$ prime ideals of K above q, the norm of each of them being q^f. Hence the left-hand side of (5.6) is $(1 - q^{-fs})^{(p-1)/f}$. Applying Lemma 5.4 with $G = (\mathbb{Z}/p\mathbb{Z})^\times$, we find that this is equal to the right-hand side of (5.6). The theorem is proved.   □

We want to adapt the residue formula (A.8) to our situation. For the pth cyclotomic field we have

$$t_1 = 0, \qquad t_2 = (p-1)/2, \qquad \omega = 2p, \qquad |\mathcal{D}_K| = p^{p-2}.$$

(see Proposition 4.4(4) and Theorem 4.6). We obtain

$$\lim_{s\downarrow 1}(s-1)\zeta_K(s) = \frac{(2\pi)^{(p-1)/2}}{2p^{p/2}}\mathcal{R}h,$$

where $\mathcal{R}$ and h are the regulator and the class number of K.

On the other hand, recall that $L(s, 1) = \zeta(s)$, which, by (A.6), implies that

$$\lim_{s\downarrow 1}(s-1)L(s, 1) = 1.$$

Since for $\chi \neq 1$ the function $s \mapsto L(s, \chi)$ is defined and continuous at $s = 1$, we have

$$\lim_{s \downarrow 1}(s - 1)\zeta_K(s) = \prod_{\chi \neq 1} L(1, \chi).$$

We obtain the fundamental identity

$$\mathcal{R}h = 2^{(3-p)/2}\pi^{(1-p)/2}p^{p/2}\prod_{\chi \neq 1} L(1, \chi). \tag{5.7}$$

An immediate consequence is the following important result of Dirichlet.

Corollary 5.5 (Dirichlet). *For $\chi \neq 1$ we have $L(1, \chi) \neq 0$.*

This result, established by Dirichlet in the course of proof of his classical theorem about primes in arithmetical progression (see Sect. 5.2.2), plays crucial role in the theory of cyclotomic fields. In particular, it is the main ingredient in the proofs of fundamental Theorems 7.18 and 10.4.

5.2.1 The Real Cyclotomic Field

To get a closer look at identity (5.7), we need an analogue of Theorem 5.3 for the real cyclotomic field K^+. For this purpose, we make one more definition. For any Dirichlet character χ we have either $\chi(-1) = 1$ or $\chi(-1) = -1$. We say that χ is *even* in the former case and χ is *odd* in the latter case.

Theorem 5.6. *Let K be the pth cyclotomic field and K^+ its maximal real subfield. Then for $s > 1$ we have*

$$\zeta_{K^+}(s) = \prod_{\chi(-1)=1} L(s, \chi),$$

where the product is over all even Dirichlet characters mod p.

Proof. Similar to the proof of Theorem 5.3, Proposition 4.8 being replaced by Proposition 4.12, and Lemma 5.4 being used with the quotient group $(\mathbb{Z}/p\,\mathbb{Z})^\times/\{\pm 1\}$ rather than with $(\mathbb{Z}/p\,\mathbb{Z})^\times$ itself. We leave the details to the reader. $\square$

For the real cyclotomic field K^+ the number t_1 of real embeddings is equal to $(p - 1)/2$, the number t_2 of pairs of complex embeddings is 0, the number ω of roots of unity is 2, and the discriminant $\mathcal{D}_{K^+}$ is $\pm p^{(p-3)/2}$ (see Proposition 4.13). Applying the residue formula (A.8) and arguing as above, we obtain the identity

$$h^+ \mathcal{R}^+ = 2^{(3-p)/2} p^{(p-3)/4} \prod_{\substack{\chi(-1)=1 \\ \chi \neq 1}} L(1, \chi), \tag{5.8}$$

where $\mathcal{R}^+$ and h^+ are the regulator and the class number of K^+.

Now recall that $\mathcal{R}/\mathcal{R}^+ = 2^{(p-3)/2}$ (see Proposition 4.14). Dividing (5.7) by (5.8), we obtain the following formula for the relative class number $h^- = h/h^+$:

$$h^- = 2^{(3-p)/2} \pi^{(1-p)/2} p^{(p+3)/4} \prod_{\chi(-1)=-1} L(1, \chi). \tag{5.9}$$

Thus, we have split identity (5.7) into two subtler identities: (5.8) for the real class number and (5.9) for the relative class number. To make them more explicit, we have to determine $L(1, \chi)$ for the nontrivial characters χ. This will be done in the next section.

5.2.2   Addendum: The Theorem of Dirichlet[2]

In this subsection we show how the inequality $L(1, \chi) \neq 0$ implies the classical theorem of Dirichlet about primes in arithmetical progressions.

Theorem 5.7 (Dirichlet). *Let m be a positive integer and let a be an integer coprime with m. Then there exist infinitely many prime numbers q satisfying $q \equiv a \bmod m$. Moreover, these primes form a regular set of Dirichlet density $1/\varphi(m)$. (See Appendix A.10 for the definition of Dirichlet density.)*

Since we defined Dirichlet characters and L-functions only in the special case of prime modulus $m = p$, we shall sketch the proof of Theorem 5.7 only for this case. The general proof requires only cosmetic changes; see Sect. 5.5.

Using the Euler product and the residue formula for the Riemann ζ-function, one finds that for $s > 1$

$$\sum_q q^{-s} = \log \frac{1}{s-1} + O(1),$$

where the sum runs over all the prime numbers q (see Appendix A.10 for the details). In a similar way, if χ is a nontrivial Dirichlet character $\bmod\ p$, then for $s > 1$ we have

$$\sum_q \chi(q) q^{-s} = \log L(s, \chi) + O(1) = O(1),$$

because $L(1, \chi) \neq 0$.

---

[2]This subsection will not be used in the sequel.

Summing up over all Dirichlet characters χ modulo p, and using the identity

$$\sum_\chi \chi(x) = \begin{cases} p-1, & \text{if } x \equiv 1 \bmod p, \\ 0, & \text{otherwise} \end{cases}$$

we obtain

$$(p-1) \sum_{q \equiv 1 \bmod p} q^{-s} = \log \frac{1}{s-1} + O(1)$$

for $s > 1$. Hence the set of prime numbers q satisfying $q \equiv 1 \bmod p$ is regular, of Dirichlet density $1/(p-1)$.

For primes $q \equiv a \bmod p$ the argument is the same, but instead of the sum $\sum_\chi \log L(s, \chi)$ one uses $\sum_\chi \overline{\chi(a)} \log L(s, \chi)$.

Let q be a prime ideal above q. Since the Frobenius element of q depends only on the residue class of q modulo p (see Sect. 4.3), the theorem of Dirichlet is a very special case of Chebotarev's density theorem (see Appendix A.7).

5.3 Calculating $L(1, \chi)$ for $\chi \neq 1$

In this section log and arg stand for the **principal branches** of the complex logarithm and argument. That is, for any nonzero complex z, we have

$$-\pi < \arg z = \operatorname{Im} \log z \leq \pi.$$

5.3.1 The Space of p-Periodic Functions

We start from afar. The set V of p-periodic functions $f : \mathbb{Z} \to \mathbb{C}$ (that is, functions satisfying $f(x + p) = f(x)$ for all $x \in \mathbb{Z}$) is a p-dimensional $\mathbb{C}$-vector space. Further, there is a natural inner product on V, defined by

$$(f, g) = \frac{1}{p} \sum_{k=0}^{p-1} f(k)\overline{g(k)}. \tag{5.10}$$

Fix a primitive complex pth root of unity ζ, and define $\psi : \mathbb{Z} \to \mathbb{C}$ by $\psi(x) = \zeta^x$. One immediately verifies that for $a, b \in \mathbb{Z}$

$$(\psi^a, \psi^b) = \begin{cases} 0 & \text{if } a \not\equiv b \bmod p, \\ 1 & \text{if } a \equiv b \bmod p. \end{cases}$$

In particular, the functions $1, \psi, \ldots, \psi^{p-1}$ form an orthogonal basis of V. It follows that for every $f \in V$ we have

$$f = (f, 1) + (f, \psi)\psi + \cdots + (f, \psi^{p-1})\psi^{p-1}.$$

For a nontrivial Dirichlet character χ we have $\chi(0) = 0$. Hence

$$(\chi, 1) = \chi(1) + \cdots + \chi(p - 1) = 0.$$

It follows that

$$\chi = (\chi, \psi)\psi + \cdots + (\chi, \psi^{p-1})\psi^{p-1}. \tag{5.11}$$

5.3.2 The General Formula for $L(1, \chi)$

We wish to calculate the infinite sum

$$L(1, \chi) = \sum_{n=1}^{\infty} \frac{\chi(n)}{n}. \tag{5.12}$$

Using (5.11), one can reduce this task to calculating the sums $\sum_{n=1}^{\infty} \psi(n)^a / n$ for $a = 1, \ldots, p - 1$.

The latter sums can be determined easily. Indeed, for $z \in \mathbb{C}$ satisfying $|z| \leq 1$ and $z \neq 1$, we have

$$\sum_{n=1}^{\infty} \frac{z^n}{n} = -\log(1 - z),$$

(recall that log stands for the principal branch of the complex logarithm). Hence for a not divisible by p we have

$$\sum_{n=1}^{\infty} \frac{\psi(n)^a}{n} = \sum_{n=1}^{\infty} \frac{\zeta^{an}}{n} = -\log(1 - \zeta^a).$$

Combining this with (5.11), we obtain the identity

$$L(1, \chi) = -\sum_{a=1}^{p-1} (\chi, \psi^a) \log(1 - \zeta^a). \tag{5.13}$$

Thus, we found a finite expression for the infinite sum (5.12). Still, identity (5.13), as it stands, is not very informative. Below we obtain much more explicit formulas for the absolute values $|L(1, \chi)|$, which will be sufficient for our purposes.

Remark 5.8. In the argument above, one must justify the change of order of summation in

$$L(1, \chi) = \sum_{n=1}^{\infty} \frac{1}{n} \sum_{a=1}^{p-1} (\chi, \psi^a) \psi(n)^a = \sum_{a=1}^{p-1} (\chi, \psi^a) \sum_{n=1}^{\infty} \frac{\psi(n)^a}{n},$$

because convergence here is not absolute. Probably, the easiest way to do this is by observing that for $s > 1$ we have

$$L(s, \chi) = \sum_{n=1}^{\infty} \frac{1}{n^s} \sum_{a=1}^{p-1} (\chi, \psi^a) \psi(n)^a = \sum_{a=1}^{p-1} (\chi, \psi^a) \sum_{n=1}^{\infty} \frac{\psi(n)^a}{n^s}$$

(now the convergence is absolute!) and taking limits[3] as $s \downarrow 1$ on both sides.

5.3.3   The Fourier Coefficients

First of all, we have to understand the "Fourier coefficients" (χ, ψ^a). For them we have the following statement.

Proposition 5.9. *Let χ be a nontrivial Dirichlet character and a an integer not divisible by p. Then $(\chi, \psi^a) = \overline{\chi(a)}(\chi, \psi)$ and $|(\chi, \psi^a)| = p^{-1/2}$.*

Proof. In the definition (5.10) of the inner product we can replace the set $\{0, 1, \ldots, p - 1\}$ by any complete system T of residues $\mod p$:

$$(f, g) = \frac{1}{p} \sum_{k \in T} f(k) \overline{g(k)}.$$

In particular, since a is not a multiple of p, we may take

$$T = \{0, a, 2a, \ldots, (p - 1)a\}.$$

---

[3] The series

$$\sum_{n=1}^{\infty} \frac{\psi(n)^a}{n^s} \qquad (a = 1, \ldots, p - 1)$$

converge for $s > 0$ to continuous (and even infinitely differentiable) functions: this can be proved in exactly the same way as Theorem 5.2.

We obtain

$$(\chi, \psi) = \frac{1}{p} \sum_{k=0}^{p-1} \chi(ak)\overline{\psi(ak)} = \frac{1}{p}\chi(a) \sum_{k=0}^{p-1} \chi(k)\overline{\psi(k)^a} = \chi(a)(\chi, \psi^a).$$

Hence

$$(\chi, \psi^a) = \chi(a)^{-1}(\chi, \psi) = \overline{\chi(a)}(\chi, \psi).$$

In particular, $|(\chi, \psi^a)| = |(\chi, \psi)|$ for every a not divisible by p. It follows that

$$(\chi, \chi) = \sum_{a=1}^{p-1} |(\chi, \psi^a)|^2 = (p-1)|(\chi, \psi)|^2.$$

Since $(\chi, \chi) = (p-1)/p$, this shows that $|(\chi, \psi)| = p^{-1/2}$. The proposition is proved. □

Remark 5.10. The expression

$$-p\,(\chi, \bar{\psi}) = -\sum_{k=0}^{p-1} \chi(k)\psi(k)$$

is called *Gauss sum*. Gauss sums play fundamental role in arithmetic; they will be studied in detail in Chap. 7.

5.3.4 *Explicit Formulas for $|L(1, \chi)|$*

We are ready to establish the main result of this section.

Theorem 5.11. *Let χ be a nontrivial Dirichlet character. Then*

$$|L(1, \chi)| = p^{-1/2}\left|\sum_{a=1}^{p-1} \chi(a) \log|1 - \zeta^a|\right| \tag{5.14}$$

if χ is even, and

$$|L(1, \chi)| = \pi p^{-3/2}\left|\sum_{a=1}^{p-1} a\chi(a)\right| \tag{5.15}$$

if χ is odd.

Recall that the character χ is *even* if $\chi(-1) = 1$ and *odd* if $\chi(-1) = -1$.

Proof. By Proposition 5.9 $\overline{(\chi, \psi^a)} = \chi(a)\overline{(\chi, \psi)}$. Hence

$$\overline{L(1, \chi)} = -\sum_{a=1}^{p-1} \overline{(\chi, \psi^a)} \, \overline{\log(1 - \zeta^a)}$$

$$= -\overline{(\chi, \psi)} \sum_{a=1}^{p-1} \chi(a) \overline{\log(1 - \zeta^a)}$$

$$= -\overline{(\chi, \psi)} \sum_{a=1}^{p-1} \chi(a) \log |1 - \zeta^a| + i\overline{(\chi, \psi)} \sum_{a=1}^{p-1} \chi(a) \arg(1 - \zeta^a).$$

$$(5.16)$$

Now observe that for any a we have $1 - \zeta^{p-a} = \overline{1 - \zeta^a}$. Therefore

$$|1 - \zeta^{p-a}| = |1 - \zeta^a|, \qquad \arg(1 - \zeta^{p-a}) = -\arg(1 - \zeta^a).$$

Also, $\chi(a) = \chi(p - a)$ for an even character χ, and $\chi(a) = -\chi(p - a)$ for an odd χ.

It follows that, for an even character χ, the second sum in (5.16) vanishes:

$$\sum_{a=1}^{p-1} \chi(a) \arg(1 - \zeta^a) = \sum_{a=1}^{(p-1)/2} (\chi(a) - \chi(p - a)) \arg(1 - \zeta^a) = 0.$$

Since $|(\chi, \psi)| = p^{-1/2}$, this proves (5.14).

Similarly, for an odd χ, the first sum in (5.16) vanishes, and we obtain

$$|L(1, \chi)| = p^{-1/2} \left| \sum_{a=1}^{p-1} \chi(a) \arg(1 - \zeta^a) \right|$$

$$= p^{-1/2} \left| \sum_{a=1}^{(p-1)/2} 2\chi(a) \arg(1 - \zeta^a) \right|.$$

$$(5.17)$$

Now specify $\zeta = e^{2\pi i/p}$. Observe that $\arg(1 + z) = \frac{1}{2} \arg z$ for any $z \in \mathbb{C}$ with $|z| = 1$ and $z \neq -1$. Hence for $1 \leq a \leq (p - 1)/2$ we have

$$\arg(1 - \zeta^a) = \frac{1}{2} \arg(-\zeta^a) = -\frac{\pi}{2} + \frac{\pi a}{p} = \frac{\pi}{2p}(a - (p - a)).$$

It follows that

$$2\chi(a)\arg(1-\zeta^a) = \frac{\pi}{p}\big(a\chi(a) + (p-a)\chi(p-a)\big)$$

for $1 \le a \le (p-1)/2$. Substituting this into (5.17), we obtain (5.15). □

Corollary 5.12. *For an even character χ we have $\sum_{a=1}^{p-1} \chi(a)\log|1-\zeta^a| \ne 0$. For an odd character χ we have $\sum_{a=1}^{p-1} a\chi(a) \ne 0$.*

Proof. For the nontrivial characters this is a direct consequence of Theorem 5.11 and Corollary 5.5. For the trivial character we have

$$\sum_{a=1}^{p-1} \chi(a)\log|1-\zeta^a| = \sum_{a=1}^{p-1}\log|1-\zeta^a| = \log\left|\prod_{a=1}^{p-1}(1-\zeta^a)\right| = \log p \ne 0.$$

□

   Mention that, though the statement of Corollary 5.12 contains no reference to L-functions, its only known proof is by deducing it from Theorem 5.11.
   The following trivial observation (already used in the proof of Theorem 5.11) is a useful complement to Corollary 5.12.

Proposition 5.13. *We have $\sum_{a=1}^{p-1} \chi(a)\log|1-\zeta^a| = 0$ for an odd character χ and $\sum_{a=1}^{p-1} a\chi(a) = 0$ for a nontrivial even character χ.*

5.4   Class Number Formulas

We are ready to give the promised formulas for the real and relative class numbers.

Theorem 5.14 (Kummer). *Let p be an odd prime number, and let h_p^+ and h_p^- be the pth real and relative class numbers. Also, let $\mathcal{R}_p^+$ be the regulator of the pth real cyclotomic field. Then*

$$h_p^+ \mathcal{R}_p^+ = 2^{(3-p)/2}\prod_{\substack{\chi(-1)=1 \\ \chi\ne 1}}\sum_{a=1}^{p-1}\chi(a)\log|1-\zeta^a|, \qquad (5.18)$$

$$h_p^- = (2p)^{(3-p)/2}\prod_{\chi(-1)=-1}\sum_{a=1}^{p-1}a\chi(a), \qquad (5.19)$$

where the product in (5.18) extends to nontrivial even Dirichlet characters mod p and the product in (5.19) extends to odd Dirichlet characters mod p.

Table 5.1 The relative class number h_p^- for $p \leq 41$

p	h_p^-	p	h_p^-	p	h_p^-	p	h_p^-
3	1	11	1	19	1	31	9
5	1	13	1	23	3	37	37
7	1	17	1	29	8	41	121

Proof. Combining Theorem 5.11 with identities (5.8) and (5.9), we obtain

$$
h_p^+ \mathcal{R}_p^+ = 2^{(3-p)/2} \left| \prod_{\substack{\chi(-1)=1 \\ \chi \neq 1}} \sum_{a=1}^{p-1} \chi(a) \log |1 - \zeta^a| \right|,
$$

$$
h_p^- = (2p)^{(3-p)/2} \left| \prod_{\chi(-1)=-1} \sum_{a=1}^{p-1} a\chi(a) \right|.
$$

On the right-hand sides of (5.18) and (5.19) every character appears together with its complex conjugate. It follows that both products are nonnegative real numbers, and the absolute value delimiters can be omitted. The theorem is proved. □

Equality (5.19) can be used for computing the relative class numbers. Kummer himself determined the pth relative class number h_p^- for $p < 100$. Table 5.1 gives the values of h_p^- for $p \leq 41$, which is sufficient for the present book. See [136, pp. 412–420] for extensive tables of relative class numbers and further references.

Unfortunately, identity (5.18) cannot be used to determine the real class number, because of the $\mathcal{R}_p^+$-factor. In fact, computing h_p^+ is a rather difficult task. At present, the following is known.

Theorem 5.15. *For all primes $p \leq 67$ we have $h_p^+ = 1$, and, thereby, $h_p = h_p^-$. In particular, in Table 5.1 one may replace h_p^- by h_p.*

This result is due to Masley [77]. The proof is rather involved and heavily relies on Odlyzko's famous lower bounds for discriminants [112, 113]. Van der Linden [72] extended Masley work; see [136, p. 421] for comments and further bibliography.

The proof of Catalan's conjecture does not use Theorem 5.15.

5.5   Composite Moduli

In this section we briefly explain how the results of this chapter extend to arbitrary cyclotomic fields. The material of this section will not be used in the sequel.

Let m be a positive integer. As in the prime case, we start from a character χ^* of the multiplicative group $(\mathbb{Z}/m\mathbb{Z})^\times$ and define the corresponding Dirichlet character $\chi : \mathbb{Z} \to \mathbb{C}$. For an integer a coprime with m we put $\chi(a) = \chi^*(a^*)$, where a^* is the image of a in $(\mathbb{Z}/m\mathbb{Z})^\times$. However, it is not clear how one should define $\chi(a)$ when a is not coprime with m. The "naive" definition "$\chi(a) = 0$ when $\gcd(a, m) > 1$" is not very convenient.

Let n be a divisor of m. The natural homomorphism $(\mathbb{Z}/m\mathbb{Z})^\times \to (\mathbb{Z}/n\mathbb{Z})^\times$ implies that every character of $(\mathbb{Z}/n\mathbb{Z})^\times$ lifts to a character of $(\mathbb{Z}/m\mathbb{Z})^\times$. A character of the multiplicative group $(\mathbb{Z}/m\mathbb{Z})^\times$ is called *primitive* if it is not a lifting of a character of $(\mathbb{Z}/n\mathbb{Z})^\times$, where $n \mid m$ and $n \neq m$.

In general, for every character of $(\mathbb{Z}/m\mathbb{Z})^\times$, there exists the smallest integer n, dividing m, such that our character is a lifting of a character of $(\mathbb{Z}/n\mathbb{Z})^\times$ (which is, obviously, a primitive character of $(\mathbb{Z}/n\mathbb{Z})^\times$). This n is called *the conductor* of our character. The conductor of a primitive character is m; the conductor of the trivial character is 1.

Now let χ^* be a character of $(\mathbb{Z}/m\mathbb{Z})^\times$. If χ^* is primitive then we define a function χ on $\mathbb{Z}$ by $\chi(a) = \chi^*(a^*)$ if $\gcd(a, m) = 1$ and $\chi(a) = 0$ if $\gcd(a, m) > 1$. The function χ, defined in this way, is called a *primitive Dirichlet character of conductor m* (associated to χ^*).

In general, let n be the conductor of χ^*. Then χ^* is the lifting of certain primitive character of $(\mathbb{Z}/n\mathbb{Z})^\times$, and we let χ be the corresponding primitive Dirichlet character of conductor n. With this definition of χ we have $\chi(a) = \chi^*(a^*)$ for a coprime with m.

A primitive Dirichlet character of conductor dividing m is called a *primitive Dirichlet character* $\bmod\, m$. Thus, we described above a one-to-one correspondence between the characters of $(\mathbb{Z}/m\mathbb{Z})^\times$ and the primitive Dirichlet characters $\bmod\, m$. In particular, there exist exactly $\varphi(m)$ primitive Dirichlet characters $\bmod\, m$.

To every Dirichlet character χ we associate the L-series as in (5.2). Using Proposition 4.27, it is not difficult to show that

$$\zeta_{K_m}(s) = \prod_\chi L(s, \chi), \qquad \zeta_{K_m^+}(s) = \prod_{\chi(-1)=1} L(s, \chi),$$

where K_m is the mth cyclotomic field, K_m^+ is its maximal totally real subfield, and χ runs through the primitive Dirichlet characters $\bmod\, m$ in the first product and through the even primitive Dirichlet characters $\bmod\, m$ in the second product. As in the prime case, we obtain the nonvanishing property $L(1, \chi) \neq 0$ for $\chi \neq 1$ and deduce from it the theorem of Dirichlet modulo m (see Sect. 5.2.2).

Now, arguing as in Sects. 5.3 and 5.4, we obtain formulas for the real and relative class numbers h_m^+ and h_m^- similar to (5.18) and (5.19). See Chap. 4 of [136] for the details.

Chapter 6
Higher Divisibility Theorems

Starting from this chapter, we shall systematically apply the theory of cyclotomic fields to Catalan's problem.

In this chapter we drastically refine Cassels' divisibility theorem: we show that $p^2 \mid y$ (and $q^2 \mid x$ by symmetry). First we do this under an additional restriction (Theorem 6.2) and then unconditionally (Theorem 6.14).

6.1 The Most Important Lemma

Let (x, y, p, q) be a solution of Catalan's equation (as defined in the beginning of Chap. 3). As above, we denote by ζ a primitive pth root of unity and put $K = \mathbb{Q}(\zeta)$.

The theory of cyclotomic fields applies to Catalan's equation through the following statement.

Lemma 6.1 (The "most important lemma"). *The number*

$$\lambda = \frac{x - \zeta}{1 - \zeta}$$

is an algebraic integer, and the principal ideal (λ) is a qth power of an ideal of the cyclotomic field K.

Proof. Recall (see Proposition 4.1) that $\mathfrak{p} = (1 - \zeta)$ is a prime ideal of K and that $(p) = \mathfrak{p}^{p-1}$. Since $p \mid (x - 1)$ by (3.4), the prime ideal $\mathfrak{p}$ divides $x - \zeta$, but $\mathfrak{p}^2$ does not. Hence λ is an algebraic integer, not divisible by $\mathfrak{p}$. The same is true for $\lambda_k = (x - \zeta^k)/(1 - \zeta^k)$, where $k = 1, \ldots, p - 1$. The identity

$$(1 - \zeta^k)\lambda_k - (1 - \zeta^m)\lambda_m = \zeta^m - \zeta^k$$

© Springer International Publishing Switzerland 2014
Y.F. Bilu et al., *The Problem of Catalan*, DOI 10.1007/978-3-319-10094-4_6

implies that, for distinct $k, m \in \{1, \ldots, p-1\}$, the greatest common divisor of λ_k and λ_m divides $(\zeta^k - \zeta^m) = \mathfrak{p}$. Hence the numbers $\lambda_1, \ldots, \lambda_{p-1}$ are pairwise coprime.

Now let

$$\Phi_p(X) = \frac{X^p - 1}{X - 1} = (X - \zeta)(X - \zeta^2) \cdots (X - \zeta^{p-1})$$

be the pth cyclotomic polynomial. Then

$$\lambda_1 \cdots \lambda_{p-1} = \frac{\Phi_p(x)}{\Phi_p(1)} = \frac{x^p - 1}{x - 1} \cdot \frac{1}{p}.$$

The second relation in (3.4) implies now $\lambda_1 \cdots \lambda_{p-1} = u^q$. Since the factors are pairwise coprime, each principal ideal (λ_k) is a qth power of an ideal. $\qquad\square$

Numerous arguments in this book have this lemma as a starting point. The first application of Lemma 6.1, called *Inkeri's divisibility theorem*, will be given already in the next section.

6.2   Inkeri's Divisibility Theorem

If (x, y, p, q) is a solution of Catalan's equation, then $q \mid x$ and $p \mid y$, by Cassels' divisibility theorem. Inkeri [50, 51] used the "most important" Lemma 6.1 to refine Cassels' theorem under certain assumptions.

Theorem 6.2 (Inkeri's divisibility theorem).   *Let (x, y, p, q) be a solution of Catalan's equation. Assume that q does not divide the class number h_p of the cyclotomic field K_p. Then $q^2 \mid x$.*

The significance of this result is explained by the following simple observation.

Proposition 6.3.   *Let (x, y, p, q) be a solution of Catalan's equation. Then $q^2 \mid x$ if and only if $p^{q-1} \equiv 1 \bmod q^2$.*

The congruence $p^{q-1} \equiv 1 \bmod q^2$ is often called *Wieferich's condition* (this term originates to the work of Wieferich [138] and Mirimanoff [95] on the Fermat equation; see Sect. 2.3.3).

Proof. Since $q \mid x$, the first equality in (3.4) implies that $p^{q-1}a^q \equiv -1 \bmod q$. Since $p^{q-1} \equiv 1 \bmod q$, we have $a^q \equiv -1 \bmod q$.

Now recall the following elementary fact[1]: if rational integers A, B and a prime number q satisfy $A^q \equiv B^q \bmod q$, then $A^q \equiv B^q \bmod q^2$. It follows that $a^q \equiv -1 \bmod q^2$.

---

[1]This is an easy consequence of Lemma 2.8: if q divides $(A^q - B^q)$, then it divides $(A^q - B^q)/(A - B)$ as well; see also Lemma 6.7 below.

Thus, $p^{q-1} \equiv 1 \bmod q^2$ is equivalent to $p^{q-1}a^q \equiv -1 \bmod q^2$, which, again by (3.4), is equivalent to $q^2 \mid x$. The proposition is proved. $\qquad\square$

Thus, Inkeri's theorem gives an efficient necessary condition for the existence of a solution (x, y, p, q) with given p and q.

Corollary 6.4. *Let (x, y, p, q) be a solution of Catalan's equation. Then either $q \mid h_p$ or $p^{q-1} \equiv 1 \bmod q^2$.*

Inkeri himself, Mignotte [86], Schwarz [125], and others suggested various versions and refinements of Corollary 6.4. For instance, Schwarz showed that the full class number h_p can be replaced by the relative class number h_p^-; see Remark 6.6.

Mignotte and Roy [90] used (a refined version of) Corollary 6.4 together with extensive electronic computations to show that Catalan's equation has no solutions with $\min\{p, q\} < 10^5$ (with a few exceptions later treated by Bugeaud and Hanrot [14]).

Inkeri's divisibility theorem has been drastically refined by Mihăilescu, who proved that $q^2 \mid x$ and $p^2 \mid y$ unconditionally (see Theorem 6.14 in Sect. 6.5). Another partial refinement, due to Bugeaud and Hanrot [14], will be proved in Sect. 8.6.

Thus, Inkeri's theorem is formally obsolete in this book. Nevertheless, we include a proof, which is simple, beautiful, and very instructive.

First of all, we apply the "most important" Lemma 6.1 to establish the following statement.

Proposition 6.5. *Let (x, y, p, q) be a solution of Catalan's equation such that q does not divide h_p. Then*

$$\mu = \frac{1 - \zeta_p x}{1 - \overline{\zeta_p} x} \in K_p$$

is a qth power in K_p.

Proof. We write $\zeta = \zeta_p$, $K = K_p$, etc. Put $\lambda = (x - \zeta)/(1 - \zeta)$. Lemma 6.1 implies that $(\lambda) = \mathfrak{a}^q$, where $\mathfrak{a}$ is an ideal of the field K. Let H be the class group of K, and let $\mathrm{cl}(\mathfrak{a}) \in H$ be the class of $\mathfrak{a}$. Since $\mathfrak{a}^q$ is a principal ideal, the order of $\mathrm{cl}(\mathfrak{a})$ divides q.

But, by the assumption, $h = |H|$ is coprime with q. It follows that $\mathrm{cl}(\mathfrak{a})$ is trivial, that is, $\mathfrak{a}$ is a principal ideal. Write $\mathfrak{a} = (\alpha)$, where $\alpha \in K$. Then $\lambda = \alpha^q \eta$, where η is a unit of K.

Theorem 4.9 implies that η is a real unit times a root of unity, the latter being a qth power in K (Corollary 4.5). Hence, redefining α, we may assume that $\eta \in \mathbb{R}$. It follows that $\lambda/\bar{\lambda} = (\alpha/\bar{\alpha})^q$ is a qth power in K. Hence

$$\mu = \frac{\zeta}{\bar{\zeta}} \cdot \frac{1 - \bar{\zeta}\,\bar{\lambda}}{1 - \zeta\,\lambda} = -\zeta \cdot \frac{\bar{\lambda}}{\lambda} \tag{6.1}$$

is a qth power as well. The proposition is proved. $\qquad\square$

Remark 6.6. In this proposition one may replace h_p by the relative class number h_p^-, which would imply the corresponding refinement of Theorem 6.2 and Corollary 6.4. (This was observed by Schwarz [125].)

Indeed, assume that q does not divide $h_p^- = [H:H^+]$, where H^+ stands for the class group of the real cyclotomic field K^+. It follows that $\mathrm{cl}(\mathfrak{a}) \in H^+$. Thus, $\mathfrak{a} = \alpha\mathfrak{b}$, where $\alpha \in K^\times$ and $\mathfrak{b}$ is an ideal of K^+. Write the principal ideal $\mathfrak{b}^q$ as (β), where $\beta \in K^+$. Then λ is equal to $\alpha^q \beta$ times a unit of K.

Theorem 4.9 implies that every unit of K is a real unit times a root of unity, the latter being a qth power in K. Hence, redefining α and β, we obtain $\lambda = \alpha^q \beta$ with $\alpha \in K^\times$ and $\beta \in K^+$. It follows that $\lambda/\bar{\lambda} = (\alpha/\bar{\alpha})^q$ is a qth power in K, and we finish the argument as in the proof of Proposition 6.5.

Further improvements are possible as well. We do not go into this, because in Sect. 6.5 we obtain a refinement of Theorem 6.2 which does not refer to class numbers at all.

We also need two simple algebraic facts. In the next two lemmas K is a number field and $\mathfrak{q}$ is a prime ideal of K. Recall that the local ring $\mathcal{O}_\mathfrak{q}$ is defined by

$$\mathcal{O}_\mathfrak{q} = \{\alpha \in K : \mathrm{Ord}_\mathfrak{q}(\alpha) \geq 0\}.$$

Recall also that for $\alpha, \beta \in \mathcal{O}_\mathfrak{q}$, we write $\alpha \equiv \beta \bmod \mathfrak{q}^N$ if $\mathrm{Ord}_\mathfrak{q}(\alpha - \beta) \geq N$.

First of all, we generalize the "elementary fact" used in the proof of Proposition 6.3.

Lemma 6.7. *Let q be the prime number below $\mathfrak{q}$. Then for any $\alpha, \beta \in \mathcal{O}_\mathfrak{q}$ satisfying* $\alpha^q \equiv \beta^q \bmod \mathfrak{q}$, *we have* $\alpha^q \equiv \beta^q \bmod \mathfrak{q}^2$.

Proof. We have $(\alpha - \beta)^q \equiv \alpha^q - \beta^q \equiv 0 \bmod \mathfrak{q}$. Since $\mathfrak{q}$ is a prime ideal, this implies $\alpha \equiv \beta \bmod \mathfrak{q}$. Write $\alpha = \beta + \gamma$ with $\gamma \in \mathfrak{q}$. Then

$$\alpha^q = \beta^q + \gamma \sum_{k=1}^{q-1} \binom{q}{k} \gamma^{k-1} + \gamma^q.$$

Since q divides the binomial coefficients $\binom{q}{1}, \binom{q}{2}, \ldots, \binom{q}{q-1}$, and since $\mathfrak{q}$ divides γ, this implies $\alpha^q \equiv \beta^q \bmod \mathfrak{q}^2$. □

Our second lemma is a particular case of Proposition 2.18.

Lemma 6.8. *For any $n \in \mathbb{Z}$ and any $\alpha \in K$ satisfying $\mathrm{Ord}_\mathfrak{q}(\alpha) > 0$ we have*

$$(1 + \alpha)^n \equiv 1 + n\alpha \bmod \mathfrak{q}^2.$$

Now we are ready to finish the proof of Theorem 6.2.

Proof of Theorem 6.2. We use the notation $\zeta = \zeta_p$, $K = K_p$, $\mathcal{O} = \mathcal{O}_K$, etc. Since q is unramified in K (see Proposition 4.4), it suffices to show that $\mathfrak{q}^2 \mid x$ for some prime ideal $\mathfrak{q}$ above q.

Thus, let q be a prime ideal of K dividing q and $\mathcal{O}_q$ its local ring. Since $q \mid x$, the number μ, defined in Proposition 6.5, satisfies $\mu \equiv 1 \bmod q$. Since μ is a qth power, Lemma 6.7 implies that $\mu \equiv 1 \bmod q^2$.

On the other hand, Lemma 6.8 implies that $(1 - \bar{\zeta}x)^{-1} \equiv 1 + \bar{\zeta}x \bmod q^2$. Hence $\mu \equiv 1 + (\bar{\zeta} - \zeta)x \bmod q^2$, which yields $1 + (\bar{\zeta} - \zeta)x \equiv 1 \bmod q^2$. Since $\mathrm{Ord}_q(\bar{\zeta} - \zeta) = 0$, this means that $\mathrm{Ord}_q(x) \geq 2$. The theorem is proved. □

6.3 A Deviation: Catalan's Problem with Exponent 3

The equation $x^3 - y^q = 1$ has been solved in 1921 by Nagell [99], who proved that it has no solution in nonzero integers x, y and odd prime q.

Since Cassels' Divisibility Theorem was not available at that time, Nagell had to solve both the equations $x^2 + x + 1 = u^q$ and $x^2 + x + 1 = 3u^q$.

By Cassels' divisibility theorem, the former equation has no solutions, so only the latter is to be treated. Inkeri [51] noticed that the argument he used to prove Theorem 6.2 in the special case $p = 3$ implies a quick proof that $x^2 + x + 1 = 3z^q$ has no nontrivial solution. We reproduce it below.

The results of this section will not be used in the sequel. Later, we shall obtain much stronger results than Theorem 6.10 below using different methods.

Inkeri's argument relies on the following curious fact: with very few exceptions, an odd power of a quadratic integer cannot have trace ± 1. More precisely, we have the following.

Lemma 6.9. *Let K be a quadratic extension of $\mathbb{Q}$ and $\alpha \in \mathcal{O}_K$. Assume that for some odd $n > 1$, we have $\mathrm{Tr}_{K/\mathbb{Q}}(\alpha^n) = \pm 1$. Then $K = \mathbb{Q}(\sqrt{-3})$ and $\alpha = \left(\pm 1 \pm \sqrt{-3}\right)/2$.*

Proof. Let $\bar{\alpha}$ be the conjugate of α, so that $\mathrm{Tr}(\alpha^n) = \alpha^n + \bar{\alpha}^n$. If m is a divisor of n, then $\mathrm{Tr}(\alpha^m) = \alpha^m + \bar{\alpha}^m$ divides $\mathrm{Tr}(\alpha^n) = \pm 1$, which implies that $\mathrm{Tr}(\alpha^m) = \pm 1$ for any such m. In particular, $\mathrm{Tr}\,\alpha = \pm 1$ and $\mathrm{Tr}(\alpha^q) = 1$, where q is a prime divisor of n (such a q always exists, because $n > 1$).

It follows that the difference $(\alpha + \bar{\alpha})^q - \alpha^q - \bar{\alpha}^q$ is equal to 0 or to ± 2. Since $(\alpha + \bar{\alpha})^q \equiv \alpha^q + \bar{\alpha}^q \bmod q$, the difference is 0 (recall that q is an odd prime number).

Now consider the polynomial

$$F(X, Y) = \frac{(X + Y)^q - X^q - Y^q}{qXY} \in \mathbb{Z}[X, Y].$$

As we have just seen, $F(\alpha, \bar{\alpha}) = 0$. Further, in the ring $\mathbb{Z}[X, Y]$, we have the congruences

$$F(X, Y) \equiv X^{q-2} + Y^{q-2} \bmod XY$$

$$\equiv (X + Y)^{q-2} \bmod XY;$$

that is, $F(X,Y) = (X + Y)^{q-2} + XYG(X,Y)$ with $G(X,Y) \in \mathbb{Z}[X,Y]$. Substituting $X = \alpha$ and $Y = \bar{\alpha}$, we obtain $0 = \pm 1 + \alpha\bar{\alpha}G(\alpha,\bar{\alpha})$.

Thus, the norm $\mathcal{N}\alpha = \alpha\bar{\alpha}$ divides ± 1, which means that $\mathcal{N}\alpha = \pm 1$. Since also $\mathrm{Tr}\,\alpha = \pm 1$, our α is a root of one of the polynomials $T^2 \pm T \pm 1$. In other words, either $\alpha = \left(\pm 1 \pm \sqrt{-3}\right)/2$ or $\alpha = \left(\pm 1 \pm \sqrt{5}\right)/2$. In the former case we are done. In the latter case

$$|\mathrm{Tr}\,(\alpha^n)| \ge \left(\frac{1 + \sqrt{5}}{2}\right)^n - \left(\frac{-1 + \sqrt{5}}{2}\right)^n > \frac{1 + \sqrt{5}}{2} - \frac{-1 + \sqrt{5}}{2} = 1,$$

a contradiction. This proves the lemma.   □

Theorem 6.10 (Nagell). *Let $q \ge 5$ be a prime number. Then the only solutions in $x, u \in \mathbb{Z}$ of the equation $x^2 + x + 1 = 3u^q$ are $x = u = 1$. Also, the equation $x^3 - y^q = 1$ has no solutions in nonzero integers x and y.*

Proof (Inkeri). The cyclotomic field K_3 is the imaginary quadratic field $K = \mathbb{Q}(\sqrt{-3})$. In particular, its unit group consists of roots of unity and is generated by $-j$, where $j = \zeta_3 = \left(1 + \sqrt{-3}\right)/2$. Also, its class number is 1.

Now let $q \ge 5$ be a prime number. Arguing as in the proof of Proposition 6.5, we show that $(x - j)/(1 - j)$ is a qth power times a unit. Dividing by j, we find that $(x - j)/(j - \bar{j})$ is a qth power times a unit as well. But all units in K are qth powers (they are 6th roots of unity, and q is coprime with 6). Hence $(x - j)/(j - \bar{j})$ is a pure qth power in K:

$$\frac{x - j}{j - \bar{j}} = \alpha^q,$$

where $\alpha \in K$. Taking the trace, we find

$$\mathrm{Tr}\,(\alpha^q) = \frac{x - j}{j - \bar{j}} + \frac{x - \bar{j}}{\bar{j} - j} = -1,$$

and Lemma 6.9 implies that α is a root of unity. Then $(x - j)/(1 - j)$ is a root of unity as well, which implies that $x = 1$. This proves the first statement on the equation $x^2 + x + 1 = 3u^q$. The second statement is an immediate consequence of the first one and of Cassels' relations.   □

6.4 The Group Ring

If a group G acts on an abelian group A, the group A (together with this action) is called a *G-module*. Let K be a Galois extension of $\mathbb{Q}$ with Galois group G. Then we have various G-modules (called in this case *Galois modules*): the *additive group* K, the *multiplicative group* $K^\times$, the *group of units* $\mathcal{U}_K$, the *class group* H_K, and so on.

Every G-module is a module (in the usual sense) over the group ring $\mathbb{Z}[G]$. It is very useful to replace the group by the group ring, which is a much more flexible and rich object.

In this book we mainly study multiplicative Galois modules $K^\times$, $\mathcal{U}_K$, H_K, and others. Hence it will be natural to make the following convention: *starting from this point, the Galois action will be written exponentially*. Say, if $\alpha \in K$ and $\sigma \in G$, then the σ-image of α is written as[2] α^σ.

Similarly, we shall write exponentially the action of the group ring $\mathbb{Z}[G]$ on our multiplicative Galois modules. Say, if $\alpha \in K^\times$ and $\Theta = \sum_{\sigma \in G} a_\sigma \sigma \in \mathbb{Z}[G]$, then

$$\alpha^\Theta = \prod_{\sigma \in G} (\alpha^\sigma)^{a_\sigma} .$$

Now return to Catalan's equation. We want to refine Inkeri's divisibility theorem, by relaxing or suppressing the assumption $q \mid h_p$. Recall that this assumption implies that the ideal $\mathfrak{a}$ from the "most important" lemma is principal.

In general, this is not guaranteed. One idea to overcome this difficulty is to find $\Theta \in \mathbb{Z}[G]$ such that the ideal $\mathfrak{a}^\Theta$ is principal and then deal with the number λ^Θ instead of λ.

In the sequel (x, y, p, q) is a solution of Catalan's equation, ζ is a primitive pth root of unity, $K = \mathbb{Q}(\zeta)$ is the pth cyclotomic field, $H = H_K$ is its class group, and $G = \mathrm{Gal}(K/\mathbb{Q})$ is its Galois group. Recall that $\iota \in G$ stands for the complex conjugation.

Proposition 6.11. *Assume that $\Theta \in \mathbb{Z}[G]$ annihilates the class group H (that is, for any ideal $\mathfrak{a}$ of K, the ideal $\mathfrak{a}^\Theta$ is principal). Then $(1 - \zeta x)^{(1-\iota)\Theta}$ is a qth power in K.*

Proof. The proof copies that of Proposition 6.5. We again have $(\lambda) = \mathfrak{a}^q$, where $\lambda = (x - \zeta)/(1 - \zeta)$ and $\mathfrak{a}$ is an ideal of K. By the assumption, $\mathfrak{a}^\Theta$ is a principal ideal; write $\mathfrak{a}^\Theta = (\alpha)$. Then $\lambda^\Theta = \eta \alpha^q$, where η is a unit of K, and we may assume that η is real. It follows that $(\lambda/\bar{\lambda})^\Theta = (\alpha/\bar{\alpha})^q$ is a qth power in K. Hence

$$(1 - \zeta x)^{(1-\iota)\Theta} = \left(\frac{1-\zeta x}{1-\bar{\zeta} x} \right)^\Theta = \left(\frac{\zeta}{\bar{\zeta}} \cdot \frac{1-\bar{\zeta}}{1-\zeta} \right)^\Theta \left(\frac{\bar{\lambda}}{\lambda} \right)^\Theta = (-\zeta)^\Theta \left(\frac{\bar{\alpha}}{\alpha} \right)^q$$

is a qth power as well. The proposition is proved. $\qquad\square$

Proposition 6.12. *Assume that there exists $\Theta \in \mathbb{Z}[G]$ such that $(1 - \zeta x)^\Theta$ is a qth power and Θ is not divisible by q (in the ring $\mathbb{Z}[G]$). Then $q^2 \mid x$.*

---

[2] In general, one should be careful here, because the usual writing $\alpha \mapsto \sigma(\alpha)$ corresponds to the *left* action, while the exponential writing $\alpha \mapsto \alpha^\sigma$ corresponds to the *right* action. However, this warning is relevant only for Chap. 12, the only chapter where non-abelian Galois groups occur.

Proof. As in the proof of Theorem 6.2, it suffices to show that $q^2 \mid x$ for some prime ideal q, dividing q.

Thus, let q be a prime ideal above q, to be specified later. Since $q \mid x$, we have $(1 - \zeta x)^\Theta \equiv 1 \bmod q$. But the number $(1 - \zeta x)^\Theta$ is, by the assumption, a qth power. Hence Lemma 6.7 implies that

$$(1 - \zeta x)^\Theta \equiv 1 \bmod q^2. \tag{6.2}$$

Now let us specify q. Write $\Theta = \sum_{\sigma \in G} a_\sigma \sigma$. Then at least one of the coefficients a_σ is not divisible by q. Since $\{\zeta^\sigma : \sigma \in G\}$ is an integral basis of the ring of integers $\mathcal{O}$ (Theorem 4.6), this implies that q does not divide $\alpha = \sum_{\sigma \in G} a_\sigma \zeta^\sigma$ in $\mathcal{O}$. Since q is unramified in K, there exists a prime ideal of K dividing q and not dividing α. In the sequel, we assume that q is this prime ideal.

Lemma 6.8 implies that for any $\sigma \in G$ and $a \in \mathbb{Z}$ we have

$$(1 - \zeta^\sigma x)^a \equiv 1 - a\zeta^\sigma x \bmod q^2.$$

Hence

$$(1 - \zeta x)^\Theta = \prod_{\sigma \in G} (1 - \zeta^\sigma x)^{a_\sigma} \equiv 1 - x \sum_{\sigma \in G} a_\sigma \zeta^\sigma \bmod q^2,$$

that is, $(1 - \zeta x)^\Theta \equiv 1 - \alpha x \bmod q^2$. Together with (6.2) this implies that $\mathrm{Ord}_q(\alpha x) \geq 2$. But $\mathrm{Ord}_q(\alpha) = 0$ by our choice of q. Hence $\mathrm{Ord}_q(x) \geq 2$. The proposition is proved. $\qquad\square$

6.5   Stickelberger, Mihăilescu, and Wieferich

To make use of Propositions 6.11 and 6.12, one should find $\Theta \in \mathbb{Z}[G]$ with the following two properties:

- Θ annihilates the class group of K, and
- $(1 - \iota)\Theta$ is not divisible by q.

It is easy to find a nonzero Θ with the first property. For instance, $\Theta = \mathcal{N}$ will do, where $\mathcal{N} = \sum_{\sigma \in G} \sigma$ is called the *norm element* of $\mathbb{Z}[G]$ (or, simply, the *norm*). However, $(1 - \iota)\mathcal{N} = 0$, so the second property is not satisfied.

An element Θ with both properties is provided by the classical theorem of Stickelberger [131]. Recall that $G = \{\sigma_1, \ldots, \sigma_{p-1}\}$, where σ_k is defined by $\zeta \mapsto \zeta^k$.

Theorem 6.13 (Stickelberger). *The element* $\Theta_S = \sum_{a=1}^{p-1} a\sigma_a^{-1} \in \mathbb{Z}[G]$ *annihilates the class group of* K.

It is easy to verify that $(1 - \iota)\Theta_S$ is not divisible by q. Indeed,

$$(1 - \iota)\Theta_S = \sum_{a=1}^{p-1} (2a - p)\sigma_a^{-1}.$$

In particular, the coefficient of $\sigma_{(p+1)/2}$ is 1. Hence q does not divide $(1 - \iota)\Theta_S$.

Theorem 6.13 will be proved in Chap. 7.

Now we shall use Theorem 6.13 to obtain the promised unconditional refinement of Inkeri's divisibility theorem, due to Mihăilescu [92].

Theorem 6.14 (Mihăilescu). *Let (x, y, p, q) be a solution of Catalan's equation. Then $q^2 \mid x$ (and $p^2 \mid y$, by symmetry).*

Proof. Applying Proposition 6.11 with $\Theta = \Theta_S$, we find that $(1 - \zeta x)^{(1-\iota)\Theta_S}$ is a qth power in K. Applying Proposition 6.12 with $\Theta = (1 - \iota)\Theta_S$, we obtain the result. $\qquad\qquad\square$

Together with Proposition 6.3 this has the following consequence.

Corollary 6.15. *Let (x, y, p, q) be a solution of Catalan's equation. Then $p^{q-1} \equiv 1 \bmod q^2$ (and $q^{p-1} \equiv 1 \bmod p^2$, by symmetry).*

This is a very strong result. To appreciate it, mention that one currently knows [1, 29, 54] only seven pairs of prime numbers (p, q) such that

$$p^{q-1} \equiv 1 \bmod q^2, \quad q^{p-1} \equiv 1 \bmod p^2$$

(called *double Wieferich pairs*). They are

$$(2, 1093), \ (3, 1006003), \ (5, 1645333507), \ (5, 188748146801),$$

$$(83, 4871), \ (911, 318917), \ (2903, 18787).$$

that no other double Wieferich pair with $\min\{p, q\} \leq 3.2 \times 10^8$ exist. However, it is unknown whether the set of double Wieferich pairs is finite or infinite.

The results of the last two sections suggest that the elements $\Theta \in \mathbb{Z}[G]$ such that $(1 - \zeta x)^{\Theta}$ is a qth power in K play an important role in the theory of Catalan's equation. In Chap. 8 we shall study them in detail.

Chapter 7
Gauss Sums and Stickelberger's Theorem

In Sect. 6.5 we already used (but did not prove) Stickelberger's theorem, which provides a nontrivial annihilator for the class group. In this chapter we prove this theorem, in a stronger form: we define an ideal of the group ring $\mathbb{Z}[G]$ (called *Stickelberger's ideal*) and show that all its elements annihilate the class group.

All known proofs of Stickelberger's theorem rely on properties of *Gauss sums*, an arithmetical object interesting by itself. We develop the theory of Gauss sums to the extent needed for the proof of Stickelberger's theorem.

In the final sections we provide deeper insight into the structure of Stickelberger's ideal. We determine its $\mathbb{Z}$-rank, find a free $\mathbb{Z}$-basis, study its real and relative parts, and prove Iwasawa's class number formula.

7.1 Stickelberger's Ideal and Stickelberger's Theorem

Let p be an odd prime number, $K_p = \mathbb{Q}(\zeta_p)$ the pth cyclotomic field, and $G = \mathrm{Gal}(K_p/\mathbb{Q})$ its Galois group. In Sect. 6.5 we stated the theorem of Stickelberger: the element $\sum_{a=1}^{p-1} a\sigma_a^{-1}$ of the group ring $\mathbb{Z}[G]$ annihilates the class group of K_p. In this chapter we shall prove it, in a stronger form.

Stickelberger's element is defined by

$$\theta = \frac{1}{p} \sum_{a=1}^{p-1} a\sigma_a^{-1} \in \mathbb{Q}[G].$$

© Springer International Publishing Switzerland 2014
Y.F. Bilu et al., *The Problem of Catalan*, DOI 10.1007/978-3-319-10094-4_7

Stickelberger's ideal is

$$\mathcal{I}_S = \theta \mathbb{Z}[G] \cap \mathbb{Z}[G].$$

The main result of this chapter is the following theorem, which refines Theorem 6.13.

Theorem 7.1. *Every* $\Theta \in \mathcal{I}_S$ *annihilates the class group of* K_p.

The proof of this theorem relies on the careful study of an important arithmetical object called *Gauss sums*. The theorem will be proved in Sect. 7.5, after we establish necessary properties of Gauss sums in Sects. 7.2–7.4.

To conclude this section, we determine a convenient system of $\mathbb{Z}$-generators of $\mathcal{I}_S$. As indicated in Appendix D, in the group ring $\mathbb{Z}[G]$, we identify 1 and σ_1, which is the neutral element of G. In particular, $\mathbb{Z}$ is a subring of $\mathbb{Z}[G]$.

Given $b \in \mathbb{Z}$ not divisible by p, we put

$$\Theta_b = (b - \sigma_b)\theta.$$

Since $\sigma_{p+1} = \sigma_1 = 1$, we have

$$\Theta_1 = 0, \qquad \Theta_{p+1} = p\theta = \sum_{a=1}^{p-1} a\sigma_a^{-1}.$$

It turns out that the elements Θ_b belong to $\mathbb{Z}[G]$ and generate $\mathcal{I}_S$. More precisely, we have the following.

Proposition 7.2.

1. *For any* $b \in \mathbb{Z}$ *not divisible by* p *we have*

$$\Theta_b = \sum_{a=1}^{p-1} \left\lfloor \frac{ba}{p} \right\rfloor \sigma_a^{-1}. \tag{7.1}$$

   In particular, $\Theta_b \in \mathcal{I}_S$ *for all such* b.
2. *Stickelberger's ideal* $\mathcal{I}_S$ *is generated over* $\mathbb{Z}$ *by the elements* $\Theta_1, \ldots, \Theta_{p-1}$ *and* $\Theta_{p+1} = p\theta$.

Proof. Let a and b be integers not divisible by p, and let c be such that $\sigma_b \sigma_a^{-1} = \sigma_c^{-1}$. Then $a \equiv bc \bmod p$. In particular, if $0 < a < p$ then

$$a = bc - p\left\lfloor \frac{bc}{p} \right\rfloor.$$

It follows that

$$\sigma_b \theta = \sum_{c=1}^{p-1} \left(\frac{bc}{p} - \left\lfloor \frac{bc}{p} \right\rfloor \right) \sigma_c^{-1}.$$

Hence

$$\Theta_b = b\theta - \sigma_b \theta = \sum_{a=1}^{p-1} \frac{ba}{p} \sigma_a^{-1} - \sum_{c=1}^{p-1} \left(\frac{bc}{p} - \left\lfloor \frac{bc}{p} \right\rfloor \right) \sigma_c^{-1} = \sum_{a=1}^{p-1} \left\lfloor \frac{ba}{p} \right\rfloor \sigma_a^{-1},$$

which proves (7.1).

Identity (7.1) implies that $\Theta_b \in \mathbb{Z}[G]$. Since $\Theta_b \in \theta \mathbb{Z}[G]$ as well, this implies that $\Theta_b \in \mathcal{I}_S$. Part (1) is proved.

Now let Θ be an element of $\mathcal{I}_S$. To prove part (2), we have to express Θ as a $\mathbb{Z}$-linear combination of $\Theta_1, \ldots, \Theta_{p-1}$ and $p\theta$.

Since $\Theta \in \theta \mathbb{Z}[G]$ and $\Theta \in \mathbb{Z}[G]$, we may write

$$\Theta = \theta \sum_{a=1}^{p-1} x_a \sigma_a = \sum_{a=1}^{p-1} y_a \sigma_a,$$

where all x_a and y_a are integers. We find from here

$$y_1 = \frac{1}{p} \sum_{a=1}^{p-1} a x_a,$$

which implies

$$\Theta = \theta \sum_{a=1}^{p-1} x_a (\sigma_a - a) + \theta \sum_{a=1}^{p-1} a x_a = -\sum_{a=1}^{p-1} x_a \Theta_a + y_1 p\theta.$$

This proves part (2). $\qquad\qquad\qquad\square$

A natural question is whether $\Theta_1, \ldots, \Theta_{p-1}$ and $p\theta$ form a *free* $\mathbb{Z}$-basis of $\mathcal{I}_S$. The answer is obviously "no" because $\Theta_1 = 0$. Less obviously, the answer is also negative[1] for $\Theta_2, \ldots, \Theta_{p-1}$ and $p\theta$: we shall see in Proposition 7.20 that $\Theta_a + \Theta_{p-a} = \Theta_b + \Theta_{p-b}$ for any a and b. However, all this is not relevant for the proof of Stickelberger's theorem, where even the weaker statement "$\mathcal{I}_S$ is generated by the elements Θ_b as an ideal" is sufficient. Nevertheless, a free basis will be needed in Chap. 8, and in Sect. 7.6 we shall determine the $\mathbb{Z}$-rank and a free $\mathbb{Z}$-basis of $\mathcal{I}_S$.

---

[1] Except when $p = 3$

7.2   Gauss Sums

In this section by a *character* of a finite abelian group G we mean a $\mathbb{C}$-character, that is, a homomorphism $G \to \mathbb{C}^\times$.

Let ℓ be a prime number, and let $\mathbb{F} = \mathbb{Z}/\ell\mathbb{Z}$ be the field of ℓ elements. We consider two types of characters on $\mathbb{F}$. An *additive character* is a character of the additive group of $\mathbb{F}$. A *multiplicative character* is a character of the multiplicative group $\mathbb{F}^\times$. An additive (respectively, multiplicative) character is called *trivial* if it is identically 1 on $\mathbb{F}$ (respectively, on $\mathbb{F}^\times$), and *nontrivial* otherwise. A multiplicative character χ is defined only on $\mathbb{F}^\times$, but we extend it to the entire $\mathbb{F}$ by setting[2] $\chi(0) = 0$. The order of a nontrivial additive character is ℓ. The order of a multiplicative character is a divisor of $\ell - 1$.

According to Proposition D.4, a nontrivial additive character ψ and a nontrivial multiplicative character χ satisfy

$$\sum_{x \in \mathbb{F}} \psi(x) = 0, \qquad \sum_{x \in \mathbb{F}^\times} \chi(x) = 0,$$

which can be also written as

$$\sum_{x \in \mathbb{F}^\times} \psi(x) = -1, \qquad \sum_{\substack{x \in \mathbb{F}^\times \\ x \neq 1}} \chi(x) = -1. \tag{7.2}$$

Now, for a nontrivial additive character ψ and for a (trivial or nontrivial) multiplicative character χ, we define the Gauss sum as

$$g(\psi, \chi) = -\sum_{x \in \mathbb{F}} \psi(x)\chi(x).$$

Gauss sums belong to the most important arithmetical objects; they are indispensable in analytic number theory, algebraic number theory, arithmetic geometry, cryptography, etc. But for us Gauss sums are nothing more than a tool for proving Stickelberger's theorem. Therefore, we establish here only the basic properties needed for this purpose.

In particular, we do not discuss here the historical and methodological reasons for putting the "−" sign before the sum. For our present purposes this is absolutely irrelevant, but we follow the tradition.

Gauss sums already implicitly appeared in this book, in Sect. 5.3. However, our notation here is very different, and we prefer to re-prove in this section several statements already made therein.

---

[2] Whenever χ is trivial or nontrivial, which is somewhat inconsistent with the conventions made in Chap. 5

The Gauss sum $g(\psi, \chi)$ is a function of two variables, a nontrivial additive character ψ and a multiplicative character χ. It turns out that the behavior of g as a function of ψ is quite simple. Indeed, if ψ is a fixed nontrivial additive character, then the complete list of nontrivial additive characters is $\psi, \psi^2, \ldots, \psi^{\ell-1}$, and we have the following statement.

Proposition 7.3. *Let ψ be a nontrivial additive character, χ a multiplicative character, and b an integer non-divisible by ℓ. Then*[3] $g(\psi^b, \chi) = \bar{\chi}(b)g(\psi, \chi)$.

Here and below we denote by $\bar{\chi}$ the complex conjugate of the character χ, defined by $\bar{\chi}(x) = \overline{\chi(x)}$ for $x \in \mathbb{F}$. Notice that for $x \neq 0$ we have

$$\bar{\chi}(x) = \chi(x)^{-1} = \chi(x^{-1}).$$

Proof. Since b is not divisible by ℓ, the product bx runs through $\mathbb{F}$ when x runs through $\mathbb{F}$. It follows that

$$g(\psi, \chi) = -\sum_{x \in \mathbb{F}} \psi(bx)\chi(bx) = -\chi(b)\sum_{x \in \mathbb{F}} \psi(x)^b \chi(x) = \chi(b)g(\psi^b, \chi).$$

Since $\bar{\chi}(b) = \chi(b)^{-1}$, the proposition follows. □

In particular, viewing the Gauss sum as an algebraic integer in some number field, the principal ideal it defines is independent of the choice of the additive character ψ.

The behavior of g as a function of χ is much more interesting, and this is what we are going to study in detail. Thus, from now on, **we fix a nontrivial additive character ψ and write $g(\chi)$ instead of $g(\psi, \chi)$**. Also, in the sequel the word *character* will mean a *multiplicative character*, unless the contrary is stated explicitly.

The only case of a "simple" relation between Gauss sums for two distinct characters is when they are complex conjugate.

Proposition 7.4. *Let χ be a (multiplicative) character and $\bar{\chi}$ its complex conjugate. Then $g(\bar{\chi}) = \chi(-1)\overline{g(\chi)}$.*

Proof. We have

$$g(\bar{\chi}) = -\sum_{x \in \mathbb{F}} \psi(x)\bar{\chi}(x) = -\overline{\sum_{x \in \mathbb{F}} \psi(-x)\chi(x)} = -\overline{\sum_{x \in \mathbb{F}} \psi(x)\chi(-x)} = \overline{\chi(-1)g(\chi)}.$$

Since $\chi(-1) \in \{\pm 1\}$, we have $\overline{\chi(-1)} = \chi(-1)$, and the proposition follows. □

---

[3] As it is commonly done, for $b \in \mathbb{Z}$ we write $\chi(b)$ instead of $\chi(b^*)$, where b^* is the image of b in $\mathbb{Z}/\ell\mathbb{Z}$.

The first identity in (7.2) implies that $g(\chi) = 1$ for the trivial character χ. For the nontrivial characters, computing the precise value of the Gauss sum is rather difficult. Much easier is to determine the *absolute value* of the Gauss sum.

Theorem 7.5. *Let χ be a nontrivial multiplicative character. Then*

$$g(\chi)\overline{g(\chi)} = \ell. \tag{7.3}$$

In particular, $g(\chi)g(\bar{\chi}) = \chi(-1)\ell$.

Proof. Since $\overline{\psi(x)} = \psi(-x)$ and $\overline{\chi(x)} = \chi(x^{-1})$, we have

$$g(\chi)\overline{g(\chi)} = \sum_{x \in \mathbb{F}^\times} \psi(x)\chi(x) \sum_{y \in \mathbb{F}^\times} \psi(-y)\chi(y^{-1})$$

$$= \sum_{x,y \in \mathbb{F}^\times} \psi(x - y)\chi(xy^{-1})$$

$$= \sum_{z,y \in \mathbb{F}^\times} \psi(zy - y)\chi(z) \qquad \text{(we put } z = xy^{-1}\text{)}$$

$$= \sum_{y \in \mathbb{F}^\times} \psi(0)\chi(1) + \sum_{\substack{z \in \mathbb{F}^\times \\ z \neq 1}} \chi(z) \sum_{y \in \mathbb{F}^\times} \psi(y(z - 1)).$$

The first sum here is $\ell - 1$. Further, when $z \neq 1$ and y runs through $\mathbb{F}^\times$, the product $y(z - 1)$ runs through $\mathbb{F}^\times$ as well. It follows that

$$\sum_{y \in \mathbb{F}^\times} \psi(y(z - 1)) = -1,$$

and we obtain

$$g(\chi)\overline{g(\chi)} = \ell - 1 - \sum_{\substack{z \in \mathbb{F}^\times \\ z \neq 1}} \chi(z) = \ell.$$

Proposition 7.4 now implies that $g(\chi)g(\bar{\chi}) = \chi(-1)\ell$. The theorem is proved. □

Another proof is indicated in Sect. 5.3.3. Let V be the space of complex functions on $\mathbb{F}$. It is a $\mathbb{C}$-vector space of dimension ℓ. Also, we have a natural inner product on V defined by

$$(u, v) = \ell^{-1} \sum_{x \in \mathbb{F}} u(x)\overline{v(x)}.$$

A straightforward verification shows that the functions $1, \psi, \ldots, \psi^{\ell-1}$ form an orthonormal basis of V. Since $(\chi, 1) = 0$, we have

$$\chi = \sum_{k=1}^{\ell-1} (\chi, \psi^k) \psi^k.$$

Proposition 7.3 implies that $(\chi, \psi^k) = \ell^{-1} \bar{\chi}(-k) g(\chi)$. In particular,

$$|(\chi, \psi^k)| = \ell^{-1} |g(\chi)| \qquad (k = 1, \ldots, \ell-1).$$

We obtain

$$\frac{\ell-1}{\ell} = (\chi, \chi) = \sum_{k=1}^{\ell-1} |(\chi, \psi^k)|^2 = (\ell-1) \frac{|g(\chi)|^2}{\ell^2},$$

whence $|g(\chi)| = \sqrt{\ell}$.

7.3   Multiplicative Combinations of Gauss Sums

The values of the additive character ψ lie in the cyclotomic field $K_\ell = \mathbb{Q}(\zeta_\ell)$, and the multiplicative character χ has values in the field $K_{\ell-1} = \mathbb{Q}(\zeta_{\ell-1})$, or, more precisely, in $K_m = \mathbb{Q}(\zeta_m)$, where m is the order of χ. It follows that the Gauss sum $g(\chi)$ belongs to the "large" cyclotomic field $K_{m\ell} = \mathbb{Q}(\zeta_m, \zeta_\ell)$, and, in general, one cannot say anything better.

It is quite remarkable that a simple multiplicative combination of several Gauss sums lies in a much smaller field.

Theorem 7.6. *For any multiplicative characters χ_1 and χ_2 the quotient $g(\chi_1) g(\chi_2)/g(\chi_1 \chi_2)$ is an algebraic integer from the field K_m, where m is the least common multiple of the orders of χ_1 and χ_2.*

Before proving the theorem, let us state two important consequences. The following assertion is proved by a simple induction in a.

Corollary 7.7. *Let χ be a multiplicative character of order m. Then for any nonnegative integer a the quotient $g(\chi)^a / g(\chi^a)$ is an algebraic integer from K_m.*

Now let m be a divisor of $\ell - 1$, and for an integer a coprime with m let σ_a be the morphism of the cyclotomic field $K_{m\ell}$ defined by $\zeta_m \mapsto \zeta_m^a$ and $\zeta_\ell \mapsto \zeta_\ell$ (see Proposition 4.26).

Corollary 7.8. *Let χ be a multiplicative character of order m. Then for any a coprime with m the number $g(\chi)^{a-\sigma_a}$ is an algebraic integer from the field K_m. In particular, $g(\chi)^m \in K_m$.*

Proof. Since for any $x \in \mathbb{F}$ the number $\chi(x)$ is an mth root of unity, we have $\chi(x)^{\sigma_a} = \chi(x)^a$. Also, since σ_a is identical on the ℓth roots of unity, we have $\psi^{\sigma_a} = \psi$. It follows that $g(\chi)^{\sigma_a} = g(\chi^a)$. Applying Corollary 7.8, we prove that $g(\chi)^{a-\sigma_a}$ is an integer from K_m. Putting $a = m + 1$, we obtain $g(\chi)^m \in K_m$. $\square$

For the proof of Theorem 7.6 we introduce an auxiliary quantity called *Jacobi sum*. Let χ_1 and χ_2 be two multiplicative characters. The Jacobi sum is

$$J(\chi_1, \chi_2) = \sum_{x \in \mathbb{F}} \chi_1(x)\chi_2(1 - x).$$

Obviously, $J(\chi_1, \chi_2) \in \mathbb{Q}(\zeta_m)$, where m is the least common multiple of the orders of χ_1 and χ_2.

Proposition 7.9. *Let χ_1 and χ_2 be multiplicative characters such that $\chi_1 \neq \overline{\chi_2}$. Then $J(\chi_1, \chi_2) = g(\chi_1)g(\chi_2)/g(\chi_1\chi_2)$.*

Proof. We have

$$g(\chi_1)g(\chi_2) = \sum_{x,y \in \mathbb{F}} \chi_1(x)\chi_2(y)\psi(x + y)$$

$$= \sum_{x,z \in \mathbb{F}} \chi_1(x)\chi_2(z - x)\psi(z)$$

$$= \sum_{\substack{x,z \in \mathbb{F} \\ z \neq 0}} \chi_1(x)\chi_2(z - x)\psi(z) + \sum_{x \in \mathbb{F}} \chi_1(x)\chi_2(-x).$$

Since $\chi_1 \neq \overline{\chi_2}$, the character $\chi_1\chi_2$ is nontrivial. It follows that

$$\sum_{x \in \mathbb{F}} \chi_1(x)\chi_2(-x) = \chi_2(-1) \sum_{x \in \mathbb{F}} (\chi_1\chi_2)(x) = 0.$$

Hence

$$g(\chi_1)g(\chi_2) = \sum_{\substack{x,z \in \mathbb{F} \\ z \neq 0}} \chi_1(x)\chi_2(z - x)\psi(z)$$

$$= \sum_{\substack{t \in \mathbb{F} \\ z \in \mathbb{F}^\times}} \chi_1(tz)\chi_2(z - tz)\psi(z)$$

$$= \sum_{t \in \mathbb{F}} \chi_1(t)\chi_2(1 - t) \sum_{z \in \mathbb{F}^\times} (\chi_1\chi_2)(z)\psi(z)$$

$$= J(\chi_1, \chi_2)g(\chi_1\chi_2),$$

as wanted. $\square$

Proof of Theorem 7.6. In the case $\chi_1 \neq \bar{\chi}_2$ the theorem is a direct consequence of Proposition 7.9. If $\chi_1 = \bar{\chi}_2$ then $\chi_1\chi_2$ is a trivial character and $g(\chi_1\chi_2) = 1$. Now, if χ_1 itself is nontrivial, then we have $g(\chi_1)g(\chi_2)/g(\chi_1\chi_2) = \chi(-1)\ell \in \mathbb{Z}$ by Theorem 7.5. If χ_1 is trivial, we have $g(\chi_1)g(\chi_2)/g(\chi_1\chi_2) = 1$. The theorem is proved. □

7.4 Prime Decomposition of a Gauss Sum

In this section we determine the prime decomposition of a Gauss sum $g(\chi)$. We shall restrict to the case when χ is a character of an odd prime order p, but with purely cosmetic changes (see Remark 7.16) our argument extends to a general χ.

The order of a multiplicative character must divide $\ell - 1$. Thus, let p be an odd prime number dividing $\ell - 1$. Since the group of multiplicative characters is cyclic, there exist precisely $p - 1$ characters of (exact) order p. On the other hand, since $\ell \equiv 1 \bmod p$, the cyclotomic field K_p has precisely $p - 1$ prime ideals above ℓ, each of degree 1 (see Proposition 4.8). We wish to establish a natural one-to-one correspondence between the two sets.

Proposition 7.10. *There is a one-to-one correspondence between the characters of order p and the prime ideals of K_p above ℓ such that, if χ is a character and $\mathfrak{l}$ the corresponding ideal, then*

$$\chi(x) \equiv x^{\frac{\ell-1}{p}} \bmod \mathfrak{l} \qquad (x \in \mathbb{Z}). \tag{7.4}$$

Proof. Let $\mathfrak{l}$ be a prime ideal of K_p above ℓ. Since $\mathfrak{l}$ is of degree 1, every residue class mod $\mathfrak{l}$ contains a rational integer. In particular, there exists $r \in \mathbb{Z}$ such that $\zeta_p \equiv r \bmod \mathfrak{l}$. Taking pth power, we obtain $r^p \equiv 1 \bmod \mathfrak{l}$. Since both sides of the last congruence are rational integers, this implies that $r^p \equiv 1 \bmod \ell$. Hence there exists a generator s of the multiplicative group $\mathbb{F}^\times$ such that $r \equiv s^{\frac{\ell-1}{p}} \bmod \ell$.

Since s generates $\mathbb{F}^\times$, there exists a (unique) character χ with $\chi(s) = \zeta_p$. It is straightforward that it satisfies (7.4).

Thus, to every prime ideal $\mathfrak{l}$ above ℓ, we associate a character χ satisfying (7.4). Such χ is unique. Indeed, let χ and χ' be two characters of order p satisfying (7.4). Then $\chi(x) \equiv \chi'(x) \bmod \mathfrak{l}$ for all $x \in \mathbb{Z}$, and, in particular, $\chi(s) \equiv \chi'(s) \bmod \mathfrak{l}$, where s is a primitive root modulo ℓ. Corollary 4.2 implies that $\chi(s) = \chi'(s)$ and thereby $\chi = \chi'$.

Since a character of order p has values in K_p, the Galois group $\mathrm{Gal}(K_p/\mathbb{Q})$ acts on the set of these characters in the natural way. Moreover, if χ is the character associated to $\mathfrak{l}$ and $\sigma \in \mathrm{Gal}(K_p/\mathbb{Q})$, then χ^σ is associated to $\mathfrak{l}^\sigma$: congruence (7.4) implies that

$$\chi^\sigma(x) \equiv x^{\frac{\ell-1}{p}} \bmod \mathfrak{l}^\sigma \qquad (x \in \mathbb{Z}),$$

and the character associated to $\mathfrak{l}^\sigma$ is unique, as we have seen in the previous paragraph. In particular, to distinct prime divisors of ℓ, we associate distinct characters.

We have defined an injective map from the set of prime divisors of ℓ into the set of characters of order p. Since both these sets have $p - 1$ elements, this map is actually bijective. This proves the proposition. $\square$

Now fix a character χ of order p. We are going to define a special numbering of all prime ideals of K_p above ℓ. For $a \in \{1, 2, \ldots, p-1\}$ we denote by $\mathfrak{l}_a$ the ideal corresponding (as in Proposition 7.10) to $\bar{\chi}^b$, where b is the inverse of a modulo p (that is, $ab \equiv 1 \bmod p$). This choice of numbering looks somewhat artificial, but it will be justified later.

The prime ideal $\mathfrak{l}_1$, corresponding to the character $\bar{\chi}$, will be denoted simply by $\mathfrak{l}$. Thus, we have

$$\bar{\chi}(x) \equiv x^{\frac{\ell-1}{p}} \bmod \mathfrak{l} \qquad (x \in \mathbb{Z}). \tag{7.5}$$

Recall that $\mathrm{Gal}(K_p/\mathbb{Q}) = \{\sigma_1, \ldots, \sigma_{p-1}\}$, where σ_a is defined by $\zeta_p \mapsto \zeta_p^a$. With our numbering we have

$$\mathfrak{l}_a^{\sigma_a} = \mathfrak{l} \qquad (a = 1, \ldots, p-1),$$

or, equivalently, $\mathfrak{l}_a = \mathfrak{l}^{\sigma_a^{-1}}$.

Unfortunately, we cannot stay in the field K_p, because the Gauss sum $g(\chi)$ lives in the wider field $K_{p\ell} = \mathbb{Q}(\zeta_p, \zeta_\ell)$. According to Proposition 4.17, every $\mathfrak{l}_a$ totally ramifies in $K_{p\ell}$. If $\mathfrak{L}_a$ is the prime ideal of $K_{p\ell}$ above $\mathfrak{l}_a$, then $\mathfrak{l}_a = \mathfrak{L}_a^{\ell-1}$. In particular, ℓ splits in $K_{p\ell}$ as

$$(\ell) = \mathfrak{L}_1^{\ell-1} \cdots \mathfrak{L}_{p-1}^{\ell-1}. \tag{7.6}$$

We also denote the ideal $\mathfrak{L}_1$ by $\mathfrak{L}$, so that $\mathfrak{l} = \mathfrak{L}^{\ell-1}$. If $\sigma_a \in \mathrm{Gal}(K_{p\ell}/\mathbb{Q})$ is defined by $\zeta_p \mapsto \zeta_p^a$ and $\zeta_\ell \mapsto \zeta_\ell$, then

$$\mathfrak{L}_a^{\sigma_a} = \mathfrak{L} \qquad (a = 1, \ldots, p-1),$$

or, equivalently, $\mathfrak{L}_a = \mathfrak{L}^{\sigma_a^{-1}}$.

We are ready to formulate the main result of this section.

Theorem 7.11 (Kummer). *Let χ be a multiplicative character of order p, and let $\mathfrak{L}_1, \ldots, \mathfrak{L}_{p-1}$ be the prime ideals of $K_{p\ell}$ defined above. Then the principal ideal $(g(\chi))$ of the field $K_{p\ell}$ decomposes as*

$$(g(\chi)) = \left(\mathfrak{L}_1 \mathfrak{L}_2^2 \ldots \mathfrak{L}_{p-1}^{p-1}\right)^{\frac{\ell-1}{p}}. \tag{7.7}$$

The proof relies on two simple lemmas.

Lemma 7.12. *We have* $\text{Ord}_{\mathcal{L}}(1 - \zeta_\ell) = 1$. *Also, for every nonnegative integer* b, *we have*

$$\frac{\zeta_\ell^b - 1}{\zeta_\ell - 1} \equiv b \bmod \mathcal{L}.$$

Proof. Since $(\ell) = (1 - \zeta_\ell)^{\ell-1}$, equality (7.6) implies that

$$(1 - \zeta_\ell) = \mathcal{L}_1 \cdots \mathcal{L}_{p-1}.$$

In particular, $\text{Ord}_{\mathcal{L}}(1 - \zeta_\ell) = 1$.

Further, $\zeta_\ell \equiv 1 \bmod \mathcal{L}$ implies that $\zeta_\ell^k \equiv 1 \bmod \mathcal{L}$ for $k = 0, 1, 2, \ldots$. It follows that

$$\frac{\zeta_\ell^b - 1}{\zeta_\ell - 1} = 1 + \zeta + \cdots + \zeta^{b-1} \equiv b \bmod \mathcal{L},$$

as wanted. $\qquad\square$

Our second lemma is just a reformulation of Proposition 7.3. For every $b \in \mathbb{Z}$ not divisible by ℓ we define the automorphism $\tau_b \in \text{Gal}(K_{p\ell}/\mathbb{Q})$ by $\zeta_\ell \mapsto \zeta_\ell^b$ and $\zeta_p \mapsto \zeta_p$. Now Proposition 7.3 can be restated as follows.

Lemma 7.13. *For an integer* b *not divisible by* ℓ *we have* $g(\chi)^{\tau_b} = \bar{\chi}(b)g(\chi)$.

Proof of Theorem 7.11. Theorem 7.5 implies that $g(\chi) \mid \ell$. Hence the only prime ideals that can occur in the decomposition of $g(\chi)$ are $\mathcal{L}_1, \ldots, \mathcal{L}_{p-1}$, and their multiplicities do not exceed $\ell - 1$:

$$(g(\chi)) = \mathcal{L}_1^{s_1} \cdots \mathcal{L}_{p-1}^{s_{p-1}},$$

where

$$0 \le s_a \le \ell - 1 \qquad (a = 1, \ldots, p - 1).$$

Since $\mathcal{L}_a^{\sigma_a} = \mathcal{L}$ and $\chi^{\sigma_a} = \chi^a$, we have

$$s_a = \text{Ord}_{\mathcal{L}_a}(g(\chi)) = \text{Ord}_{\mathcal{L}}(g(\chi^a)).$$

It follows that the algebraic number

$$\beta = \frac{(1 - \zeta_\ell)^{s_a}}{g(\chi^a)}$$

is an $\mathcal{L}$-adic unit; that is, $\text{Ord}_{\mathcal{L}}(\beta) = 0$.

Now we apply the beautiful argument of Kummer to determine s_a. Let b be an integer non-divisible by ℓ. Using Lemmas 7.12 and 7.13, we obtain

$$\beta^{\tau_b} = \frac{(1 - \zeta_\ell^b)^{s_a}}{\bar{\chi}(b)^a g(\chi^a)} = \frac{1}{\bar{\chi}(b)^a} \left(\frac{1 - \zeta_\ell^b}{1 - \zeta_\ell} \right)^{s_a} \beta \equiv \frac{b^{s_a}}{\bar{\chi}(b)^a} \beta \bmod \mathfrak{L}.$$

On the other hand, Proposition 4.19 implies that $\beta^{\tau_b} \equiv \beta \bmod \mathfrak{L}$. It follows that the congruence $\bar{\chi}(b)^a \equiv b^{s_a} \bmod \mathfrak{L}$ holds for every b. Comparing this with (7.5), we obtain

$$b^{a\frac{\ell-1}{p}} \equiv b^{s_a} \bmod \mathfrak{L} \tag{7.8}$$

for all integers b. Hence the same congruence holds $\bmod\, \ell$ as well, which implies that

$$s_a \equiv a \frac{\ell - 1}{p} \bmod (\ell - 1).$$

But we have seen in the very beginning that $0 \le s_a \le \ell - 1$. Hence $s_a = a\frac{\ell-1}{p}$. The theorem is proved.   □

7.5 Proof of Stickelberger's Theorem

We are ready now to prove Theorem 7.1. We begin with the following general statement.

Proposition 7.14. *Let K be a number field. Then every ideal class of K contains a prime ideal (and even infinitely many prime ideals) of degree 1 (over $\mathbb{Q}$).*

Proof. Fix an ideal class $C \in \mathcal{H}_K$. Let L be the Hilbert Class Field of K (see Appendix A.11), and let $\sigma \in \mathrm{Gal}(L/K)$ be the element of the Galois group corresponding, via the Artin map, to the fixed class C. Then for an ideal $\mathfrak{a}$ of K we have

$$\mathfrak{a} \in C \iff \left[\frac{\mathfrak{a}}{L/K} \right] = \sigma.$$

By the Chebotarev density theorem, there exists infinitely many prime ideals $\mathfrak{l}$ of K of degree 1 such that $\left[\frac{\mathfrak{l}}{L/K} \right] = \sigma$. This proves the proposition.   □

Since Stickelberger's ideal is generated by the elements Θ_b (see Proposition 7.2), Stickelberger's theorem is a consequence of Proposition 7.14 and the following assertion.

Proposition 7.15. *Let $\mathfrak{l}$ be a prime ideal of K_p of degree 1. Then for every positive integer b the ideal $\mathfrak{l}^{\Theta_b}$ is principal.*

Proof. Denote by ℓ the prime below $\mathfrak{l}$. Since $\mathfrak{l}$ is a prime ideal of degree 1, we have $\ell \equiv 1 \bmod p$ (see Proposition 4.8), and we can use the setup and the results of Sect. 7.4.

Thus, let χ be the character satisfying (7.5), and let $\mathfrak{l}_1 = \mathfrak{l}, \dots, \mathfrak{l}_{p-1}$ and $\mathfrak{L}_1 = \mathfrak{L}, \dots, \mathfrak{L}_{p-1}$ have the same meaning as in Sect. 7.4. Since

$$\mathfrak{l}_a = \mathfrak{l}^{\sigma_a^{-1}}, \qquad \mathfrak{L}_a = \mathfrak{L}^{\sigma_a^{-1}} \qquad (a = 1, \dots, p-1),$$

the prime decomposition (7.7) can be rewritten as

$$(g(\chi)) = \mathfrak{L}^{\frac{\ell-1}{p}\left(\sigma_1^{-1} + 2\sigma_2^{-1} + \cdots + (p-1)\sigma_{p-1}^{-1}\right)}.$$

In other words, we have

$$\mathfrak{L}^{(\ell-1)\theta} = (g(\chi)). \tag{7.9}$$

Since $\Theta_b = (b - \sigma_b)\theta$, this implies

$$\mathfrak{L}^{(\ell-1)\Theta_b} = \left(g(\chi)^{b-\sigma_b}\right). \tag{7.10}$$

Now, while in (7.9) we had ideals of the field $K_{p\ell}$, in (7.10) we already deal with ideals of K_p. Indeed, the ideal on the left of (7.10) is $\mathfrak{l}^{\Theta_b}$, and on the right we have a principal ideal of K_p, as follows from Theorem 7.6. Thus, $\mathfrak{l}^{\Theta_b}$ is a principal ideal of the field K_p. The proposition is proved, and this completes the proof of Stickelberger's theorem.   □

Remark 7.16. As a careful reader might have noticed, the present proof of Stickelberger's theorem did not make much use of the primality of p. Indeed, one can formulate and prove, using the same argument, Stickelberger's theorem for general cyclotomic fields K_m; one just has to replace everywhere in this chapter p by m and the set $\{1, \dots, p-1\}$ by the set

$$\{a : 0 < a < m, \ (a, m) = 1\}.$$

(By the way, we already did it in Sect. 7.3.) We leave the details to the reader.

7.6 Kummer's Basis

In Proposition 7.2 we found a system of $\mathbb{Z}$-generators of Stickelberger's ideal. In this section we determine its $\mathbb{Z}$-rank and find a free $\mathbb{Z}$-basis.

If χ is a character of G, then $\chi(\iota)$ can be equal to 1 or -1. (Recall that ι stands for the complex conjugation.) A character of G will be called *even* (respectively, *odd*) if $\chi(\iota) = 1$ (respectively, $\chi(\iota) = -1$).

We start from the following statement.

Proposition 7.17. *Let χ be a character of G. Then $\chi(\theta) = 0$ if and only if χ is a nontrivial even character. Moreover,*

$$\prod_{\chi(\iota)=-1} \chi(\theta) = 2^{(p-3)/2} p^{-1} h_p^-, \tag{7.11}$$

where the product extends to the odd characters of G and h_p^- is the relative pth class number.

In this section we need only the first statement. Identity (7.11) is to be used in Sect. 7.8.

Proof. Let χ be a nontrivial character. Then the map $\mathbb{Z} \to \mathbb{C}$ defined by

$$a \mapsto \begin{cases} \chi(\sigma_a) & \text{if } a \text{ is coprime with } p, \\ 0 & \text{if } a \text{ is divisible by } p \end{cases} \tag{7.12}$$

is a nontrivial Dirichlet character mod p, as defined in Sect. 5.1. Moreover, if χ is an even (respectively, odd) character of G then the Dirichlet character (7.12) is even (respectively, odd) as well. Since

$$\chi(\theta) = \frac{1}{p} \sum_{a=1}^{p-1} a \bar{\chi}(\sigma_a) = \frac{1}{p} \overline{\sum_{a=1}^{p-1} a \chi(\sigma_a)}, \tag{7.13}$$

Corollary 5.12 implies that $\chi(\theta) \neq 0$ for an odd χ, and Proposition 5.13 implies that $\chi(\theta) = 0$ for a nontrivial even χ. It remains to notice that $\chi(\theta) = (p-1)/2 \neq 0$ for the trivial character χ.

Finally, identity (7.11) is a direct consequence of the class number formula (5.19).   □

Theorem 7.18. *The $\mathbb{Z}$-rank of Stickelberger's ideal is $(p+1)/2$.*

Proof. Observe that $\theta \mathbb{Q}[G]$ is the ideal of the group ring $\mathbb{Q}[G]$ generated by $\mathcal{I}_S$. It follows that the $\mathbb{Z}$-rank of $\mathcal{I}_S$ is equal to the $\mathbb{Q}$-dimension of $\theta \mathbb{Q}[G]$. Proposition D.13 implies that the latter dimension is the number of characters of G satisfying $\chi(\theta) \neq 0$.

As we have seen in Proposition 7.17, exactly $(p+1)/2$ characters of G do not vanish at θ (all the odd characters and the trivial even character). This proves the theorem.   □

Remark 7.19. Theorem 7.18 can be viewed as an "algebraic reformulation" of Dirichlet's nonvanishing relation $L(1, \chi) \neq 0$ (Corollary 5.5) for the *odd* characters. The similar statement for the *even* characters will be obtained in Chap. 10, see Theorem 10.4.

Recall that for b non-divisible by p we put $\Theta_b = (b - \sigma_b)\theta$. In Proposition 7.2 we showed that every Θ_b belongs to $\mathcal{I}_S$ and that $\Theta_1, \ldots, \Theta_{p-1}$ and $p\theta = \Theta_{p+1}$ generate $\mathcal{I}_S$.

Now for any $b \not\equiv 0, -1 \bmod p$ we put

$$\Psi_b = \Theta_{b+1} - \Theta_b.$$

Also, we consider the *norm element* (see Appendix D.1)

$$\mathcal{N} = \sum_{\sigma \in G} \sigma = \sigma_1 + \cdots + \sigma_{p-1}.$$

One verifies that

$$\mathcal{N} = (1 + \iota)\theta,$$

which implies that $\mathcal{N} \in \mathcal{I}_S$.

We need some simple properties of the elements Θ_b and Ψ_b.

Proposition 7.20. *For any b non-divisible by p we have*

$$\Theta_b + \Theta_{p-b} = p\theta - \mathcal{N}. \tag{7.14}$$

For any $b \not\equiv 0, -1 \bmod p$ we have

$$\Psi_b = \Psi_{p-1-b}. \tag{7.15}$$

Proof. Since $\sigma_{p-b} = \iota \sigma_b$ we have

$$\Theta_b + \Theta_{p-b} = (p - \sigma_b - \iota \sigma_b)\theta = p\theta - \sigma_b \mathcal{N}.$$

This proves (7.14) because $\sigma_b \mathcal{N} = \mathcal{N}$. Equality (7.15) is an immediate consequence of (7.14). $\qquad\square$

Theorem 7.21 (Kummer). *The elements*

$$\Psi_1, \ldots, \Psi_{(p-1)/2}, \mathcal{N} \tag{7.16}$$

form a free $\mathbb{Z}$-basis of $\mathcal{I}_S$.

Proof. Since the $\mathbb{Z}$-rank of $\mathcal{I}_S$ is $(p + 1)/2$, we only have to prove that elements (7.16) generate $\mathcal{I}_S$ over $\mathbb{Z}$. Let $\mathcal{M}$ be the $\mathbb{Z}$-module generated by (7.16).

By Proposition 7.2 it suffices to show that each of $\Theta_1, \ldots, \Theta_{p-1}$ and $p\theta$ belongs to $\mathcal{M}$.

Equality (7.15) implies that $\Psi_1, \ldots, \Psi_{p-2} \in \mathcal{M}$. Further, $\Theta_1 = 0 \in \mathcal{M}$, and by the definition of Ψ_b we have

$$\Theta_b = \Psi_1 + \ldots + \Psi_{b-1} \in \mathcal{M} \qquad (b = 2, \ldots, p-1).$$

Finally, (7.14) implies that $p\theta = \Theta_1 + \Theta_{p-1} + \mathcal{N} \in \mathcal{M}$. The theorem is proved.

□

In Chap. 8 we shall deal mainly with the ideal $(1 - \iota)\mathcal{I}_S$ rather than $\mathcal{I}_S$. In the next proposition we determine the rank and a free basis of this ideal.

Proposition 7.22. *The $\mathbb{Z}$-rank of $(1 - \iota)\mathcal{I}_S$ is $(p-1)/2$. The elements*

$$(1 - \iota)\Psi_b \qquad (b = 1, \ldots, (p-1)/2) \qquad\qquad (7.17)$$

form a free $\mathbb{Z}$-basis of $(1 - \iota)\mathcal{I}_S$.

Proof. As in the proof of Theorem 7.18, the rank is equal to the number of characters χ with $\chi((1 - \iota)\theta) \neq 0$. If $\chi(\iota) = 1$ then, obviously, $\chi((1 - \iota)\theta) = 0$. If $\chi(\iota) = -1$ then $\chi((1 - \iota)\theta) = 2\chi(\theta) \neq 0$ by Proposition 7.17. Thus, there are exactly $(p-1)/2$ characters with the required property, which proves that the rank is $(p-1)/2$.

Further, elements (7.17) generate $(1 - \iota)\mathcal{I}_S$, because elements (7.16) generate $\mathcal{I}_S$ and $(1 - \iota)\mathcal{N} = 0$. Since the number of elements (7.17) is equal to the rank of $(1 - \iota)\mathcal{I}_S$, they form a free basis.   □

7.7 The Real and the Relative Part of Stickelberger's Ideal[4]

The *real part*, or *plus-part*, of the ring $R = \mathbb{Z}[G]$ is the ideal $R^+ := (1 + \iota)R$; the *relative part*, or *minus-part*, is the ideal $R^- := (1 - \iota)R$.

Equivalently: $\Theta \in R^+$ (respectively, $\Theta \in R^-$) if $\iota\Theta = \Theta$ (respectively, if $\iota\Theta = -\Theta$). One more equivalent definition: $\Theta = \alpha_1\sigma_1 + \cdots + \alpha_{p-1}\sigma_{p-1}$ belongs to R^+ (respectively, to R^-) if $\alpha_k = \alpha_{p-k}$ (respectively, $\alpha_k = -\alpha_{p-k}$) for $k = 1, \ldots, p-1$.

The real and relative parts of an ideal $\mathcal{I}$ of R are defined by $\mathcal{I}^+ = \mathcal{I} \cap R^+$ and $\mathcal{I}^- = \mathcal{I} \cap R^-$. Obviously,

$$\mathcal{I}^+ \supseteq (1 + \iota)\mathcal{I}, \qquad \mathcal{I}^{\text{aug}} \supseteq \mathcal{I}^- \supseteq (1 - \iota)\mathcal{I}.$$

(Recall that $\mathcal{I}^{\text{aug}}$ consists of elements of $\mathcal{I}$ of weight 0.)

---

[4]The results of this (and the next) section will not be used in the rest of the book.

In this (and the next) section we study in detail the real part $\mathcal{I}_S^+$ and, mainly, the relative part $\mathcal{I}_S^-$ of Stickelberger's ideal. In particular, we prove the remarkable relation $[R^-:\mathcal{I}_S^-] = h^-$, known as *Iwasawa's class number formula*. This result is not used in the proof of Catalan's conjecture, but we have everything needed for its proof, and it would be unwise to miss such an opportunity.

We start from the real part. Its theory is very simple.

Proposition 7.23. *We have* $\mathcal{I}_S^+ = (1 + \iota)\mathcal{I}_S = \mathcal{N}\mathbb{Z}$. *In particular, the* $\mathbb{Z}$*-rank of* $\mathcal{I}_S^+$ *is 1.*

Proof. Since $(1 + \iota)\theta = \mathcal{N}$, we have $\mathcal{I}_S^+ \supseteq (1 + \iota)\mathcal{I}_S \supseteq \mathcal{N}\mathbb{Z}$, and we have to show that $\mathcal{I}_S^+ \subseteq \mathcal{N}\mathbb{Z}$. Moreover, it suffices to verify that $\mathcal{I}_S^+ \subseteq \mathcal{N}\mathbb{Q}$, because $\mathcal{N}\mathbb{Z} = \mathcal{N}\mathbb{Q} \cap R$.

For any $\Theta \in R^+$, we have $\iota\Theta = \Theta$, which implies $(1 + \iota)\Theta = 2\Theta$. We obtain

$$2\mathcal{I}_S^+ = (1 + \iota)\mathcal{I}_S^+ \subseteq (1 + \iota)\mathcal{I}_S \subseteq (1 + \iota)\theta R = \mathcal{N}R = \mathcal{N}\mathbb{Z}.$$

(Recall that $\mathcal{N}R = \mathcal{N}\mathbb{Z}$ by Proposition D.2.) Thus, $\mathcal{I}_S^+ \subseteq \mathcal{N}\mathbb{Q}$, as wanted. □

Much more substantial is the theory of the relative part $\mathcal{I}_S^-$. We have

$$R^- \supset \mathcal{I}_S^- \supset (1 - \iota)\mathcal{I}_S,$$

and since the rank of both R^- and $(1 - \iota)\mathcal{I}_S$ is $(p - 1)/2$, so is the rank of $\mathcal{I}_S^-$. In particular, both indices $[R^-:\mathcal{I}_S^-]$ and $[\mathcal{I}_S^-:(1 - \iota)\mathcal{I}_S]$ are finite. We are going to determine them. We follow the beautiful exposition of Chapman [22] with insignificant changes.

The basic object that we work with in this (and the next) section is the ideal

$$\mathcal{J} := \{\Phi \in R : \theta\Phi \in R\} \tag{7.18}$$

so that

$$\mathcal{I}_S = \theta\mathcal{J}.$$

In particular, for any $\Theta \in (1 - \iota)\mathcal{I}_S$, there exists $\Phi \in \mathcal{J}$ such that

$$\Theta = (1 - \iota)\theta\Phi.$$

Of course, this Φ is not well defined, but, as we are going to show, the parity of its weight is well defined.

Proposition 7.24. *Let* $\Phi_1, \Phi_2 \in \mathcal{J}$ *be such that* $(1 - \iota)\theta\Phi_1 = (1 - \iota)\theta\Phi_2$. *Then* $w(\Phi_1) \equiv w(\Phi_2) \bmod 2$.

Proof. It suffices to show that

$$(1 - \iota)\theta\Phi = 0 \tag{7.19}$$

implies $2|w(\Phi)$. Equality (7.19) means that $\theta\Phi \in \mathcal{I}_S^+$. Hence $\theta\Phi \in \mathcal{N}\mathbb{Z}$. Since $w(\mathcal{N}) = p - 1$, this implies that $(p - 1) \mid w(\theta\Phi)$. Since $w(\theta) = (p - 1)/2$, the weight of Φ must be even. $\qquad\square$

An element $\Theta = (1 - \iota)\theta\Phi \in (1 - \iota)\mathcal{I}_S$ will be called *even*[5] (respectively, *odd*) if $w(\Phi)$ is even (respectively, odd). Even elements form a subgroup of index 1 or 2 in $(1 - \iota)\mathcal{I}_S$. Since odd elements do exist (for instance, $(1 - \iota)p\theta$ is odd), we obtain the following statement.

Proposition 7.25. *Even elements form an index 2 subgroup of* $(1 - \iota)\mathcal{I}_S$.

After this preparation we are ready to determine the index $\left[\mathcal{I}_S^- : (1 - \iota)\mathcal{I}_S\right]$. Actually, we prove slightly more.

Proposition 7.26. *We have* $\mathcal{I}_S^{\mathrm{aug}} = \mathcal{I}_S^-$ *and* $\left[\mathcal{I}_S^- : (1 - \iota)\mathcal{I}_S\right] = 2^{(p-3)/2}$.

Proof. For any $\Theta \in R^-$ we have $\iota\Theta = -\Theta$, which implies $(1 - \iota)\Theta = 2\Theta$. Hence, for any ideal $\mathcal{I}$ of R, we have $2\mathcal{I}^- = (1 - \iota)\mathcal{I}^-$. In particular,

$$2\mathcal{I}_S^- = (1 - \iota)\mathcal{I}_S^- \subseteq (1 - \iota)\mathcal{I}_S.$$

Since the rank of $\mathcal{I}_S^-$ is $(p - 1)/2$, we have $[\mathcal{I}_S^- : 2\mathcal{I}_S^-] = 2^{(p-1)/2}$. The proposition would follow if we show that $2\mathcal{I}_S^- = 2\mathcal{I}_S^{\mathrm{aug}}$ and $[(1 - \iota)\mathcal{I}_S : 2\mathcal{I}_S^-] = 2$. In view of Proposition 7.25, it suffices to prove that

$$2\mathcal{I}_S^- = 2\mathcal{I}_S^{\mathrm{aug}} = \{\text{the even elements of } (1 - \iota)\mathcal{I}_S\}. \tag{7.20}$$

Obviously, $2\mathcal{I}_S^- \subseteq \mathcal{I}_S^{\mathrm{aug}}$. It remains to prove that $2\mathcal{I}_S^{\mathrm{aug}} \subseteq \mathcal{I}_0$ and $\mathcal{I}_0 \subseteq 2\mathcal{I}_S^-$, where we denote by $\mathcal{I}_0$ the ideal of even elements.

Fix $\Theta \in \mathcal{I}_S^{\mathrm{aug}}$ and write $\Theta = \theta\Phi$ with $\Phi \in \mathcal{J}$. Since $w(\Theta) = 0$ and $w(\theta) \neq 0$, we have $w(\Phi) = 0$. It follows that

$$
\begin{aligned}
2\Theta = 2\theta\Phi &= (1 - \iota)\theta\Phi + (1 + \iota)\theta\Phi \\
&= (1 - \iota)\theta\Phi + \mathcal{N}\Phi \\
&= (1 - \iota)\theta\Phi + w(\Phi)\mathcal{N} \\
&= (1 - \iota)\theta\Phi.
\end{aligned}
$$

---

[5]This notion of parity will be used only in this section. It has nothing to do with the parity of the characters of G.

(Recall that $\Phi\mathcal{N} = \mathrm{w}(\Phi)\mathcal{N}$ by Proposition D.2.) Thus, 2Θ is an even element. This proves the inclusion $2\mathcal{I}_S^{\mathrm{aug}} \subseteq \mathcal{I}_0$.

Further, let $\Theta = (1 - \iota)\theta\Phi$ be an even element, and write $\mathrm{w}(\Phi) = 2m$. Then

$$\Theta + 2m\mathcal{N} = (1 - \iota)\theta\Phi + \mathcal{N}\Phi = (1 - \iota)\theta\Phi + (1 + \iota)\theta\Phi = 2\theta\Phi,$$

which implies that $\Theta + 2m\mathcal{N}$ belongs to $2\mathcal{I}_S$. Hence Θ itself belongs to $2\mathcal{I}_S$. Since $\Theta \in R^-$, we obtain $\Theta \in 2\mathcal{I}_S^-$. This proves the inclusion $\mathcal{I}_0 \subseteq 2\mathcal{I}_S^-$.   □

The index $[R^-:\mathcal{I}_S^-]$ was determined by Iwasawa [53].

Theorem 7.27 (Iwasawa). *The index* $\left[R^-:\mathcal{I}_S^-\right]$ *is equal to* h_p^-, *the relative class number.*

This beautiful result, known as *Iwasawa's class number formula*, is proved in Sect. 7.8.

7.8 Proof of Iwasawa's Class Number Formula

To begin with, we establish an "invariant characterization" of the ideal $\mathcal{J}$, defined in (7.18). Since $\sigma_a\sigma_b = \sigma_{ab}$, the map $\sigma_a \mapsto a$ defines a ring homomorphism $u : R \to \mathbb{Z}/p\mathbb{Z}$.

Proposition 7.28. *We have* $\mathcal{J} = \ker u$ *and* $\mathcal{J}^- = (1 - \iota)\mathcal{J}$. *Also,*

$$[R:\mathcal{J}] = [R^-:(1 - \iota)\mathcal{J}] = p. \tag{7.21}$$

Proof. As follows from Proposition 7.2, the ideal $\mathcal{J}$ is generated by the elements of the form $b - \sigma_b$. Hence $\mathcal{J} = \ker u$. Also, $\mathcal{J}^- = R^- \cap \ker u$, that is, $\mathcal{J}^-$ is the kernel of the restriction $u\,|_{R^-}$. Since u is surjective on both R and R^-, we obtain $R/\mathcal{J} \cong R^-/\mathcal{J}^- \cong \mathbb{Z}/p\mathbb{Z}$. In particular,

$$[R:\mathcal{J}] = [R^-:\mathcal{J}^-] = p.$$

Further, if $\Theta \in \mathcal{J}^-$ then $\Theta = (1 - \iota)\Phi$, where $\Phi \in R$. Since $u(\Theta) = 0$ and $u(1 - \iota) = 2 \neq 0$, we have $u(\Phi) = 0$, that is, $\Phi \in \mathcal{J}$. Thus, $\mathcal{J}^- = (1 - \iota)\mathcal{J}$. This completes the proof of the proposition.   □

For the proof of Theorem 7.27 we shall need a notion of *index* more general than commonly used. Let V be a $\mathbb{Q}$-vector space of finite dimension n. A *lattice* in V is a free abelian subgroup of V of rank n. Any two lattices A and B in V are "commeasurable" in the sense that the intersection $A \cap B$ is of finite index in both A and B. Now we define the index $[A:B]$ by

$$[A:B] := \frac{[A:A \cap B]}{[B:A \cap B]}.$$

Thus, the index is a positive rational number, which is equal to the usual index when $B \subseteq A$. It has the standard properties of the usual index, collected in the following proposition.

Proposition 7.29. *1. The index is multiplicative: if A, B, and C are lattices in V, then $[A{:}C] = [A{:}B] \cdot [B{:}C]$. In particular, $[B{:}A] = [A{:}B]^{-1}$.*
2. *If $f : V \to V$ is a non-singular linear transformation, then for any lattice A we have $[A{:}f(A)] = |\det f|$.*

Proof. In part (1), put $D = A \cap B \cap C$. Then

$$[A{:}B] = \frac{[A{:}A \cap B] \cdot [A \cap B{:}D]}{[B{:}A \cap B] \cdot [A \cap B{:}D]} = \frac{[A{:}D]}{[B{:}D]},$$

and, similarly,

$$[B{:}C] = \frac{[B{:}D]}{[C{:}D]}, \qquad [A{:}C] = \frac{[A{:}D]}{[C{:}D]}.$$

We obtain

$$[A{:}B] \cdot [B{:}C] = \frac{[A{:}D]}{[B{:}D]} \cdot \frac{[B{:}D]}{[C{:}D]} = \frac{[A{:}D]}{[C{:}D]} = [A{:}C],$$

which proves part (1).

The statement of part (2) is obvious if $f(A) \subseteq A$. In the general case, there exists a positive integer λ such that $\lambda f(A) \subseteq A$. Put $g = \lambda f$. Since $g(A) \subseteq A$, we have $[A{:}g(A)] = |\det g| = \lambda^n |\det f|$. Now, using part (1), we obtain

$$[A{:}f(A)] = \frac{[A{:}g(A)]}{[f(A){:}g(A)]} = \frac{\lambda^n |\det f|}{\lambda^n} = |\det f|,$$

as wanted.   □

Proof of Theorem 7.27. We shall apply the previously given definition of index with $V = \mathbb{Q}[G]^-$. We wish to compute the index $[(1 - \iota)\mathcal{J}{:}(1 - \iota)\mathcal{I}_S]$. Since $(1 - \iota)\mathcal{I}_S = (1 - \iota)\theta\mathcal{J}$, the index is equal to $|\det f|$, where f is the multiplication by θ.

To compute the determinant of a linear map we may extend the base field as we please and choose the most convenient basis. Thus, let us extend the base field to $\mathbb{C}$. The ideal $\mathbb{C}[G]^-$ of the group ring $\mathbb{C}[G]$ is the common kernel of the *even* characters (that is, the characters satisfying $\chi(\iota) = 1$). According to Appendix D.5, the ideal $\mathbb{C}[G]^-$ has a $\mathbb{C}$-basis consisting of idempotents ε_χ, where χ runs over the *odd* characters (those with $\chi(\iota) = -1$). Proposition D.15 implies that

$$f(\varepsilon_\chi) = \theta\varepsilon_\chi = \chi(\theta)\varepsilon_\chi.$$

Hence

$$| \det f | = \prod_{\chi(\iota)=-1} \chi(\theta) = 2^{(p-3)/2} p^{-1} h_p^-$$

by Proposition 7.17.

We have proved that

$$[(1-\iota)\mathcal{J}:(1-\iota)\mathcal{I}_S] = 2^{(p-3)/2} p^{-1} h_p^-.$$

Earlier, in Propositions 7.26 and 7.28, we showed that

$$[\mathcal{I}_S^-:(1-\iota)\mathcal{I}_S] = 2^{(p-3)/2}, \qquad [R^-:(1-\iota)\mathcal{J}] = p.$$

Putting all this together, we find

$$[R^-:\mathcal{I}_S^-] = \frac{[R^-:(1-\iota)\mathcal{J}] \cdot [(1-\iota)\mathcal{J}:(1-\iota)\mathcal{I}_S]}{[\mathcal{I}_S^-:(1-\iota)\mathcal{I}_S]} = h_p^-,$$

as wanted.   □

We learned this argument from Chapman [22]. A very similar proof can be found in Lemmermeyer's book [69] ; see Theorem 11.25. Chapter 11 of [69] contains, among other things, a very good historical account of Stickelberger's theory and many useful references.

Theorem 7.27 suggests a natural question: does equality $\left[R^-:\mathcal{I}_S^-\right] = h_p^-$ extend to an isomorphism of abelian groups $R^-/\mathcal{I}_S^-$ and H^-, the relative class group? The answer is "no"; see [136, end of Sect. 6.4] and [69, pages 382–383].

Note in conclusion that Iwasawa's class number formula extends, with the same proof, to the cyclotomic field K_{p^k}. For the general field K_m the situation is more complicated; see Kučera [57] and Sinnott [129].

Chapter 8
Mihăilescu's Ideal

Let (x, y, p, q) be a solution of Catalan's equation. Arguments from Sects. 6.4 and 6.5 illustrate the important role of the elements $\Theta \in \mathbb{Z}[G]$ such that $(x - \zeta)^{\Theta}$ (or, equivalently, $(1 - \zeta x)^{\Theta}$) is a qth power in K.

Elements Θ with this property form an ideal of the group ring $\mathbb{Z}[G]$, called *Mihăilescu's ideal*. It is convenient to study Mihăilescu's ideal on its own, without any reference to Catalan's equation.

Thus, in this chapter, we fix, once and for all, distinct odd prime numbers p and q and a nonzero integer x. We stress that, **unless the contrary is stated explicitly, we do not assume that our x, p, and q come from a solution of Catalan's equation**.

Besides this, we employ in this chapter our standard notation: ζ is a primitive pth root of unity, $K = \mathbb{Q}(\zeta)$ is the pth cyclotomic field, and $G = \mathrm{Gal}(K/\mathbb{Q})$ is its Galois group. We also **fix, once and for all, an embedding $K \hookrightarrow \mathbb{C}$** and view the elements of K as complex algebraic numbers.

In this chapter we use the notion of the height of an algebraic number. The reader is advised to look through Appendix B before reading the chapter.

And the final convention: in this chapter log and arg stand for the **principal branches** of the complex logarithm and argument. That is, for any nonzero complex z, we have

$$-\pi < \arg z = \mathrm{Im} \log z \leq \pi .$$

8.1 Definitions and Main Theorems

We start from the basic definition.

Definition 8.1. Mihăilescu's ideal $\mathcal{I}_M$ is the ideal of the group ring $\mathbb{Z}[G]$ consisting of $\Theta \in \mathbb{Z}[G]$ such that $(x - \zeta)^{\Theta} \in (K^{\times})^q$.

© Springer International Publishing Switzerland 2014
Y.F. Bilu et al., *The Problem of Catalan*, DOI 10.1007/978-3-319-10094-4_8

To formulate the main result, we also need some definitions concerning the group ring $\mathbb{Z}[G]$. Recall (see Appendix D) that the *weight homomorphism* $w : \mathbb{Z}[G] \to \mathbb{Z}$ is defined by

$$ w\left(\sum_{\sigma \in G} a_\sigma \sigma \right) = \sum_{\sigma \in G} a_\sigma . $$

Its kernel, consisting of elements of weight 0, is called the *augmentation ideal* of the group ring $\mathbb{Z}[G]$. Given an ideal $\mathcal{I}$ of $\mathbb{Z}[G]$, we define the *augmented part of* $\mathcal{I}$ as the intersection of $\mathcal{I}$ with the augmentation ideal:

$$ \mathcal{I}^{\mathrm{aug}} = \{ \Theta \in \mathcal{I} : w(\Theta) = 0 \} . $$

In addition to the weight function, we define the *size function* $\| \cdot \|$ by

$$ \left\| \sum_{\sigma \in G} a_\sigma \sigma \right\| = \sum_{\sigma \in G} |a_\sigma| . $$

One immediately verifies the inequalities

$$ \| \Theta_1 \Theta_2 \| \leq \| \Theta_1 \| \cdot \| \Theta_2 \|, \quad \| \Theta_1 + \Theta_2 \| \leq \| \Theta_1 \| + \| \Theta_2 \| . $$

Let $\mathcal{I}$ be an ideal of the ring $\mathbb{Z}[G]$ and r a positive real number. We define the *r-ball* of $\mathcal{I}$ by

$$ \mathcal{I}(r) := \{ \Theta \in \mathcal{I} : \| \Theta \| \leq r \}. $$

Now we are ready to state the first main theorem of this chapter, which is due to Mihăilescu [94] (see also [10]). Roughly speaking, it asserts that, for sufficiently large $|x|$, Mihăilescu's ideal cannot have many elements of zero weight and small size.

Theorem 8.2 (Mihăilescu). *Let ε be a real number satisfying $0 < \varepsilon \leq 1$, and assume that*

$$ |x| \geq \max \left\{ \left(\frac{20 \cdot 2^{p-1}}{(p-1)^2} \right)^{1/\varepsilon}, \frac{4}{\pi} \frac{q}{p-1} + 1 \right\} . \tag{8.1} $$

Put $r = (2 - \varepsilon)q/(p-1)$. Then $\left| \mathcal{I}_M^{\mathrm{aug}}(r) \right| \leq q$.

Obviously, $\mathcal{I}(r) = \mathcal{I}(\lfloor r \rfloor)$ for any $\mathcal{I}$ and r. Since

$$ \left\lfloor \frac{2q}{p-1} \right\rfloor \geq \frac{2q-2}{p-1} $$

(except the case $p = 3$), we obtain, specifying $\varepsilon = 2/q$, the following consequence: for $p \geq 5$ and

$$x \geq \left(\frac{20 \cdot 2^{p-1}}{(p-1)^2} \right)^{q/2} \tag{8.2}$$

Mihăilescu's ideal has at most q elements of weight 0 and of size not exceeding $2q/(p-1)$. Similarly, specifying $\varepsilon = 1/q$, we deduce that for $p = 3$ and $|x| \geq 20^q$, Mihăilescu's ideal has at most q elements of weight 0 and of size not exceeding $q - 1$.

Condition (8.2) is too strong for applications to Catalan's problem. Fortunately, the following consequence, obtained by taking $\varepsilon = 1$, is sufficient for us.

Theorem 8.3. *Assume that* $|x| \geq \max\{2^{p+2}, q\}$. *Put* $r = q/(p-1)$. *Then* $\left| \mathcal{I}_M^{\mathrm{aug}}(r) \right| \leq q$.

To deduce it from Theorem 8.2, just observe that $20 \cdot 2^{p-1}/(p-1)^2 \leq 2^{p+2}$ and $(4/\pi)q/(p-1) + 1 \leq q$.

When $p \leq (1 - \varepsilon/2)q + 1$ and (8.1) is satisfied, Theorem 8.2 implies that $|\mathcal{I}_M^{\mathrm{aug}}(2)| \leq q$; that is, $\mathcal{I}_M$ has at most $q - 1$ elements of weight 0 and size 2. This can be refined when (8.1) is replaced by a slightly stronger assumption.

Theorem 8.4. *Let* ε *be a real number satisfying* $0 < \varepsilon \leq 1$, *and assume that* $p \leq (2 - \varepsilon)q + 1$. *Assume further that*

$$|x| \geq \max \left\{ \left(\frac{20 \cdot 2^{p-1}}{(p-1)^2} \right)^{1/\varepsilon}, 8q^q \right\}, \tag{8.3}$$

Then $\mathcal{I}_M^{\mathrm{aug}}(2) = \{0\}$.

Thus, when (8.3) is satisfied, Mihăilescu's ideal cannot have elements of weight 0 and size 2.

Again, it is useful to state separately the particular case of Theorem 8.4 corresponding to $\varepsilon = 1$.

Theorem 8.5. *Assume that* $p < q$ *and that* $|x| \geq 8q^q$. *Then* $\mathcal{I}_M^{\mathrm{aug}}(2) = \{0\}$.

To deduce Theorem 8.5 from Theorem 8.4, observe that $q > 2$ and $16 > 20/(p-1)^2$. Hence $8q^q > 16 \cdot 2^{q-1} > 20 \cdot 2^{p-1}/(p-1)^2$ whenever $q > p$.

Theorems 8.2 and 8.4 will be proved in Sects. 8.4 and 8.5, respectively, after some preparation in Sects. 8.2 and 8.3.

8.2   The Algebraic Number $(x - \zeta)^\Theta$

In this section, we investigate the number $(x - \zeta)^\Theta$. First of all, we estimate its height.

Obviously, $(x - \zeta)^\Theta$ is a product of $\|\Theta\|$ factors of type $(x - \zeta^\sigma)^{\pm 1}$. We have $h(\zeta^\sigma) = 0$, because ζ^σ is a root of unity, and $h(x) = \log |x|$ by Proposition B.2(5). Hence, using Proposition B.2(7), we obtain

$$h(x - \zeta^\sigma) \le \log |x| + \log 2 = \log |2x|$$

and $h\left((x - \zeta)^\Theta\right) \le \|\Theta\| \log |2x|$. With slightly more effort one can show that $h(x - \zeta^\sigma) \le \log(|x| + 1)$ and thereby $h\left((x - \zeta)^\Theta\right) \le \|\Theta\| \log(|x| + 1)$.

This is already a reasonable estimate, but one can drastically improve on it in the case $w(\Theta) = 0$ using Proposition B.2(6).

Proposition 8.6. *Let* $\Theta \in \mathbb{Z}[G]$ *satisfy* $w(\Theta) = 0$. *Then*

$$h\left((x - \zeta)^\Theta\right) \le \frac{1}{2}\|\Theta\| \log(|x| + 1).$$

Before proving the proposition, we make a simple remark. We say that $\Theta = \sum_{\sigma \in G} a_\sigma \sigma$ is *nonnegative* (notation: $\Theta \ge 0$) if $a_\sigma \ge 0$ for all $\sigma \in G$. For such Θ we have $w(\Theta) = \|\Theta\|$.

Any Θ can be presented as a difference $\Theta_+ - \Theta_-$, where $\Theta_+, \Theta_- \ge 0$ and $\|\Theta\| = \|\Theta_+\| + \|\Theta_-\|$. Indeed, write $\Theta = \sum_{\sigma \in G} a_\sigma \sigma$ and put

$$\Theta_+ = \sum_\sigma \max\{a_\sigma, 0\}\, \sigma, \quad \Theta_- = -\sum_\sigma \min\{a_\sigma, 0\}\, \sigma.$$

Proof of Proposition 8.6. Write $\Theta = \Theta_+ - \Theta_-$ as above. Since $w(\Theta) = 0$, we have $w(\Theta_+) = w(\Theta_-)$. Since $\Theta_+, \Theta_- \ge 0$, this means that $\|\Theta_+\| = \|\Theta_-\|$. We denote this number by m:

$$m = \|\Theta_+\| = \|\Theta_-\| = \frac{1}{2}\|\Theta\|.$$

We have $(x - \zeta)^\Theta = \alpha/\beta$, where $\alpha = (x - \zeta)^{\Theta_+}$ and $\beta = (x - \zeta)^{\Theta_-}$ are algebraic integers. Since α is a product of m terms of the type $x - \zeta^\sigma$, we have $|\alpha| \le (|x| + 1)^m$. More generally, for any $\sigma \in G$ we have $|\alpha^\sigma| \le (|x| + 1)^m$, and, similarly, $|\beta^\sigma| \le (|x| + 1)^m$.

Since α and β are algebraic integers, we may use (B.5). We obtain

$$\mathrm{h}\left((x - \zeta)^\Theta\right) \leq \frac{1}{[K:\mathbb{Q}]} \sum_{\sigma \in G} \log \max \{|\alpha^\sigma|, |\beta^\sigma|\}$$

$$\leq \frac{1}{[K:\mathbb{Q}]} \sum_{\sigma \in G} m \log(|x| + 1)$$

$$= m \log(|x| + 1),$$

as wanted.   □

Next, we observe that, for large x, the algebraic number $(x - \zeta)^\Theta$ is "very close" to 1 if $\mathrm{w}(\Theta) = 0$.

Proposition 8.7. *If* $|x| > 1$ *and* $\mathrm{w}(\Theta) = 0$ *then*

$$\left|\log(x - \zeta)^\Theta\right| \leq \frac{\|\Theta\|}{|x| - 1}.$$

(Recall that log stands for the principal branch of the complex logarithm.)

Proof. For any complex z satisfying $|z| < 1$ we have

$$|\log(1 + z)| \leq \frac{|z|}{1 - |z|}.$$

In particular,

$$\left|\log\left(1 - \frac{\zeta^\sigma}{x}\right)\right| \leq \frac{1}{|x| - 1}.$$

Since $(x - \zeta)^\Theta = (1 - \zeta/x)^\Theta$ when $\mathrm{w}(\Theta) = 0$, the result follows.   □

Finally, we show that $(x - \zeta)^\Theta$ is distinct from 1 for $|x| \geq 3$.

Proposition 8.8. *Assume that* $|x| \geq 3$. *Then, for a nonzero* $\Theta \in \mathbb{Z}[G]$, *we have* $(x - \zeta)^\Theta \neq 1$.

Proof. Let $\mathfrak{p}$ be the prime ideal of K lying over p. Then $\mathfrak{p}^{p-1} = (p)$ and $\mathfrak{p} = (\zeta^\sigma - \zeta^\tau)$ for any distinct $\sigma, \tau \in G$. In particular, for distinct σ and τ we have

$$(x - \zeta^\sigma, x - \zeta^\tau) | \mathfrak{p}. \tag{8.4}$$

If $x - \zeta$ has no prime divisors other than $\mathfrak{p}$, then $(x - \zeta) = \mathfrak{p}^k$. In this case $(x - \zeta^\sigma) = \mathfrak{p}^k$ for any $\sigma \in G$, because $\mathfrak{p}$ is stable under the Galois action. Now (8.4) implies that $k \leq 1$. It follows that the norm of $x - \zeta$ is either ± 1 or $\pm p$.

On the other hand, the assumption $|x| \geq 3$ implies that

$$|\mathcal{N}(x - \zeta)| = \prod_{\sigma \in G} |x - \zeta^{\sigma}| \geq 2^{p-1} > p \, .$$

This shows that $x - \zeta$ has a prime divisor $\tilde{\mathfrak{p}}$ distinct from $\mathfrak{p}$.
Put $\ell = \mathrm{Ord}_{\tilde{\mathfrak{p}}}(x - \zeta)$. Then (8.4) implies that

$$\mathrm{Ord}_{\tilde{\mathfrak{p}}^{\sigma}}(x - \zeta^{\tau}) = \begin{cases} \ell, & \text{if } \sigma = \tau, \\ 0, & \text{if } \sigma \neq \tau. \end{cases}$$

Therefore, writing $\Theta = \sum_{\sigma \in G} a_{\sigma} \sigma$, we obtain

$$\mathrm{Ord}_{\tilde{\mathfrak{p}}^{\sigma}}\left((x - \zeta)^{\Theta}\right) = \ell a_{\sigma} \quad (\sigma \in G).$$

Now, if $(x - \zeta)^{\Theta} = 1$ then $\ell a_{\sigma} = 0$ for all $\sigma \in G$. Since $\ell \neq 0$, we conclude that $a_{\sigma} = 0$ for all σ. Hence $\Theta = 0$. $\qquad \square$

8.3 The qth Root of $(x - \zeta)^{\Theta}$

By the definition of Mihăilescu's ideal, for every $\Theta \in \mathcal{I}_M$, the algebraic number $(x - \zeta)^{\Theta}$ has a qth root in K. Actually, this root is unique and has some nice properties.

Proposition 8.9. *1. For any $\Theta \in \mathcal{I}_M$ there exists a unique $\alpha(\Theta) \in K^{\times}$ such that $\alpha(\Theta)^q = (x - \zeta)^{\Theta}$.*
2. For any $\Theta_1, \Theta_2 \in \mathcal{I}_M$ we have

$$\alpha(\Theta_1 + \Theta_2) = \alpha(\Theta_1)\alpha(\Theta_2). \tag{8.5}$$

In other words, the map $\alpha : \mathcal{I}_M \to K^{\times}$ is a group homomorphism.
3. For $\sigma \in G$ and $\Theta \in \mathcal{I}_M$ we have $\alpha(\Theta\sigma) = \alpha(\Theta)^{\sigma}$.
4. If $|x| \geq 3$ then for a nonzero $\Theta \in \mathcal{I}_M$ we have $\alpha(\Theta) \neq 1$. In other words, the homomorphism $\alpha : \mathcal{I}_M \to K^{\times}$ is injective.
5. For any $\Theta \in \mathcal{I}_M^{\mathrm{aug}}$ we have

$$\mathrm{h}(\alpha(\Theta)) \leq \frac{1}{2q}\|\Theta\|\log(|x| + 1). \tag{8.6}$$

Proof. In part (1), only uniqueness is to be proved. Thus, let $\alpha_1, \alpha_2 \in K$ satisfy $\alpha_1^q = \alpha_2^q = (x - \zeta)^{\Theta}$. Then $\alpha_1/\alpha_2 \in K$ is a qth root of unity. Since K does not contain qth roots of unity other than 1 (Proposition 4.4(4)), we obtain $\alpha_1 = \alpha_2$.

To prove part (2), observe that both parts of (8.5) belong to K, and their qth powers are equal to $(x - \zeta)^{\Theta_1 + \Theta_2}$. By the uniqueness, they should be equal.

Similarly one proves part (3): the qth powers of $\alpha(\Theta)^\sigma$ and of $\alpha(\sigma\Theta)$ are both equal to $(x - \zeta)^{\sigma\Theta}$. By the uniqueness, $\alpha(\Theta)^\sigma = \alpha(\sigma\Theta)$.

Part (4) follows from Proposition 8.8, and part (5) is a consequence of Proposition 8.6 and (B.9).   □

In the sequel, we shall exploit the notion of the *nearest qth root of unity*. Let z be a nonzero complex number. Then there exists a unique qth root of unity ξ (called the *nearest qth root of unity* for z) such that

$$-\pi/q < \arg\left(z\xi^{-1}\right) \le \pi/q \,.$$

Some simple properties of the nearest qth root of unity are collected in the following proposition. The proofs are left to the reader.

Proposition 8.10. *1. Let z be a nonzero complex number and ξ its nearest qth root of unity. Then*

$$\log\left(z\xi^{-1}\right) = \frac{1}{q}\log\left(z^q\right).$$

2. *Let z be a nonzero complex number and ξ its nearest qth root of unity. Assume that $\left|\arg\left(z^q\right)\right| < \pi$ (that is, z^q is not a negative real number). Then ξ^{-1} is the nearest qth root of unity of z^{-1}.*
3. *Let z_1, z_2 be nonzero complex numbers and ξ_1, ξ_2 their nearest qth roots of unity, respectively. Assume that $\left|\arg z_1^q\right|, \left|\arg z_2^q\right| < \pi/2$. Then $\xi_1\xi_2$ is the nearest qth root of unity of $z_1 z_2$.*

Now we define a new map $\xi : \mathcal{I}_M \to \boldsymbol{\mu}_q$, where $\boldsymbol{\mu}_q$ stands for the group of qth roots of unity. For $\Theta \in \mathcal{I}_M$ we let $\xi(\Theta)$ be the nearest qth root of unity for $\alpha(\Theta)$. As we have seen in Proposition 8.7, when $\mathrm{w}(\Theta) = 0$, the number $\alpha(\Theta)^q$ is "very close" to 1 for large x. Hence $\alpha(\Theta)$ should, under the same assumptions, be "very close" to $\xi(\Theta)$. Since $\alpha : \mathcal{I}_M \to K^\times$ is a group homomorphism, we may expect from the map $\xi : \mathcal{I}_M^{\mathrm{aug}} \to \boldsymbol{\mu}_q$ certain "homomorphism-like" behavior. All this is realized in the following proposition.

Proposition 8.11. *1. For any $\Theta \in \mathcal{I}_M^{\mathrm{aug}}$ we have*

$$\left|\log\left(\alpha(\Theta)\xi(\Theta)^{-1}\right)\right| \le \frac{1}{q}\frac{\|\Theta\|}{|x| - 1} \,. \tag{8.7}$$

2. *If $\Theta \in \mathcal{I}_M^{\mathrm{aug}}$ satisfies $\|\Theta\| < \pi(|x| - 1)$ then $\xi(-\Theta) = \xi(\Theta)^{-1}$.*

3. *If* $\Theta_1, \Theta_2 \in \mathcal{I}_M^{\mathrm{aug}}$ *satisfy*

$$\|\Theta_1\|, \|\Theta_2\| < \frac{\pi}{2}(|x| - 1),$$

then

$$\xi(\Theta_1 + \Theta_2) = \xi(\Theta_1)\xi(\Theta_2).$$

Proof. Apply Propositions 8.7 and 8.10.   □

Let A and B be two groups and S a subset of A. We say that a map $f : S \to B$ is a *quasi-homomorphism* if for any $x, y \in S$ we have

$$f(x^{-1}) = f(x)^{-1}, \qquad f(xy) = f(x)f(y).$$

In these terms, parts (2) and (3) of Proposition 8.11 can be reformulated as follows: *for any $r < \frac{\pi}{2}(|x| - 1)$ the map $\xi : \mathcal{I}_M^{\mathrm{aug}}(r) \to \mu_q$ is a quasi-homomorphism.*

8.4   Proof of Theorem 8.2

After all this preparation, we are ready to prove Theorem 8.2. Recall that we fix a positive $\varepsilon \le 1$ and define

$$r = (2 - \varepsilon)\frac{q}{p - 1}.$$

Let us rewrite (8.1) in the form we are going to use in the proof. First of all,

$$|x| \ge \left(\frac{20 \cdot 2^{p-1}}{(p-1)^2}\right)^{1/\varepsilon}. \qquad (8.8)$$

Second, we have $|x| \ge (4/\pi)q/(p - 1) + 1$, which implies

$$r < \frac{\pi}{2}(|x| - 1). \qquad (8.9)$$

Theorem 8.2 is an easy consequence of the following statement.

Proposition 8.12. *Assume (8.8) and (8.9), and let $\Theta \in \mathcal{I}_M^{\mathrm{aug}}(2r)$ satisfy $\xi(\Theta) = 1$. Then $\Theta = 0$.*

Indeed, assume that Proposition 8.12 is true. Let $\Theta_1, \Theta_2 \in \mathcal{I}_M^{\mathrm{aug}}(r)$ satisfy $\xi(\Theta_1) = \xi(\Theta_2)$. Inequality (8.9) implies that

$$\|\Theta_1\|, \|\Theta_2\| < \frac{\pi}{2}(|x| - 1).$$

Then $\xi(\Theta_1 - \Theta_2) = 1$ by Proposition 8.11, and Proposition 8.12 implies that $\Theta_1 - \Theta_2 = 0$. We have shown that the map $\xi : \mathcal{I}_M^{\mathrm{aug}}(r) \to \mu_q$ is injective, which proves Theorem 8.2.

For the proof of Proposition 8.12, we need a simple lemma.

Lemma 8.13. *Let a be a positive real number. Then for any complex w, satisfying $|w| \leq a$, we have*

$$|e^w - 1| \leq \frac{e^a - 1}{a}|w|.$$

Proof. For $|z| \leq 1$ we have

$$|e^{az} - 1| = \left| az + \frac{(az)^2}{2!} + \frac{(az)^3}{3!} + \ldots \right| \leq a + \frac{a^2}{2!} + \frac{a^3}{3!} + \ldots = e^a - 1.$$

Hence the complex function $f(z) = (e^{az} - 1)/(e^a - 1)$ satisfies $f(0) = 0$ and $|f(z)| \leq 1$ for $|z| \leq 1$. By the Schwarz lemma, $|f(z)| \leq |z|$ for $|z| \leq 1$. Putting $z = w/a$, we obtain the result.   □

Proof of Proposition 8.12. Fix a nonzero $\Theta \in \mathcal{I}_M^{\mathrm{aug}}(2r)$ with $\xi(\Theta) = 1$ and put $\alpha = \alpha(\Theta)$. Inequality (8.7) becomes

$$|\log \alpha| \leq \frac{1}{q}\frac{\|\Theta\|}{|x| - 1}. \tag{8.10}$$

We shall obtain an upper estimate for $|\alpha - 1|$, using (8.10), and a lower estimate, using "Liouville's inequality" (Proposition B.3). Then we shall see that the two estimates are contradictory when (8.8) holds.

Throughout the proof we shall use the lower bound $|x| \geq 20$, which follows from (8.8).

We start with the upper estimate. Since $\|\Theta\| \leq 2r < 4q/(p - 1)$ by the assumption, we have

$$|\log \alpha| \leq \frac{4}{(p - 1)(|x| - 1)} \leq \frac{2}{19},$$

because $p \geq 3$ and $|x| \geq 20$. Using Lemma 8.13, we obtain

$$|\alpha - 1| \leq \frac{e^{2/19} - 1}{2/19}|\log \alpha| \leq \frac{1.06}{q}\frac{\|\Theta\|}{|x| - 1}. \tag{8.11}$$

On the other hand, $\alpha \neq 1$ by Proposition 8.9(4). Hence we may estimate $|\alpha - 1|$ from below, using "Liouville's inequality" (B.13) (with $\alpha - 1$ instead of α):

$$|\alpha - 1|^2 \geq e^{-[K:\mathbb{Q}]h(\alpha-1)} = e^{-(p-1)h(\alpha-1)}. \tag{8.12}$$

To make use of this, we have to estimate $h(\alpha - 1)$. From (8.6) and (B.6) we obtain

$$h(\alpha - 1) \leq \frac{1}{2q}\|\Theta\| \log(|x| + 1) + \log 2. \tag{8.13}$$

Now, combining the upper estimate (8.11) with the lower estimate (8.12), and taking into account (8.13), we obtain

$$2\left(\log(|x| - 1) - \log \frac{1.06\|\Theta\|}{q}\right) \leq (p-1)\left(\frac{\|\Theta\|}{2q} \log(|x| + 1) + \log 2\right).$$

Since

$$\|\Theta\| \leq 2r = 2(2 - \varepsilon)\frac{q}{p-1},$$

the inequality remains valid when we replace $\|\Theta\|$ by $2(2-\varepsilon)q/(p-1)$ in the right-hand side and by $4q/(p-1)$ in the left-hand side. We obtain

$$2\log(|x| - 1) - 2\log \frac{4.24}{p-1} \leq (2-\varepsilon)\log(|x| + 1) + (p-1)\log 2,$$

which can be rewritten as

$$\varepsilon \log(|x| + 1) \leq 2\log 4.24 + 2\log \frac{|x| + 1}{|x| - 1} + \log \frac{2^{p-1}}{(p-1)^2}.$$

Using again the inequality $|x| \geq 20$, we obtain

$$\varepsilon \log(|x| + 1) \leq 2\log 4.24 + 2\log \frac{21}{19} + \log \frac{2^{p-1}}{(p-1)^2} < \log \frac{20 \cdot 2^{p-1}}{(p-1)^2},$$

which contradicts (8.8). This proves Proposition 8.12 and Theorem 8.2.   □

8.5   Proof of Theorem 8.4

In this section, we prove Theorem 8.4. We start with a simple lemma, which gives a quantitative version of the following obvious fact: if a complex number is "close" to a qth root of unity, distinct from 1, then it cannot be close to 1.

Lemma 8.14. *Let α be a nonzero complex number and ξ a qth root of unity, distinct from 1. Assume that $\left|\log(\alpha\xi^{-1})\right| \leq 0.1/q$. Then $|\alpha - 1| \geq 5/q$.*

Proof. Since $\xi \neq 1$, we have $\xi = e^{2\pi i k/q}$, where $k \in \{1, 2, \ldots, q-1\}$. Hence $|\xi - 1| = 2\left|\sin(\pi k/q)\right| \geq 2\sin(\pi/q)$. Since the function $x \mapsto (\sin x)/x$ decreases on $[0, \pi/2]$, we have

$$\frac{\sin(\pi/q)}{\pi/q} \geq \frac{\sin(\pi/3)}{\pi/3},$$

which implies

$$|\xi - 1| \geq 2\sin(\pi/q) \geq 2 \cdot \frac{\sin(\pi/3)}{\pi/3} \cdot \frac{\pi}{q} \geq \frac{5.19}{q}.$$

Further, we use Lemma 8.13 to obtain

$$|\alpha - \xi| = \left|\alpha\xi^{-1} - 1\right| \leq \frac{e^{0.1} - 1}{0.1} \cdot \frac{0.1}{q} \leq \frac{0.11}{q}.$$

Combining the two estimates, we obtain $|\alpha - 1| \geq 5.19/q - 0.11/q > 5/q$, as wanted.   □

Proof of Theorem 8.4. First of all, recall our assumptions:

$$p \leq (2 - \varepsilon)q + 1, \tag{8.14}$$

$$|x| \geq \left(\frac{20 \cdot 2^{p-1}}{(p-1)^2}\right)^{1/\varepsilon}, \tag{8.15}$$

$$x \geq 8q^q. \tag{8.16}$$

Inequality (8.16) implies that $r = (2 - \varepsilon)q/(p - 1)$ satisfies

$$r \leq \frac{\pi}{2}(|x| - 1).$$

Together with (8.15) this implies that the assumptions of Proposition 8.12 are satisfied.

Assume that $\mathcal{I}_M^{\mathrm{aug}} \ni \Theta$ with $\|\Theta\| = 2$, and put $\alpha = \alpha(\Theta)$. We are going to estimate from below $|\alpha - 1|_v$ for all valuations $v \in M_K$. (For Archimedean v this is equivalent to estimating $|\alpha^\sigma - 1|$ for all $\sigma \in G$, which will be done using Proposition 8.12 and Lemma 8.14.) Lower estimates for $|\alpha - 1|_v$ would imply an upper estimate for the height $\mathrm{h}\left((\alpha - 1)^{-1}\right)$, which is equal to $\mathrm{h}(\alpha - 1)$. This, in turn, would imply an upper estimate for $\log|x| = \mathrm{h}(x)$, contradicting the assumption (8.16).

Now to the proof. Inequality (8.14) can be rewritten as $2r \geq 2$. Hence, Proposition 8.12 applies to Θ and we obtain $\xi = \xi(\Theta) \neq 1$. On the other hand, inequality (8.16) implies that $|x| \geq 8 \cdot 3^3 = 216$. Together with (8.7) this yields $\left| \log \left(\alpha \xi^{-1} \right) \right| < 0.1/q$ (we are very generous!), and Lemma 8.14 implies that $|\alpha - 1| \geq 5/q$.

We may apply the same argument for $\sigma \Theta$, where $\sigma \in G$. Since $\alpha(\sigma \Theta) = \alpha^\sigma$, we obtain $|\alpha^\sigma - 1| > 5/q$ for any $\sigma \in G$. Hence, denoting by M_K^∞ the subset of M_K consisting of the Archimedean valuations, we obtain

$$\frac{1}{[K:\mathbb{Q}]} \sum_{v \in M_K^\infty} \log \max \left\{ 1, \left| (\alpha - 1)^{-1} \right|_v \right\}$$

$$= \frac{1}{[K:\mathbb{Q}]} \sum_{\sigma \in G} \log \max \left\{ 1, \left| (\alpha^\sigma - 1)^{-1} \right| \right\} \leq \log \frac{q}{5}. \tag{8.17}$$

Now write $\Theta = \sigma_1 - \sigma_2$, where σ_1 and σ_2 are distinct elements of G, and put $\zeta_i = \zeta^{\sigma_i}$. Assume that $|\alpha - 1|_v < 1$ for a non-Archimedean v. Then $|\alpha^q - 1|_v \leq |\alpha - 1|_v < 1$. However,

$$\alpha^q - 1 = (x - \zeta)^\Theta - 1 = \frac{\zeta_2 - \zeta_1}{x - \zeta_2}. \tag{8.18}$$

It follows that, for a non-Archimedean v, we have

$$|\alpha - 1|_v \geq |\zeta_2 - \zeta_1|_v = \begin{cases} p^{-1} & \text{if } v \mid p, \\ 1 & \text{otherwise.} \end{cases}$$

Hence, denoting by M_K^f the subset of M_K consisting of the non-Archimedean valuations, and recalling that $[K:\mathbb{Q}] = p - 1$, we obtain

$$\frac{1}{[K:\mathbb{Q}]} \sum_{v \in M_K^f} \log \max \left\{ 1, \left| (\alpha - 1)^{-1} \right|_v \right\} \leq \frac{\log p}{p - 1} \leq \frac{\log 3}{2}.$$

Together with (8.17) this implies the estimate

$$h(\alpha - 1) = h \left((\alpha - 1)^{-1} \right) \leq \log q - \log \frac{5}{\sqrt{3}} < \log q - \log 2.$$

This and (B.6) imply that

$$h(\alpha) \leq h(\alpha - 1) + \log 2 < \log q.$$

Now, rewriting (8.18) as

$$x = \frac{\zeta_2 - \zeta_1}{\alpha^q - 1} + \zeta_2,$$

and repeatedly using (B.6) and (B.9), we obtain

$$\log |x| = h(x) = h\left(\frac{\zeta_2 - \zeta_1}{\alpha^q - 1} + \zeta_2\right) \le q h(\alpha) + 3\log 2 < q \log q + \log 8,$$

which contradicts (8.16). The theorem is proved.   □

Remark 8.15. With a slight additional effort $8q^q$ can be replaced by $3q^q$ and even by q^q for $q \ge 5$.

8.6 Application to Catalan's Problem I: Divisibility of the Class Number

In this section we apply Theorem 8.5 to Catalan's equation. The following result is due to Bugeaud and Hanrot [14].

Theorem 8.16. *Let (x, y, p, q) be a solution of Catalan's equation with $q > p$. Then q divides the class number h_p.*

Proof. Since $p \ge 3$ and $q \ge 5$, Hyyrö's bound (3.18) implies that

$$|x| \ge p^{q-1}(q-1)^q = \frac{q-1}{q}\left(\frac{p(q-1)}{q}\right)^{q-1} q^q \ge \frac{4}{5}\left(\frac{12}{5}\right)^4 q^q > 8q^q.$$

Hence Theorem 8.5 applies in our case.

On the other hand, Proposition 6.5 can be formulated as follows: if q does not divide h_p then for any $\sigma \in G$ we have $\sigma - \bar\sigma \in \mathcal{I}_M$. But $\sigma - \bar\sigma$ cannot belong to $\mathcal{I}_M$ by Theorem 8.5. Hence $q \mid h_p$.   □

A fundamental consequence of this theorem is that **for every p there exist only finitely many possible q.** (Indeed, if p is given, then any possible q should satisfy either $q < p$ or $q \mid h_p$.) Originally, this was proved by Schinzel and Tijdeman [122] using Baker's method. See Theorem 13.18 for a more general result.

One more estimate for q in terms of p will be given in Theorem 8.19.

Besides this, Theorem 8.16 allows one to exclude many pairs of exponents p and q. For this purpose, it is more practical to use the following refinement of Theorem 8.16, also from [14].

Theorem 8.17. *Let (x, y, p, q) be a solution of Catalan's equation with $q > p$. Then q divides the relative class number h_p^-.*

The proof is the same as for Theorem 8.16, but Proposition 6.5 should be replaced by the refinement indicated in Remark 6.6.

Corollary 8.18. *Let* (x, y, p, q) *be a solution of Catalan's equation. Then* $p, q \geq 43$.

Proof. We may assume that $q > p$. Theorem 8.17 implies that $q \mid h_p^-$. However, according to Table 5.1, for $p \leq 41$ the prime divisors of the relative class number h_p^- do not exceed p. This proves the corollary. $\square$

Using Corollary 6.15, one can go further. For instance, since $h_{43}^- = 211$, we have $q = 211$ for $p = 43$. However,

$$211^{42} \not\equiv 1 \bmod 43^2,$$

which contradicts Corollary 6.15. It follows that $p \geq 47$ and so on.

Let us mention that Mihăilescu [94] showed, by a curious p-adic argument, that Theorem 8.17 holds without the assumption $q > p$. We do not use this result in the present book.

8.7   Application to Catalan's Problem II: Mihăilescu's Ideal vs Stickelberger's Ideal

In this section we apply Theorem 8.2 to Catalan's problem. Roughly speaking, this theorem tells us that Mihăilescu's ideal contains "few" elements of small size and weight 0. On the other hand, Stickelberger's Theorem 7.1, together with Proposition 6.11, implies that

$$\mathcal{I}_M \supset (1 - \iota)\mathcal{I}_S \tag{8.19}$$

whenever (x, y, p, q) is a solution of Catalan's equation. From this one can deduce that Mihăilescu's ideal contains "many" small elements of weight 0. Elaborating this idea, we prove the following theorem.

Theorem 8.19 (Mihăilescu). *Let* (x, y, p, q) *be a solution of Catalan's equation. Then* $q < 3(p - 1)^2$.

This theorem and Mihăilescu's "double Wieferich" criterion (Corollary 6.15) have the following consequence, which will be crucial in Chap. 11.

Corollary 8.20. *Let* (x, y, p, q) *be a solution of Catalan's equation. Then* $q \not\equiv 1 \bmod p$ *(and, by symmetry, $p \not\equiv 1 \bmod q$).*

Proof. Assume that $q \equiv 1 \bmod p$. In addition to this, we have $q^{p-1} \equiv 1 \bmod p^2$, by Corollary 6.15. The two congruences together imply that $q \equiv 1 \bmod p^2$.

Now observe that equalities $q = 1 + p^2$ and $1 + 3p^2$ are impossible because both p and q are odd. Also, $q = 1 + 2p^2$ is impossible, because $1 + 2p^2$ is divisible by 3 when $p > 3$. (We can dismiss the case $p = 3$ by Corollary 8.18.) Thus, $q \geq 1 + 4p^2$, which contradicts Theorem 8.19. □

The proof of Theorem 8.19 requires a lower bound for the cardinality of the r-ball of the ideal $(1 - \iota)\mathcal{I}_S$. First of all, let us estimate the size of the elements of the basis (7.17). We use the notation

$$\mathcal{I}^* = (1 - \iota)\mathcal{I}_S, \qquad \Psi_b^* = (1 - \iota)\Psi_b.$$

Proposition 8.21. *For $b = 1, \ldots, p - 2$ we have $\Psi_b \geq 0$. Also,*

$$\|\Psi_b\| = \mathrm{w}(\Psi_b) = \frac{p - 1}{2}, \qquad \|\Psi_b^*\| \leq p - 1.$$

Proof. Using (7.1), we obtain

$$\Psi_b = \sum_{a=1}^{p-1} \left(\left\lfloor \frac{(b+1)a}{p} \right\rfloor - \left\lfloor \frac{ba}{p} \right\rfloor \right) \sigma_a^{-1} \geq 0.$$

Further, a straightforward calculation shows that $\mathrm{w}(\theta) = (p - 1)/2$. It follows that

$$\mathrm{w}(\Theta_b) = \mathrm{w}(b - \sigma_b)\mathrm{w}(\theta) = (b - 1)\frac{p - 1}{2}.$$

Since $\Psi_b \geq 0$, we obtain

$$\|\Psi_b\| = \mathrm{w}(\Psi_b) = \mathrm{w}(\Theta_{b+1}) - \mathrm{w}(\Theta_b) = \frac{p - 1}{2}.$$

Finally,

$$\|\Psi_b^*\| \leq \|1 - \iota\| \cdot \|\Psi_b\| = p - 1. \tag{8.20}$$

The proposition is proved. □

Next, we estimate the number of integral solutions of the inequality

$$|x_1| + \cdots + |x_n| \leq r. \tag{8.21}$$

Let n be a positive integer and r a positive real number. Denote by $S(n, r)$ the number of points $(x_1, \ldots, x_n) \in \mathbb{Z}^n$ satisfying (8.21).

Proposition 8.22. *1. Let n and m be positive integers. Then*

$$S(n,m) = \sum_{k=0}^{\min\{m,n\}} 2^k \binom{n}{k}\binom{m}{k}. \tag{8.22}$$

2. Let n be an integer satisfying $n \geq 15$ and r a real number satisfying $r \geq 3$. Then

$$S(n,r) > 4n^2 r. \tag{8.23}$$

Proof. Part (1) is a standard exercise in combinatorics. There are exactly $\binom{m}{k}$ solutions of the inequality $y_1 + \cdots + y_k \leq m$ in positive integers $y_1, \ldots, y_k$. Indeed, the map

$$(y_1, \ldots, y_k) \mapsto \{y_1, y_1 + y_2, \ldots, y_1 + y_2 + \cdots + y_k\}$$

is a one-to-one correspondence between the solutions of this inequality and the subsets of $\{1, 2, \ldots, m\}$.

It follows that $y_1 + \cdots + y_n \leq m$ has exactly $\binom{n}{k}\binom{m}{k}$ solutions in nonnegative integers $y_1, \ldots, y_n$ among which exactly k are positive. Hence there are exactly $2^k \binom{n}{k}\binom{m}{k}$ solutions of (8.21) (with $r = m$) in integers $x_1, \ldots, x_n$ among which exactly k are nonzero. Summing up, we obtain (8.22).

To prove part (2), put $m = \lfloor r \rfloor$. Since $n \geq 15$ and $m \geq 3$, we have

$$S(n,r) = S(n,m) \geq 8\binom{n}{3}\binom{m}{3}$$

$$= \frac{2}{9}n(n-1)(n-2)m(m-1)(m-2)$$

$$\geq \frac{2}{9} \cdot \frac{14}{15} \cdot n^2 \cdot 13 \cdot \frac{3}{4} \cdot (m+1) \cdot 2 \cdot 1$$

$$= \frac{182}{45}n^2(m+1) > 4n^2 r,$$

as wanted.   □

In part (2), the assumption $n \geq 15$ is not really important and can be easily relaxed. However, the assumption $r \geq 3$ is substantial. Indeed, for $2 \leq r < 3$ we have $S(n,r) = 2n^2 + 2n + 1$, and (8.23) cannot hold.

It is curious that (8.22) implies the equality $S(n,m) = S(m,n)$. It would be interesting to find a geometric interpretation of this identity.

Now we are ready to state and prove the promised lower estimate for $|\mathcal{I}^*(r)|$.

Proposition 8.23. *Assume that $p \geq 31$ and $r \geq 3(p-1)$. Then*

$$|\mathcal{I}^*(r)| > (p-1)r.$$

Proof. Put $n = (p-1)/2$ and $\rho = r/(p-1)$. Let $(x_1, \ldots, x_n)$ be an integral vector such that $|x_1| + \cdots + |x_n| \leq \rho$. Inequality (8.20) implies that

$$x_1 \Psi_1^* + \cdots + x_n \Psi_n^* \in \mathcal{I}^*(r).$$

It follows that $|\mathcal{I}^*(r)| \geq S(n, \rho)$.

By the assumption, $n \geq 15$ and $\rho \geq 3$. Hence we may use Proposition 8.22(2). We obtain

$$|\mathcal{I}^*(r)| > 4n^2 \rho = (p-1)r,$$

as wanted. $\square$

It is not difficult to show that $|\mathcal{I}^*(r)| \sim c(p)r^{(p-1)/2}$ when $r \to \infty$. (Here $c(p) > 0$ depends only on p.) Hence for large r Proposition 8.23 can be substantially refined. However, the estimate from Proposition 8.23 is sufficient for us.

Proof of Theorem 8.19. As indicated in the beginning of this section, Stickelberger's Theorem 7.1 and Proposition 6.11 jointly imply that $\mathcal{I}^* = (1 - \iota)\mathcal{I}_S$ is contained in Mihăilescu's ideal $\mathcal{I}_M$. Moreover, since all the elements of $\mathcal{I}^*$ are of weight 0, we obtain $\mathcal{I}_M^{\mathrm{aug}} \supseteq \mathcal{I}^*$.

Put $r = q/(p-1)$. Hyyrö's inequality (3.18) implies that the hypothesis $|x| \geq \max\{2^{p+2}, q\}$ of Theorem 8.3 is satisfied. This theorem tells us that

$$\left|\mathcal{I}_M^{\mathrm{aug}}(r)\right| \leq q. \tag{8.24}$$

Now suppose that $q \geq 3(p-1)^2$. Then $r \geq 3(p-1)$. Also, we may assume that $p \geq 31$ by Corollary 8.18. The hypothesis of Proposition 8.23 being satisfied, we have

$$\left|\mathcal{I}_M^{\mathrm{aug}}(r)\right| \geq |\mathcal{I}^*(r)| > (p-1)r = q,$$

which contradicts (8.24). The theorem is proved. $\square$

8.8   On the Real Part of Mihăilescu's Ideal

Recall (see Sect. 7.7) that the real part of the ring $R = \mathbb{Z}[G]$ is the ideal $R^+ := (1 + \iota)R$ and the relative part is the ideal $R^- := (1 - \iota)R$. The real and relative parts of an ideal $\mathcal{I}$ of R are defined by $\mathcal{I}^+ = \mathcal{I} \cap R^+$ and $\mathcal{I}^- = \mathcal{I} \cap R^-$.

As we have seen before, when x is large enough, Mihăilescu's ideal has at most q elements of weight 0 and size not exceeding $2q/(p-1)$ and no elements of weight 0 and size 2. In this section we obtain a stronger assertion for the real part of Mihăilescu's ideal.

Theorem 8.24. *Assume that $0 < \varepsilon \leq 1$ and $|x| \geq 22^{1/\varepsilon}$. Then $\mathcal{I}_M^+$ has no nonzero elements of weight 0 and size not exceeding $(2 - \varepsilon)q$.*

The weight of elements of R^+ being even, the inequality $\|\Theta\| < 2q$ is equivalent to $\|\Theta\| \leq 2q - 2$. Hence, specifying $\varepsilon = 2/q$, we obtain the following statement: if $|x| \geq 5^q$ then $\mathcal{I}_M^+$ has no nonzero elements Θ with $\mathrm{w}(\Theta) = 0$ and $\|\Theta\| < 2q$.

Theorem 8.24 is formally obsolete, because in Chap. 9 we shall obtain a much stronger statement: for large x the real part of Mihăilescu's ideal has no nontrivial[1] elements of weight 0. However, we include the proof here, to illustrate the principles used in the "real case," and let the reader appreciate the more powerful method of Chap. 9 as compared with the method of this section.

The proof of Theorem 8.24 relies on the following simple but absolutely crucial property. We use the notation of Sect. 8.3.

Proposition 8.25. *Let Θ belong to the real part of Mihăilescu's ideal. Then $\alpha(\Theta)$ is a positive real number. In particular, $\xi(\Theta) = 1$.*

Proof. Observe first of all that for $\Theta \in R^+$ the number $(x - \zeta)^\Theta$ is real and positive and, in particular, belongs to the real cyclotomic field K^+. Indeed, writing $\Theta = (1 + \iota)\Theta'$, we obtain $(x - \zeta)^\Theta = (x - \zeta)^{\Theta'}\overline{(x - \zeta)^{\Theta'}} > 0$.

Now assume that $\Theta \in \mathcal{I}_M^+$. Since $(x - \zeta)^\Theta \in K^+$, the complex conjugate $\overline{\alpha(\Theta)}$ is a qth root of $(x - \zeta)^\Theta$. Since K contains only one qth root of $(x - \zeta)^\Theta$, we have $\alpha(\Theta) = \overline{\alpha(\Theta)}$. Hence $\alpha(\Theta) \in \mathbb{R}$. Since $(x - \zeta)^\Theta$ is positive, so is $\alpha(\Theta)$. $\qquad\square$

Proof of Theorem 8.24. Let $\Theta \in \mathcal{I}_M^+$ satisfy $\mathrm{w}(\Theta) = 0$ and $\|\Theta\| \leq (2 - \varepsilon)q$. By Proposition 8.25, the nearest root of unity for $\alpha(\Theta)$ is 1. Moreover, applying this proposition with $\Theta\sigma$ instead of Θ, we see that the same is true for all conjugates $\alpha(\Theta)^\sigma$. In the sequel we write α instead of $\alpha(\Theta)$.

Now we argue as in the proof of Proposition 8.12. The only difference is that this time not only α is close to 1, but all its conjugates are as well. We are going to profit from this.

As in the proof of Proposition 8.12, we start with the inequality

$$|\log \alpha^\sigma| \leq \frac{1}{q}\frac{\|\Theta\|}{|x| - 1},$$

which holds for every $\sigma \in G$. Since $\|\Theta\| \leq 2q$ and $|x| \geq 22$, this implies $|\log \alpha^\sigma| \leq 2/21$. Using Lemma 8.13, we obtain

$$|\alpha^\sigma - 1| \leq \frac{e^{2/21} - 1}{2/21}|\log \alpha^\sigma| \leq \frac{1.05}{q}\frac{\|\Theta\|}{|x| - 1}$$

---

[1]that is, non-divisible by q

for all $\sigma \in G$. In other words,

$$|\alpha - 1|_v \leq \left(\frac{1.05}{q} \frac{\|\Theta\|}{|x| - 1} \right)^2 \tag{8.25}$$

for every Archimedean valuation v of the field K.

Using Liouville's inequality (B.12) with $S = M_K^\infty$, the set of all Archimedean valuations of K, we obtain

$$\prod_{v \in M_K^\infty} |\alpha - 1|_v \geq e^{-(p-1)h(\alpha-1)}.$$

Together with (8.13) and (8.25), this implies

$$\left(\frac{1.05}{q} \frac{\|\Theta\|}{|x| - 1} \right)^{p-1} < \exp\left(-(p-1)\left(\frac{\|\Theta\|}{2q} \log(|x| + 1) + \log 2 \right) \right),$$

that is

$$\log(|x| - 1) - \log \frac{1.05\|\Theta\|}{q} \leq \frac{\|\Theta\|}{2q} \log(|x| + 1) + \log 2.$$

Replacing $\|\Theta\|$ by $(2 - \varepsilon)q$ in the right-hand side and by $2q$ in the left-hand side, we obtain

$$\log(|x| - 1) - \log 2.1 \leq \left(1 - \frac{\varepsilon}{2} \right) \log(|x| + 1) + \log 2,$$

which easily transforms into

$$\varepsilon \log(|x| + 1) < 2 \log \frac{|x| + 1}{|x| - 1} + 2 \log 4.2.$$

Since $x \geq 22$, this implies

$$\varepsilon \log(|x| + 1) < 2 \log \frac{23}{21} + 2 \log 4.2 < \log 22,$$

which contradicts the assumption $|x| \geq 22^{1/\varepsilon}$. $\qquad\qquad\qquad\qquad\qquad\square$

Chapter 9
The Real Part of Mihăilescu's Ideal

In this chapter we continue our study of Mihăilescu's ideal. As follows from the definition of this ideal, given in Chap. 8, it contains the ideal $q\mathbb{Z}[G]$ of the elements divisible by q. A basic question is whether Mihăilescu's ideal has nontrivial (that is, not divisible by q) elements. In Chap. 8 we showed that, when x is large, Mihăilescu's ideal contains at most q elements Θ with $\mathrm{w}(\Theta) = 0$ and $\|\Theta\| < 2q/(p-1)$. Moreover, in Sect. 8.8 we showed that, when x is large, the *real part* $\mathcal{I}_M^+$ of Mihăilescu's ideal contains no $\Theta \neq 0$ with $\mathrm{w}(\Theta) = 0$ and $\|\Theta\| < 2q$.

In this chapter we go further and prove that (for large x) the real part $\mathcal{I}_M^+$ contains **no nontrivial elements of weight** 0 and even of any weight divisible by q.

On the other hand, in Chap. 11 we shall see that a solution to Catalan's equation implies a nontrivial element of $\mathcal{I}_M^+$ of weight divisible by q. This contradiction would prove Catalan's conjecture.

As usual, p and q are distinct odd primes[1], ζ is a primitive pth root of unity, $K = \mathbb{Q}(\zeta)$ is the pth cyclotomic Field, and $G = \mathrm{Gal}(K/\mathbb{Q})$ is its Galois group over $\mathbb{Q}$.

9.1 The Main Theorem

As in Chap. 8, we fix a nonzero integer x and define *Mihăilescu's ideal* $\mathcal{I}_M$ as the set of all elements Θ of the group ring $R = \mathbb{Z}[G]$ such that $(x - \zeta)^\Theta$ is a qth power in K. Let $\mathcal{I}_M^+ = \mathcal{I}_M \cap R^+$ be the *real part* of Mihăilescu's ideal. An element of Mihăilescu's ideal is called *trivial* if it is divisible by q in the ring R.

In this chapter we prove the following theorem.

Theorem 9.1 (Mihăilescu). *Assume that* $|x| \geq (8q)^{(p-1)/2}$. *Then* $\mathcal{I}_M^+$ *has no nontrivial elements of weight* 0.

---

[1]Like in Chap. 8, we do not assume that p and q come out of a solution of Catalan's equation.

© Springer International Publishing Switzerland 2014
Y.F. Bilu et al., *The Problem of Catalan*, DOI 10.1007/978-3-319-10094-4_9

In other words, for large x we have $\mathcal{I}_M^+ \cap R^{\text{aug}} \subseteq qR^{\text{aug}}$, where R^{aug} is the augmentation ideal of R.

If (x, y, p, q) is a solution of Catalan's equation, then $q \geq 43$ by Corollary 8.18, and Hyyrö's bound (3.18) implies that $|x| \geq q^{p-1} \geq (43q)^{(p-1)/2}$. Hence Theorem 9.1 applies in this case.

To give an idea of the proof, recall that in the proof of Theorem 8.24, we approximated the qth root α, together with all its conjugates, by 1. When the size of Θ is not too large, this approximation is so good that it implies an equality, which is impossible.

Unfortunately, this argument fails if the size of Θ exceeds (or is equal to) $2q$. This resembles the situation with the proof of Cassels' divisibility theorem in Sect. 3.3: the simple approximation $y \approx a^p$ was sufficient for the "easy" case $p < q$ but insufficient when $p > q$.

In the proof of Theorem 9.1 we argue as in the "difficult" case of Cassels' theorem. We approximate the qth root by a partial sum of a certain power series. Next, we show that the approximation is so good that it implies the equality of the corresponding quantities. However, the equality is impossible, because one of the numbers is an algebraic integer, but the other is not.

In brief, Theorem 9.1 relates to Theorem 8.24 in the same way as the "difficult" case of Cassels' theorem (Proposition 3.10) relates to the "easy" case (Proposition 3.8).

We conclude this section with a few technical comments concerning the proof of Theorem 9.1. First of all, it is more convenient to prove (and to use) the following equivalent statement.

Theorem 9.2. *Assume that $|x| \geq (8q)^{(p-1)/2}$. Then $\mathcal{I}_M^+$ has no nontrivial elements of weight divisible by q.*

Again, as we have seen above, Theorem 9.2 applies in the case when (x, y, p, q) is a solution of Catalan's equation.

Though Theorem 9.2 looks formally stronger than Theorem 9.1, the two are, in fact, equivalent. Indeed, let Θ be a nontrivial element of $\mathcal{I}_M^+$ of weight mq, where $m \in \mathbb{Z}$. Since $\Theta \in R^+$, the weight of Θ is even. Hence m is even; write $m = 2n$. Then $\Theta - nq(1 + \iota)$ is a nontrivial element of $\mathcal{I}_M^+$ of weight 0.

Further, we may assume Θ nonnegative[2] and of bounded size. Indeed, write $\Theta = \sum_{\sigma \in G} a_\sigma \sigma$, and for each $\sigma \in G$ let a'_σ be the remainder of the Euclidean division of a_σ by q; that is,

$$0 \leq a'_\sigma \leq q - 1, \qquad a'_\sigma \equiv a_\sigma \bmod q. \tag{9.1}$$

Put $\Theta' = \sum_{\sigma \in G} a'_\sigma \sigma$. Then Θ' is a nontrivial element of $\mathcal{I}_M^+$ if and only if Θ is. Thus, we shall not restrict generality if we assume in Theorem 9.2 that $\Theta \geq 0$ and that $w(\Theta) \leq (q - 1)(p - 1)$. In fact, we can do even better.

---

[2]Recall that $\Theta = \sum_{\sigma \in G} a_\sigma \sigma$ is called *nonnegative* (notation: $\Theta \geq 0$) if $a_\sigma \geq 0$ for all $\sigma \in G$.

Proposition 9.3. *For every* $\Theta \in R$ *there exists a nonnegative* $\Theta' \in R$ *satisfying*

$$\|\Theta'\| = \mathrm{w}(\Theta') \leq \frac{q(p-1)}{2}$$

and congruent either to Θ *or to* $-\Theta$ *modulo* qR. *Also, if* $\Theta \in R^+$ *then* $\Theta' \in R^+$ *as well. In particular,* Θ' *is a nontrivial element of* $\mathcal{I}_M^+$ *if and only if* Θ *is.*

Proof. We again write $\Theta = \sum_{\sigma \in G} a_\sigma \sigma$ and define a'_σ as in (9.1). Put

$$\Theta'_1 = \sum_{\sigma \in G} a'_\sigma \sigma, \qquad \Theta'_2 = \sum_{\sigma \in G} (q - a'_\sigma) \sigma = q\mathcal{N} - \Theta'_1.$$

Obviously, both Θ'_1 and Θ'_2 are nonnegative and satisfy

$$\Theta'_1 \equiv \Theta \bmod qR, \qquad \Theta'_2 \equiv -\Theta \bmod qR.$$

Also, both Θ'_1 and Θ'_2 belong to R^+ if Θ does.

Finally, since $\Theta'_1 + \Theta'_2 = q\mathcal{N}$, we have $\mathrm{w}(\Theta'_1) + \mathrm{w}(\Theta'_2) = q(p-1)$. Hence at least one of the two weights does not exceed $q(p-1)/2$. $\square$

Thus, in the proof of Theorem 9.2 we may assume that $\Theta \geq 0$ and that $\mathrm{w}(\Theta) \leq q(p-1)/2$.

9.2 Products of Binomial Power Series

The proof of Theorem 3.3 relied on simple arithmetical and analytic properties of the binomial power series $(1 + T)^\nu$. For the proof of Theorem 9.2 we need similar properties of more general power series. In this section we extend Lemma 3.6 to series of the form

$$(1 + \xi_1 T)^{\nu_1} \cdots (1 + \xi_r T)^{\nu_r}.$$

In the sequel, the capital letter T will stand for an independent variable, and small t will be a complex number. If $A(T) = a_0 + a_1 T + a_2 T^2 + \cdots$ is a formal power series with complex coefficients, then we denote by $A(t)$ its sum at $T = t$ (of course, provided the numerical series $a_0 + a_1 t + a_2 t^2 + \cdots$ converges).

For a complex number ξ and a real number ν we consider the *binomial power series*

$$(1 + \xi T)^\nu = \sum_{k=0}^{\infty} \binom{\nu}{k} \xi^k T^k.$$

Now fix complex numbers $\xi_1, \ldots, \xi_r$ and real numbers $v_1, \ldots, v_r$ and consider the power series

$$F(T) = (1 + \xi_1 T)^{v_1} \cdots (1 + \xi_r T)^{v_r} = f_0 + f_1 T + f_2 T^2 + \cdots,$$

which is obtained by formal multiplication of the corresponding binomial series; in particular,

$$f_0 = 1, \qquad f_1 = v_1 \xi_1 + \cdots + v_r \xi_r,$$

and so on.

For every nonnegative integer m we define the mth partial sum by

$$F_m(T) = f_0 + f_1 T + \cdots + f_m T^m.$$

We want to estimate the difference $F(t) - F_m(t)$. We may assume that

$$\max \{|\xi_1|, \ldots, |\xi_r|\} \le 1, \tag{9.2}$$

because the general case reduces to (9.2) by an obvious change of variables.

Proposition 9.4. *Assume that (9.2) holds. Then for any complex t satisfying $|t| < 1$ we have*

$$|F(t) - F_m(t)| \le (1 - |t|)^{-\mu - m - 1} \left| \binom{\mu + m}{m + 1} \right| |t|^{m+1}, \tag{9.3}$$

where $\mu = |v_1| + \cdots + |v_r|$.

The proof requires the notion of *dominance* of power series. Let $A(t) = \sum_{k=0}^{\infty} a_k T^k$ be a series with complex coefficients, and let $\tilde{A}(t) = \sum_{k=0}^{\infty} \tilde{a}_k T^k$ be a series with nonnegative real coefficients. We say that $A(T)$ is *dominated* by $\tilde{A}(t)$ (notation: $A(T) \prec \tilde{A}(T)$) if $|a_k| \le \tilde{a}_k$ for $k = 0, 1, 2 \ldots$. The following properties are immediate.

Proposition 9.5. *1. The relation of dominance is preserved by addition and multiplication of power series; that is, if $A(T) \prec \tilde{A}(T)$ and $B(T) \prec \tilde{B}(T)$, then $A(T) + B(T) \prec \tilde{A}(T) + \tilde{B}(T)$ and $A(T) B(T) \prec \tilde{A}(T) \tilde{B}(T)$.*
2. If $A(T) \prec \tilde{A}(T)$ and t is a complex number such that $\tilde{A}(T)$ converges at $T = |t|$, then $A(T)$ converges at $T = t$, and $|A(t)| \le \tilde{A}(|t|)$. Moreover, for any nonnegative integer m, we have

$$|A(t) - A_m(t)| \le \left| \tilde{A}(|t|) - \tilde{A}_m(|t|) \right|,$$

where $A_m(T)$ and $\tilde{A}_m(T)$ are the mth partial sums of the corresponding series.

Proof of Proposition 9.4. For any real number v and any nonnegative integer k we have

$$\left| \binom{v}{k} \right| \leq \frac{|v|(|v|+1)\cdots(|v|+|k-1|)}{k!} = (-1)^k \binom{-|v|}{k}.$$

It follows that the series $(1 + \xi T)^v$ is dominated by $(1 - |\xi|T)^{-|v|}$, and, when $|\xi| \leq 1$, it is dominated by $(1 - T)^{-|v|}$. Hence our series $F(T)$ is dominated by $\tilde{F}(T) = (1 - T)^{-\mu}$, where $\mu = |v_1| + \cdots + |v_r|$.

Now, using Taylor's formula as in the proof of Lemma 3.6, we obtain

$$|F(t) - F_m(t)| \leq |\tilde{F}(|t|) - \tilde{F}_m(|t|)|$$

$$\leq \sup_{0 \leq \theta \leq 1} \left| \left(\frac{d^{m+1}(1-T)^{-\mu}}{dT^{m+1}} \right)\Big|_{T=\theta|t|} \right| \frac{|t|^{m+1}}{(m+1)!}$$

$$= (1 - |t|)^{-\mu-m-1} \left| \binom{-\mu}{m+1} \right| |t|^{m+1}.$$

Since

$$\left| \binom{-\mu}{m+1} \right| = \left| \binom{\mu+m}{m+1} \right|,$$

the proposition follows. □

9.3 Mihăilescu's Series $(1 + \zeta T)^{\Theta/q}$

In this section we fix

$$\Theta = \sum_{\sigma \in G} a_\sigma \sigma \in \mathbb{Z}[G],$$

and investigate the power series $(1 + \zeta T)^{\Theta/q}$, which is defined by

$$(1 + \zeta T)^{\Theta/q} = \prod_{\sigma \in G} (1 + \zeta^\sigma T)^{a_\sigma/q}. \tag{9.4}$$

Write

$$(1 + \zeta T)^{\Theta/q} = f_0(\Theta) + f_1(\Theta)T + f_2(\Theta)T^2 + \cdots,$$

so that

$$f_0(\Theta) = 1, \qquad f_1(\theta) = \frac{1}{q} \sum_{\sigma \in G} a_\sigma \zeta^\sigma,$$

and so on.

First of all, we state few simple properties of Mihăilescu's series that follow immediately from the definition. Observe that the Galois group $G = \mathrm{Gal}(K/\mathbb{Q})$ acts on the ring $K[[T]]$ of formal power series coefficient wise: if

$$A(T) = a_0 + a_1 T + a_2 T^2 + \dots$$

is a formal power series with coefficients in K, then we put

$$A^\sigma(T) := a_0^\sigma + a_1^\sigma T + a_2^\sigma T^2 + \dots$$

An obvious verification shows that this action is compatible with the arithmetic operations on power series: we have $(A(T) + B(T))^\sigma = A^\sigma(T) + B^\sigma(T)$ and $(A(T)B(T))^\sigma = A^\sigma(T)B^\sigma(T)$.

The series $(1 + \zeta T)^{\Theta/q}$ converges at $T = t$ when t is a complex number with $|t| < 1$, and we denote the sum by $(1 + \zeta t)^{\Theta/q}$.

Proposition 9.6. *1. For any $\Theta \in R$ and $\sigma \in G$ we have*

$$\left((1 + \zeta T)^{\Theta/q}\right)^\sigma = (1 + \zeta T)^{\Theta\sigma/q}$$

In other words,

$$f_k(\Theta)^\sigma = f_k(\Theta\sigma) \qquad (k = 0, 1, 2, \dots).$$

2. If $\Theta \in R^+$ then Mihăilescu's series $(1 + \zeta T)^{\Theta/q}$ has real coefficients. In particular, for any real t with $|t| < 1$, we have $(1 + \zeta t)^{\Theta/q} \in \mathbb{R}$.

Proof. Put $F(T) = \left((1 + \zeta T)^{\Theta/q}\right)^\sigma$ and $G(T) = (1 + \zeta T)^{\Theta\sigma/q}$. Then

$$F(T)^q = G(T)^q = (1 + \zeta T)^{\Theta\sigma}.$$

It follows that $F(T) = \xi G(T)$, where ξ is a qth root of unity. Since

$$F(0) = G(0) = 1$$

we have $\xi = 1$. This proves part (1).

Now assume that $\Theta \in R^+$, and write $\Theta = \Theta'(1 + \iota)$. Then, using part (1), we obtain

$$(1 + \zeta T)^{\Theta/q} = (1 + \zeta T)^{\Theta'/q}(1 + \zeta T)^{\Theta'\iota/q}$$

$$= (1 + \zeta T)^{\Theta'/q}\left((1 + \zeta T)^{\Theta'/q}\right)^{\iota}$$

$$= (1 + \zeta T)^{\Theta'/q}\overline{(1 + \zeta T)^{\Theta'/q}}.$$

Hence $(1 + \zeta T)^{\Theta/q}$ has real coefficients as a product of two complex conjugate series. This proves part (2).   □

We want to study the arithmetic of the coefficients $f_k(\Theta)$, as we did in Lemma 3.5 for the usual binomial power series. Sometimes we shall write f_k instead of $f_k(\Theta)$ when this does not lead to a confusion. In the sequel the number

$$\phi(\Theta) = \sum_{\sigma \in G} a_\sigma \zeta^\sigma = q f_1(\Theta)$$

will play the central role.

Theorem 9.7. *1. The number $q^{k+\mathrm{Ord}_q(k!)} f_k(\Theta)$ is an algebraic integer.*
2. Let $\mathfrak{q}$ be a prime ideal of the cyclotomic field K, dividing q but not dividing the number $\phi(\Theta)$ defined above. Then

$$\mathrm{Ord}_{\mathfrak{q}}(f_k) = -k - \mathrm{Ord}_q(k!) \qquad (k = 0, 1, 2, \ldots).$$

In particular, for such $\mathfrak{q}$ we have

$$0 = \mathrm{Ord}_{\mathfrak{q}}(f_0) > \mathrm{Ord}_{\mathfrak{q}}(f_1) > \mathrm{Ord}_{\mathfrak{q}}(f_2) > \ldots$$

Put $\phi_k = \phi_k(\Theta) = k! q^k f_k(\Theta)$, so that

$$(1 + \zeta q T)^{\Theta/q} = \sum_{k=0}^{\infty} \frac{\phi_k(\Theta)}{k!} T^k.$$

Theorem 9.7 is an easy consequence of the following statement.

Proposition 9.8. *The numbers $\phi_k(\Theta)$ are algebraic integers satisfying the congruence*

$$\phi_k(\Theta) \equiv \phi(\Theta)^k \mod q$$

for $k = 0, 1, 2, \ldots$

Proof of Theorem 9.7 (assuming Proposition 9.8). Let a be an integer. Lemma 3.5 implies that for every k the binomial coefficient $\binom{a/q}{k}$ is an integer divided by a power of q. Hence every of the coefficients $f_k(\Theta)$ is an algebraic integer divided by a power of q; in other words, for each k there exists an integer N such that $q^N f_k(\Theta)$ is an algebraic integer.

On the other hands, Proposition 9.8 implies that $q^k k! f_k(\Theta)$ is an algebraic integer. Hence $q^{k+\mathrm{Ord}_q(k!)} f_k(\Theta)$ is an algebraic integer. This proves part (1).

Next, let $\mathfrak{q}$ be a prime ideal dividing q but not dividing $\phi(\Theta)$. Since $\phi_k(\Theta) \equiv \phi(\Theta)^k \bmod \mathfrak{q}$, we have $\mathrm{Ord}_\mathfrak{q}(\phi_k) = 0$. Hence

$$\mathrm{Ord}_\mathfrak{q}(f_k) = \mathrm{Ord}_\mathfrak{q}\left(\frac{\phi_k}{q^k k!}\right) = -e(k + \mathrm{Ord}_q(k!)),$$

where $e = \mathrm{Ord}_\mathfrak{q}(q)$. Since q is unramified in K (Proposition 4.8), we have $e = 1$. This proves part (2).   □

The proof of Proposition 9.8 relies on the following purely algebraic lemma.

Lemma 9.9. *Let R be an integral domain of characteristic 0, let $A(T)$, $B(T)$, and $C(T) = A(T)B(T)$ be formal power series over its field of quotients, and let $\mathfrak{a}$ be an ideal of R. Write*

$$A(T) = \sum_{k=0}^{\infty} \frac{\alpha_k}{k!}T^k, \qquad B(T) = \sum_{k=0}^{\infty} \frac{\beta_k}{k!}T^k, \qquad C(T) = \sum_{k=0}^{\infty} \frac{\gamma_k}{k!}T^k$$

and assume that all the coefficients α_k and β_k belong to R. Further, assume that there exist $\alpha, \beta \in R$ such that

$$\alpha_k \equiv \alpha^k \bmod \mathfrak{a}, \qquad \beta_k \equiv \beta^k \bmod \mathfrak{a} \qquad (k = 0, 1, 2, \ldots).$$

Then the coefficients γ_k also belong to R and satisfy

$$\gamma_k \equiv (\alpha + \beta)^k \bmod \mathfrak{a} \qquad (k = 0, 1, 2, \ldots).$$

Proof. We have $\gamma_k = \sum_{j=0}^{k} \binom{k}{j}\alpha_j \beta_{k-j}$. Hence $\gamma_k \in R$ and

$$\gamma_k \equiv \sum_{j=0}^{k} \binom{k}{j}\alpha^j \beta^{k-j} \equiv (\alpha + \beta)^k \bmod \mathfrak{a},$$

as wanted.   □

Proof of Proposition 9.8. Lemma 9.9 implies that the statement of the proposition is additive in Θ. That is, if $\Theta_1, \Theta_2 \in \mathbb{Z}[G]$ are such that for all k the numbers $\phi_k(\Theta_1)$ and $\phi_k(\Theta_2)$ are algebraic integers satisfying

$$\phi_k(\Theta_1) \equiv \phi(\Theta_1)^k \bmod \mathfrak{q}, \qquad \phi_k(\Theta_2) \equiv \phi(\Theta_2)^k \bmod \mathfrak{q},$$

then the numbers $\phi_k(\Theta_1 + \Theta_2)$ are algebraic integers satisfying

$$\phi_k(\Theta_1 + \Theta_2) \equiv \phi(\Theta_1 + \Theta_2)^k \bmod q.$$

Thus, it remains to verify the statement of the proposition for $\Theta = \pm\sigma$. For $\Theta = \sigma$ we have

$$f_k(\sigma) = \binom{1/q}{k}(\zeta^\sigma)^k = \frac{1 \cdot (1 - q) \cdots (1 - (k - 1)q)}{q^k k!}(\zeta^\sigma)^k.$$

Hence $\phi_k(\sigma) = 1 \cdot (1 - q) \cdots (1 - (k - 1)q)(\zeta^\sigma)^k \equiv (\zeta^\sigma)^k \bmod q$. Similarly, one shows that $\phi_k(-\sigma) \equiv (-\zeta^\sigma)^k \bmod q$. The proposition is proved. □

Finally, we wish to adapt Proposition 9.4 to Mihăilescu's series. Recall that we denote by $(1 + \zeta t)^{\Theta/q}$ the sum of $(1 + \zeta T)^{\Theta/q}$ at $T = t$.

Proposition 9.10. *Let m be a nonnegative integer and t a complex number with $|t| < 1$. Then*

$$\left| (1 + \zeta t)^{\Theta/q} - \sum_{k=0}^{m} f_k(\Theta) t^k \right| \leq (1 - |t|)^{-\mu-m-1} \left| \binom{\mu + m}{m + 1} \right| |t|^{m+1},$$

where $\mu = \frac{\|\Theta\|}{q}$.

Proof. This is a direct application of Proposition 9.4. □

9.4 Proof of Theorem 9.2

Let Θ be a nontrivial element of the real part $\mathcal{I}_M^+$ of Stickelberger's ideal, and assume that the weight of Θ is divisible by q. As follows from Proposition 9.3, we may assume that $\Theta \geq 0$ and that

$$\|\Theta\| = w(\Theta) \leq \frac{q(p - 1)}{2}.$$

We shall use this later in the proof. Also, the assumption $|x| \geq (8q)^{(p-1)/2}$ implies that $x \geq 40$, which will be used in the proof as well.

Denote by $\alpha = \alpha(\Theta)$ the qth root of $(x - \zeta)^\Theta$ belonging to K. As indicated in the introduction, we wish to approximate α by a partial sum of a certain power series. To do this, we have to express α as the sum of a certain power series. In the proof of Cassels' theorem we used binomial power series; this time we need more complicated Mihăilescu series.

Write $\mathrm{w}(\Theta) = \mu q$, where, by the assumption, $\mu \in \mathbb{Z}$. Moreover, since Θ is nonnegative and nonzero, μ is a positive integer. Analytically, a qth root of $(x - \zeta)^{\Theta}$ can be given by $x^{\mu}(1 - \zeta/x)^{\Theta/q}$, where $(1 - \zeta/x)^{\Theta/q}$ is the sum of Mihăilescu's series $(1 + \zeta T)^{\Theta/q}$ at $T = -1/x$. A priori, there is no reason for α to be equal to this quantity; all we can assert, without an additional assumption about Θ, is that α is equal to $x^{\mu}(1 - \zeta/x)^{\Theta/q}$ times a qth root of unity.

It is crucial that, in our special case $\Theta \in \mathcal{I}_M^+$, we do have the equality

$$\alpha = x^{\mu}(1 - \zeta/x)^{\Theta/q}. \tag{9.5}$$

Indeed, both parts of (9.5) are real numbers: $\alpha \in \mathbb{R}$ by Proposition 8.25 and $(1 - \zeta/x)^{\Theta/q} \in \mathbb{R}$ by Proposition 9.6. Thus, both parts of (9.5) are equal to the real qth root of $(x - \zeta)^{\Theta}$.

It follows that α can be approximated by $x^{\mu}F_m(-1/x)$, where

$$F_m(T) = f_0(\Theta) + f_1(\Theta)T + \cdots + f_m(\Theta)T^m$$

is the mth partial sum of Mihăilescu's series $(1 + \zeta T)^{\Theta/q}$. More precisely, Proposition 9.10 implies that

$$\left| \alpha - x^{\mu}F_m\left(-\frac{1}{x}\right) \right| \leq \left(1 - \frac{1}{|x|}\right)^{-\mu-m-1} \binom{\mu + m}{m + 1} |x|^{\mu-m-1}.$$

Since $x \geq 40$ and[3] $\binom{\mu+m}{m+1} \leq 2^{m+\mu-1}$, we have

$$\left| \alpha - x^{\mu}F_m\left(-\frac{1}{x}\right) \right| \leq 2.1^{m+\mu}|x|^{\mu-m-1}.$$

From now on we put $m = \mu$. It will be convenient to multiply the difference $\alpha - x^{\mu}F_{\mu}(-1/x)$ by $q^{\mu + \mathrm{Ord}_q(\mu!)}$, to get an algebraic integer (see Theorem 9.7). Thus, put

$$\beta = \beta(\Theta) = q^{\mu + \mathrm{Ord}_q(\mu!)}\left(\alpha(\Theta) - x^{\mu}F_{\mu}\left(-\frac{1}{x}\right)\right)$$

$$= q^{\mu + \mathrm{Ord}_q(\mu!)}\left(\alpha(\Theta) - \sum_{k=0}^{\mu}(-1)^k f_k(\Theta)x^{\mu-k}\right).$$

---

[3] It is easy to see that, for integers $a > 0$ and $b \in \{0, \ldots, a\}$, the binomial coefficient $\binom{a}{b}$ satisfies $\binom{a}{b} \leq 2^{a-1}$. Indeed, the cases $b = 0$ and $b = a$ are obvious, and for $0 < b < a$ we have $\binom{a}{b} = \binom{a-1}{b-1} + \binom{a-1}{b} \leq \sum_{k=0}^{a-1}\binom{a-1}{k} = 2^{a-1}$.

Then β is an algebraic integer, satisfying

$$|\beta| \le q^{\mu + \mathrm{Ord}_q(\mu!)} \cdot 2.1^{2\mu} |x|^{-1}.$$

Since $\mathrm{Ord}_q(\mu!) \le \mu/(q-1)$, we have $q^{\mathrm{Ord}_q(\mu!)} \le \left(q^{1/(q-1)}\right)^{\mu} \le \left(\sqrt{3}\right)^{\mu}$, which implies

$$|\beta| \le \left(2.1^2 \cdot \sqrt{3}q\right)^{\mu} |x|^{-1} < (8q)^{\mu} |x|^{-1}.$$

Now recall that $\mathrm{w}(\Theta) \le q(p-1)/2$, which means that $\mu \le (p-1)/2$. Using our assumption $|x| \ge (8q)^{(p-1)/2}$, we obtain $|\beta| < 1$.

If β were a rational integer, this would have been sufficient to conclude that $\beta = 0$. However, it is merely an algebraic integer, so we have to estimate all its conjugates to make a similar conclusion. Recall that, for $\sigma \in G$, we have $\alpha(\Theta)^{\sigma} = \alpha(\Theta\sigma)$ (Proposition 8.9(3)) and $f_k(\Theta)^{\sigma} = f_k(\Theta\sigma)$ (Proposition 9.6(1)). It follows that $\beta(\Theta)^{\sigma} = \beta(\Theta\sigma)$. Hence, applying the previous estimates with $\Theta\sigma$ instead of Θ, we obtain $|\beta^{\sigma}| < 1$ for every $\sigma \in G$.

Thus, β is algebraic integer with all conjugates strictly smaller than 1 in absolute value. Hence $\beta = 0$, that is, $\alpha = x^{\mu} F_{\mu}(-1/x)$. We are going to show that this equality is impossible.

Since $\Theta \ge 0$, the number $(x - \zeta)^{\Theta}$ is an algebraic integer. Hence so is $\alpha(\Theta)$. We shall see that the number $x^{\mu} F_{\mu}(-1/x)$ is not an algebraic integer. By the assumption, Θ is not divisible by q. That is, if we write $\Theta = \sum_{\sigma \in G} a_{\sigma}\sigma$, then at least one of the coefficients a_{σ} is not a multiple of q. Hence the algebraic number $\phi(\Theta) = \sum_{\sigma \in G} a_{\sigma}\zeta^{\sigma}$ is not divisible by q either[4]. Let q be a prime ideal above q which does not divide $\phi(\Theta)$. Theorem 9.7:2 implies that

$$0 = \mathrm{Ord}_q(f_0) > \mathrm{Ord}_q(f_1 x^{-1}) > \ldots > \mathrm{Ord}_q(f_{\mu} x^{-\mu})$$

(we write f_k instead of $f_k(\Theta)$). Hence $\mathrm{Ord}_q\left(F_{\mu}(-1/x)\right) = \mathrm{Ord}_q\left(f_{\mu} x^{-\mu}\right)$. It follows that

$$\mathrm{Ord}_q\left(x^{\mu} F_{\mu}(-1/x)\right) = \mathrm{Ord}_q\left(f_{\mu}\right) < 0.$$

This proves that $x^{\mu} F_{\mu}(-1/x)$ is not an algebraic integer. In particular, $\alpha \ne x^{\mu} F_m(-1/x)$. The theorem is proved.

We conclude this section with a general remark. In the course of our argument we have implicitly proved (and used) the following statement: *let Θ belong to the real part of Mihăilescu's ideal, and let $F(T) = (1 + \zeta T)^{\Theta/q}$ be the corresponding Mihăilescu series; then $F^{\sigma}(-1/x) = F(-1/x)^{\sigma}$ for any $\sigma \in G$.* That is, when we apply σ to the coefficients of the power series $F(T)$, then the sum at $T = -1/x$ of the new series is the σ-conjugate of the original sum.

---

[4]Because $\mathcal{O}_K = \mathbb{Z}[\zeta]$, see Theorem 4.6.

It is important that a similar general statement is not true. Indeed, let

$$A(T) = a_0 + a_1 T + a_2 T^2 + \cdots$$

be a formal power series with coefficients in some number field. Also, let t be a rational number; assume that the series $A(T)$ converges at $T = t$ and that the sum $A(t)$ is an algebraic number as well.

Let σ be an automorphism of a number field containing the coefficients $a_0, a_1, a_2, \ldots$ and the sum $A(t)$. Then it is not true, in general, that the series $A^\sigma(T) = a_0^\sigma + a_1^\sigma T + a_2^\sigma T^2 + \cdots$ converges at $T = t$. Moreover, even if it does converge, it is not true, in general, that $A^\sigma(t) = A(t)^\sigma$.

Example 9.11. 1. The series $\sum_{k=0}^\infty (1 - \sqrt{2})^k T^k$ converges at $T = 1$, but the series $\sum_{k=0}^\infty (1 + \sqrt{2})^k T^k$ does not.
2. The series

$$A(T) = \sqrt{3}(1 + T)^{1/2} = \sum_{k=0}^\infty \sqrt{3}\binom{1/2}{k} T^k.$$

converges at $T = 1/3$, and $A(1/3) = \sqrt{3}(1 + 1/3)^{1/2} = 2$. Let σ be the non-trivial automorphism of $\mathbb{Q}(\sqrt{3})$. Then $A^\sigma(T) = -A(T)$ also converges at $T = 1/3$, but $A^\sigma(1/3) = -2 \neq A(1/3)^\sigma$.

The reason for this unfortunate fact is simple: Galois automorphisms are, in general, not continuous in the complex topology (the only exception is the complex conjugation). Therefore one always has to be careful applying Galois action to convergent power series.

Chapter 10
Cyclotomic Units

In this short chapter we give a brief introduction to the beautiful theory of cyclotomic units, which can be viewed as a "real analogue" of Stickelberger's ideal. In particular, we prove the real class number formula (10.6).

10.1 The Circulant Determinant

This section is preparatory. We calculate a special determinant needed for the theory of cyclotomic units. We closely follow Lang [60, Sect. 3.6].

The famous "circulant determinant" identity is

$$\begin{vmatrix} x_1 & x_2 & \cdots & x_m \\ x_m & x_1 & \cdots & x_{m-1} \\ \vdots & \vdots & \ddots & \vdots \\ x_2 & x_3 & \cdots & x_1 \end{vmatrix} = P(1)P(\xi)\cdots P(\xi^{m-1})$$

where ξ is a primitive mth root of unity and

$$P(T) = x_1 + x_2 T + \cdots + x_m T^{m-1}.$$

In this section we prove a generalized version of this identity.

© Springer International Publishing Switzerland 2014
Y.F. Bilu et al., *The Problem of Catalan*, DOI 10.1007/978-3-319-10094-4_10

Let G be a finite abelian group, and let $f : G \to \mathbb{C}$ be a complex-valued function on G. We want to calculate the $|G| \times |G|$-determinant[1]

$$\det[f(a^{-1}b)]_{a,b\in G}.$$

When G is a cyclic group, this is the usual circulant determinant displayed above.

Let $\hat{G}$ be the group of $\mathbb{C}$-characters of G. The following result is credited to Dedekind, see [46].

Theorem 10.1. *We have*

$$\det[f(a^{-1}b)]_{a,b\in G} = \prod_{\chi\in\hat{G}} \sum_{a\in G} f(a)\chi(a). \qquad (10.1)$$

Proof. Let V be the space of complex functions on G. It is a $|G|$-dimensional $\mathbb{C}$-vector space. It has two natural bases. The first one is the "δ-basis" $\{\delta_a : a \in G\}$, where the function δ_a is defined by $\delta_a(a) = 1$ and $\delta_a(x) = 0$ for $x \neq a$. The second one is the "χ-basis" $\hat{G}$, consisting of the characters of G.

For $a \in G$ let $T_a : V \to V$ be the "translation by a" map, defined by $T_a g(x) = g(a^{-1}x)$ for $g \in V$. Then $T_a(\delta_b) = \delta_{ab}$ and $T_a(\chi) = \chi(a^{-1})\chi$ for a $\chi \in \hat{G}$. In particular, every χ is an eigenvector of T_a.

Now let $T : V \to V$ be the linear operator defined by

$$T = \sum_{a\in G} f(a^{-1})T_a.$$

Then

$$T\delta_b = \sum_{a\in G} f(a^{-1})\delta_{ab} = \sum_{a\in G} f(ba^{-1})\delta_a,$$

which means that the matrix of T in the δ-basis is exactly $[f(ba^{-1})]_{a,b\in G}$. Clearly, this matrix has the same determinant as $[f(a^{-1}b)]_{a,b\in G}$.

On the other hand,

$$T\chi = \left(\sum_{a\in G} f(a^{-1})\chi(a^{-1})\right)\chi = \left(\sum_{a\in G} f(a)\chi(a)\right)\chi$$

---

[1]To be precise, we fixed an ordering $G = \{a_1,\ldots,a_m\}$, and consider the determinant $\det\left[f(a_i^{-1}a_j)\right]_{1\le i,j\le m}$. Obviously, the value of the determinant does not depend on the fixed ordering.

which means that the matrix of T in the χ-basis is diagonal, with the elements $\sum_{a \in G} f(a)\chi(a)$. Equating the determinants of the two matrices, we obtain (10.1). The theorem is proved.   □

Actually, we need a variation of this result.

Corollary 10.2. *Let G be a finite abelian group, and let f be a complex function on G. Then*

$$\det[f(ab)]_{a,b \in G} = \pm \prod_{\chi \in \hat{G}} \sum_{a \in G} f(a)\chi(a), \tag{10.2}$$

$$\det[f(ab) - f(a)]_{\substack{a,b \in G \\ a,b \neq 1}} = \pm \prod_{\substack{\chi \in \hat{G} \\ \chi \neq 1}} \sum_{a \in G} f(a)\chi(a), \tag{10.3}$$

where the product in (10.3) extends to the nontrivial characters of G.

Proof. The matrix $[f(ab)]$ can be obtained from $[f(a^{-1}b)]$ by a permutation of lines, which proves (10.2). (The reader can verify that $\pm$ can be specified as $(-1)^{|G|-|G[2]|}$, where $G[2]$ is the 2-torsion subgroup of G.)

Now write $G = \{a_1 = 1, a_2, \ldots, a_m\}$. Then the determinant in (10.2) is

$$\begin{vmatrix} f(a_1) & f(a_1 a_2) & \cdots & f(a_1 a_m) \\ f(a_2) & f(a_2 a_2) & \cdots & f(a_2 a_m) \\ \vdots & \vdots & \ddots & \vdots \\ f(a_m) & f(a_m a_2) & \cdots & f(a_m a_m) \end{vmatrix}.$$

If we add all the rows to the first row, the elements of the new first row will all become equal to $\sum_{a \in G} f(a)$. It follows that

$$\det[f(ab)]_{a,b \in G} = \left(\sum_{a \in G} f(a) \right) \begin{vmatrix} 1 & 1 & \cdots & 1 \\ f(a_2) & f(a_2 a_2) & \cdots & f(a_2 a_m) \\ \vdots & \vdots & \ddots & \vdots \\ f(a_m) & f(a_m a_2) & \cdots & f(a_m a_m) \end{vmatrix}.$$

Subtracting now the first column from every other, we obtain

$$\det[f(ab)]_{a,b \in G} = \left(\sum_{a \in G} f(a) \right) \det[f(ab) - f(a)]_{\substack{a,b \in G \\ a,b \neq 1}},$$

which proves (10.3) for all f satisfying $\sum_{a \in G} f(a) \neq 0$.

To complete the proof, observe that for every f there is a sequence $\{f_k\}_{k=0}^{\infty}$ converging to f and such that $\sum_{a \in G} f_k(a) \neq 0$ for all k. As we have just proved, (10.3) holds with f replaced by f_k. Taking the limit as $k \to \infty$, we obtain (10.3) with f itself. □

10.2 Cyclotomic Units

As usual, p is an odd prime number, $\zeta = \zeta_p$ is a primitive pth root of unity, $K = \mathbb{Q}(\zeta)$ is the pth cyclotomic field, and G is the Galois group of K over $\mathbb{Q}$.

As we have seen in Proposition 4.1, the numbers

$$\frac{1 - \zeta^k}{1 - \zeta^\ell} \qquad (k, \ell \not\equiv 0 \bmod p)$$

are units of the cyclotomic field K. It follows that for any integers $a_1, \ldots, a_{p-1}$ satisfying $a_1 + \cdots + a_{p-1} = 0$ the number

$$(1 - \zeta)^{a_1} (1 - \zeta^2)^{a_2} \cdots (1 - \zeta^{p-1})^{a_{p-1}}$$

is a unit of K as well. In other words, for any Θ from the augmentation ideal R^{aug} of the group ring $R = \mathbb{Z}[G]$, the number $(1 - \zeta)^\Theta$ is a unit of K. These units are called *cyclotomic*, or *circular*, units; they form a multiplicative group, which is denoted by $\mathcal{C}_K$ or simply by $\mathcal{C}$ if there is no risk of confusion.

To begin with, we determine a system of generators for $\mathcal{C}$. Obviously, the group $\mathcal{C}$ is generated by the units

$$\theta_k = \frac{1 - \zeta^k}{1 - \zeta} \qquad (k = 2, 3, \ldots, p - 1).$$

Since $\theta_{p-1} = -\zeta$ and $\theta_{p-k} = -\zeta^{-k}\theta_k$, we obtain the following statement.

Proposition 10.3. *The group $\mathcal{C}$ of cyclotomic units is generated by $-\zeta$ and by the units $\theta_2, \ldots, \theta_{(p-1)/2}$.* □

The group $\mathcal{C}$ is a subgroup of the full unit group $\mathcal{U}$, which, by Dirichlet's unit theorem, is a finitely generated abelian group of rank $r = (p - 3)/2$. It follows that $\mathcal{C}$ is a finitely generated abelian group of rank not exceeding r. It is fundamental that the rank of $\mathcal{C}$ is equal to r (and, consequently, the index $[\mathcal{U} : \mathcal{C}]$ is finite).

Theorem 10.4. *The cyclotomic units $\theta_2, \ldots, \theta_{(p-1)/2}$ are multiplicatively independent. In particular, the rank of the group of cyclotomic units is $(p - 3)/2$.*

The proof of this theorem requires some preparation. Recall that Theorem 7.18 was an "algebraic reformulation" of Dirichlet's nonvanishing relation $L(1, \chi) \neq 0$

for the *odd* characters (see Remark 7.19). Theorem 10.4 is the corresponding statement for the (non-trivial) *even* characters. In particular, its proof is quite analogous to the proof of Theorem 7.18.

Let us translate into a more suitable language the part of Corollary 5.12 related to the even Dirichlet's characters. Let $f : G \to \mathbb{R}$ be the function defined by $f(\sigma) = \log|1 - \zeta^{\sigma}|$. Then the "even" part of Corollary 5.12 can be restated as follows: for any even[2] character χ of G, we have $\sum_{\sigma \in G} \chi(\sigma) f(\sigma) \neq 0$. Also, the class number formula (5.18) can be rewritten as

$$h^+ \mathcal{R}^+ = 2^{(3-p)/2} \prod_{\substack{\chi(\iota)=1 \\ \chi \neq 1}} \sum_{\sigma \in G} \chi(\sigma) f(\sigma),$$

where h^+ and $\mathcal{R}^+$ are the class number and the regulator of the real cyclotomic field K^+, and the product runs over all nontrivial even characters of G.

Obviously, $f(\sigma) = f(\sigma\iota)$ for any $\sigma \in G$. It follows that f defines a function on the group $G^+ = G/\{1, \iota\}$ (which is the Galois group of K^+ over $\mathbb{Q}$). This function on G^+ will be denoted by f as well. Also, there is a one-to-one correspondence between the even characters of G and the characters of G^+. With this correspondence in mind, we have the relation

$$\sum_{\sigma \in G} \chi(\sigma) f(\sigma) = 2 \sum_{\sigma \in G^+} \chi(\sigma) f(\sigma),$$

where the χ on the left is an even character of G and the χ on the right is the corresponding character of G^+.

We have proved the following statement.

Proposition 10.5. *For any character χ of G^+ we have $\sum_{\sigma \in G^+} \chi(\sigma) f(\sigma) \neq 0$. Moreover,*

$$h^+ \mathcal{R}^+ = \prod_{\chi \neq 1} \sum_{\sigma \in G^+} \chi(\sigma) f(\sigma), \tag{10.4}$$

where the product extends to the nontrivial characters of G^+.

Proof of Theorem 10.4. Put $m = (p-1)/2 = r + 1$. Assume that the units $\theta_2, \ldots, \theta_m$ satisfy a nontrivial multiplicative relation $\theta_2^{a_2} \cdots \theta_m^{a_m} = 1$. Then for any $\sigma \in G$ the conjugates $\theta_2^{\sigma}, \ldots, \theta_m^{\sigma}$ satisfy the same relation. Hence the rows of the $r \times r$ matrix $A = \left[\log|\theta_a^{\sigma_b}|\right]_{2 \leq a, b \leq m}$ (where σ_b is the automorphism of K defined by $\sigma_b(\zeta) = \zeta^b$) are linearly dependent; in particular, the determinant of this matrix is 0.

---

[2]That is, satisfying $\chi(\iota) = 1$, where ι is the complex conjugation

On the other hand, a quick reflection shows that matrix A coincides (up to a row/column permutation) with the matrix $[f(\sigma\tau) - f(\tau)]_{\sigma,\tau\in G^+, \atop \sigma,\tau\neq 1}$, where $f : G^+ \to \mathbb{R}$ is the function defined above. By Corollary 10.2,

$$\det A = \pm \prod_{\chi\neq 1} \sum_{\sigma\in G^+} \chi(\sigma) f(\sigma), \tag{10.5}$$

which is nonzero by Proposition 10.5. The theorem is proved.   □

For the proof of Theorem 10.4, we needed only the "qualitative" part (that is, the nonvanishing statement) of Proposition 10.5. Using the "quantitative" part (that is, equality (10.4)), we can go further and determine the index $[\mathcal{U} : C]$. Below we sketch the proof of the beautiful relation

$$[\mathcal{U} : C] = h^+. \tag{10.6}$$

(This result will not be used in the sequel.)

As we have seen in Sect. 4.4, the full unit group $\mathcal{U}$ is the direct product $\Omega \times \mathcal{U}_+$, where $\Omega = \langle -\zeta \rangle$ is the torsion subgroup of $\mathcal{U}$ and $\mathcal{U}_+$ is the group of positive real units, which is a free abelian group of rank r. Since $-\zeta \in C$, the group C contains Ω, which means that $C = \Omega \times C_+$, where $C_+ = C \cap \mathcal{U}_+$ is the group of real positive cyclotomic units (again a free abelian group of rank r). This shows that $[\mathcal{U} : C] = [\mathcal{U}_+ : C_+]$.

Now consider the map $\lambda : \mathcal{U} \to \mathbb{R}^r$, which associates to each $\eta \in \mathcal{U}$ the vector $(\log|\eta^{\sigma_2}|, \ldots, \log|\eta^{\sigma_m}|)$. The kernel of λ is Ω_K and the image $\lambda(\mathcal{U})$ is a lattice in $\mathbb{R}^r$ of fundamental volume $\mathcal{R}^+$. Also, $\lambda(C)$ is the lattice generated by $\lambda(\theta_2), \ldots, \lambda(\theta_m)$, and its fundamental volume is $|\det A|$, which is equal to $h^+\mathcal{R}^+$ by (10.4) and (10.5). Since λ is injective on $\mathcal{U}_+$, we have

$$[\mathcal{U} : C] = [\mathcal{U}_+ : C_+] = [\lambda(\mathcal{U}) : \lambda(C)] = \frac{h^+\mathcal{R}^+}{\mathcal{R}^+} = h^+,$$

which proves (10.6).

Remark 10.6. Equality (10.6) is a "real analogue" of Iwasawa's class number formula (see Theorem 7.27). As in that case, it is believed that the equality of orders does not correspond, in general, to isomorphism of the groups $\mathcal{U}/C$ and $\mathcal{H}^+$. See Sect. 8.1 in Washington's book [136] for a discussion on this, as well as for extending the results of this section to general cyclotomic fields.

Chapter 11
Selmer Group and Proof of Catalan's Conjecture

As follows from Theorem 9.2, Catalan's problem would be solved if we show that a solution of Catalan's equation implies a nontrivial element in the real part of Mihăilescu's ideal. It is natural to look for such elements in the annihilator of the class group. (More precisely, we want to annihilate a related group, called here the qth *Selmer group*.) Unfortunately, Stickelberger's theorem is not suitable for this purpose, because the real part of Stickelberger's ideal is uninteresting (see Sect. 7.7). In 1988 Thaine [132] discovered a partial "real" analogue of Stickelberger's theorem, and Mihăilescu showed that Thaine's theorem is sufficient for solving Catalan's problem.

11.1 Selmer Group

In this section we define the notion of qth Selmer group and reduce Catalan's problem to a certain property of this group.

Let K be any number field (not just the cyclotomic field) and let q be any prime number. Consider the subgroup $\tilde{\Sigma}$ of the multiplicative group $K^\times$ consisting of the elements $\alpha \in K^\times$ such that the principal ideal (α) is a qth power of an ideal. In other words,

$$\tilde{\Sigma} = \{\alpha \in K^\times : q \text{ divides } \mathrm{Ord}_\mathfrak{l}(\alpha) \text{ for any prime ideal } \mathfrak{l} \text{ of } K\}.$$

It contains the group $(K^\times)^q$ of qth powers, and it is easy to show that the quotient $\Sigma = \tilde{\Sigma}/(K^\times)^q$ is finite.

Indeed, let $\mathcal{H}$ be the class group of K, and consider the homomorphism $\tilde{\Sigma} \to \mathcal{H}$ which maps $\alpha \in \tilde{\Sigma}$ to the class of the ideal $\mathfrak{a}$ satisfying $\mathfrak{a}^q = (\alpha)$. The kernel of this homomorphism is $\mathcal{U}(K^\times)^q$, where $\mathcal{U}$ is the unit group of K. This homomorphism induces a homomorphism $\Sigma \to \mathcal{H}$, whose kernel $\mathcal{U}(K^\times)^q/(K^\times)^q = \mathcal{U}/\mathcal{U}^q$ is finite

© Springer International Publishing Switzerland 2014
Y.F. Bilu et al., *The Problem of Catalan*, DOI 10.1007/978-3-319-10094-4_11

(because $\mathcal{U}$ is of finite rank). Since $\mathcal{H}$ is finite as well, the finiteness of Σ follows. One can add to this that Σ is a q-torsion group; that is, the order of each element of Σ is q or 1.

If K is a Galois extension of $\mathbb{Q}$, then the Galois group G acts on both $\tilde{\Sigma}$ and $(K^{\times})^q$; in other words, both these groups are G-modules. Hence so is the quotient group Σ. Moreover, since Σ is a q-torsion group, it is annihilated by $q\mathbb{Z}[G]$. It follows that Σ has a natural structure of $\mathbb{Z}[G]/q\mathbb{Z}[G] = \mathbb{F}_q[G]$-module, where $\mathbb{F}_q = \mathbb{Z}/q\mathbb{Z}$ is the field of q elements.

Remark 11.1. The ring $\mathbb{F}_q[G]$ has an important advantage compared to $\mathbb{Z}[G]$: it is *semi-simple* when q does not divide $|G|$; see Theorem D.11. In particular, $\mathbb{F}_q[G]$ is semi-simple if K is the pth cyclotomic field and q does not divide $p-1$. We shall repeatedly use this property in this chapter.

The significance of this construction for Catalan's problem stems from the following observation. Let (x, y, p, q) be a solution of Catalan's equation, ζ a primitive pth root of unity, $K = \mathbb{Q}(\zeta)$ the pth cyclotomic field, and $G = \mathrm{Gal}(K/\mathbb{Q})$.

Proposition 11.2. *In the above setup, for any $\Theta \in \mathbb{Z}[G]$ of weight divisible by q, we have $(x - \zeta)^{\Theta} \in \tilde{\Sigma}$.*

Proof. By the "most important" Lemma 6.1 the principal ideal $(x - \zeta)$ is equal to $\mathfrak{p}\mathfrak{a}^q$, where $\mathfrak{p} = (1 - \zeta)$ is the prime ideal above p and $\mathfrak{a}$ is an ideal of K. Since $\mathfrak{p}^{\sigma} = \mathfrak{p}$ for any $\sigma \in G$, we have $\mathfrak{p}^{\Theta} = \mathfrak{p}^{w(\Theta)}$ for any $\Theta \in \mathbb{Z}[G]$. Hence if $q \mid w(\Theta)$ then, writing $w(\Theta) = mq$, we find that the principal ideal $\left((x - \zeta)^{\Theta}\right)$ is $\left(\mathfrak{p}^m \mathfrak{a}^{\Theta}\right)^q$, a qth power of an ideal.   □

We continue to assume that K is the pth cyclotomic field and G its Galois group. As we did for $\mathbb{Z}[G]$, we consider the *real part*, or the *plus-part* $\mathbb{F}_q[G]^+$, and the *relative part*, or the *minus part* $\mathbb{F}_q[G]^-$, the former being the ideal $(1 + \iota)$ and the latter being the ideal $(1 - \iota)$ (where ι, as usual, is the complex conjugation).

Assume that, in the setup of Proposition 11.2, the group Σ has a nonzero annihilator $\Theta \in \mathbb{F}_q[G]^+$ with $w(\Theta) = 0$. Then, lifting Θ to $\mathbb{Z}[G]$, it is easy to produce (see the end of this section) a nontrivial element of the real part of Mihăilescu's ideal, of weight divisible by q, contradicting Theorem 9.2.

Thus, Catalan's problem will be resolved if we find in the plus- part of $\mathbb{F}_q[G]$ a nonzero annihilator of Σ of weight 0. Unfortunately, the theory of cyclotomic fields in its present state does not guarantee the existence of such an annihilator. Mention that we cannot use Stickelberger's ideal to produce it, because the real part of Stickelberger's ideal is very small and uninteresting (see Proposition 7.23).

Mihăilescu's bright idea was to replace Σ by a smaller group, for which a required annihilator exists (under suitable assumptions). As it was for Σ, the definition applies to any number field K and any prime q. Thus, we return to the general case: now K is any number field.

Let $\alpha \in \mathcal{O} = \mathcal{O}_K$ be coprime with q. We say that α is q-primary if α is congruent to a qth power modulo q^2:

$$\alpha \equiv \beta^q \bmod q^2$$

for some $\beta \in \mathcal{O}$.

More generally, let α be a nonzero element of K. We say that α is q-primary if $\alpha = \alpha_1 \alpha_2^{-1} \gamma^q$, where $\alpha_1, \alpha_2 \in \mathcal{O}$ are coprime with q and q-primary as defined above and $\gamma \in K^{\times}$.

Remark 11.3. Assume that q is an odd prime number unramified in K (which is the case when K is the pth cyclotomic field and $q \neq p$). A reader familiar with the notion of q-adic completion can easily verify that in this case $\alpha \in K^{\times}$ is q-primary if and only if it is a "q-adic qth power" (that is, a q-adic qth power for any prime ideal q | q). Perhaps, it would have been more "conceptual" to take this property as the definition of q-primary numbers. However, our definition is more convenient to use and more suitable for an unsophisticated reader.

The q-primary elements of K form a multiplicative group, containing the group $(K^{\times})^q$ of qth powers. Now let $\tilde{S}$ be the subgroup of $\tilde{\Sigma}$, consisting of q-primary elements, and put $S = \tilde{S}/(K^{\times})^q$. Since S is a subgroup of Σ, it is also a finite q-torsion group. We call it the *qth Selmer group*, or (when q is fixed) simply the Selmer group of the field K, because its definition is inspired by the definition of the Selmer group in the theory of elliptic curves.[1]

If K is a Galois extension of $\mathbb{Q}$ and $G = \mathrm{Gal}(K/\mathbb{Q})$ then the multiplicative group of q-primary numbers is a G-module. It follows that $\tilde{S}$ is a G-submodule of $\tilde{\Sigma}$ and S is an $\mathbb{F}_q[G]$-submodule of Σ.

Mihăilescu's divisibility theorem (Theorem 6.14) implies that in Proposition 11.2 one can replace $\tilde{\Sigma}$ by $\tilde{S}$.

Proposition 11.4. *In the setup of Proposition 11.2, we have $(x - \zeta)^{\Theta} \in \tilde{S}$.*

Proof. It suffices to show that $x - \zeta$ is q-primary, which is true because $-\zeta$ is a qth power and x is divisible by q^2 by Theorem 6.14.   □

Now we are ready to state the main result of this chapter. Let p and q be distinct odd prime numbers, let K be the pth cyclotomic field, let G be the Galois group of $K/\mathbb{Q}$ and let S be the qth Selmer group of K.

Theorem 11.5. *Assume that $p > q$ and that $p \not\equiv 1 \bmod q$. Then there exists a nonzero $\Theta \in \mathbb{F}_q[G]^+$ with $\mathrm{w}(\Theta) = 0$ which annihilates S.*

This theorem, together with the previously established results, solves Catalan's problem.

---

[1] We do not assume from our reader any knowledge of the theory of elliptic curves.

Proof of Theorem 3.1 (assuming Theorem 11.5). Let (x, y, p, q) be a solution of Catalan's equation. By symmetry, we can assume that $p > q$, and Corollary 8.20 implies that $p \not\equiv 1 \bmod q$.

Let ζ, K, and G be as above. Since q does not divide $|G| = p - 1$, the ring $\mathbb{F}_q[G]$ is semi-simple (see Remark 11.1).

Applying Theorem 11.5, we find $\Theta \in \mathbb{F}_q[G]^+$ with $\mathrm{w}(\Theta) = 0$ which annihilates S. Write $\Theta = (1 + \iota)\Theta_0$, and let $\tilde{\Theta}_0 \in \mathbb{Z}[G]$ be an arbitrary lifting of Θ_0 to $\mathbb{Z}[G]$. Then the element $\tilde{\Theta} := (1 + \iota)\tilde{\Theta}_0 \in \mathbb{Z}[G]^+$ is a lifting of Θ. It has the following properties:

- for any $\alpha \in \tilde{S}$ we have $\alpha^{\tilde{\Theta}} \in (K^\times)^q$;
- the weight of $\tilde{\Theta}$ is divisible by q (because $\mathrm{w}(\Theta) = 0$);
- the element $\tilde{\Theta}^2$ is not divisible by q (indeed, if $q \mid \tilde{\Theta}^2$ then $\Theta^2 = 0$ which is impossible because $\Theta \neq 0$ and $\mathbb{F}_q[G]$ is semi-simple).

Since $q \mid \mathrm{w}(\tilde{\Theta})$, Proposition 11.4 implies that $(x - \zeta)^{\tilde{\Theta}} \in \tilde{S}$. Hence

$$(x - \zeta)^{\tilde{\Theta}^2} \in (K^\times)^q.$$

Thus, $\tilde{\Theta}^2 \in \mathbb{Z}[G]^+$ is a nontrivial element of Mihăilescu's ideal of weight divisible by q. This contradicts Theorem 9.2.   □

Now we can forget about Catalan's problem and concentrate on the proof of Theorem 11.5.

11.2   Selmer Group as Galois Module

In this section we reduce Theorem 11.5, which is a statement on the somewhat "mysterious" Selmer group, to a statement on more "familiar" objects: units and ideal classes.

As usual, let p and q be distinct odd prime numbers; let K be the pth cyclotomic field, G its Galois group, and $\mathcal{O} = \mathcal{O}_K$ its ring of integers. Further, let $\mathcal{U}$ be the unit group and $\mathcal{H}$ the class group of K. Finally, we denote by $\mathcal{U}_q$ the subgroup of $\mathcal{U}$ consisting of the q-primary units.

Proposition 11.6. *Assume that q does not divide $p - 1$. Then the Selmer group S is $\mathbb{F}_q[G]$-isomorphic to a submodule of $\mathcal{U}_q / \mathcal{U}^q \oplus \mathcal{H}/\mathcal{H}^q$.*

Proof. Recall the homomorphism $\tilde{\Sigma} \to \mathcal{H}$ already used in the beginning of Sect. 11.1: to each $\alpha \in \tilde{\Sigma}$ we associate the class of the ideal $\mathfrak{a}$ such that $\mathfrak{a}^q = (\alpha)$. We restrict this homomorphism to $\tilde{S}$. The kernel of this restriction is

$$\mathcal{U}(K^\times)^q \cap \tilde{S} = \mathcal{U}_q (K^\times)^q.$$

Further, since $\mathfrak{a}^q = (\alpha)$ is the principal ideal, the image of our homomorphism is contained in the q-torsion subgroup[2] $\mathcal{H}[q]$. Also, all groups above are G-modules and our homomorphism is a G-homomorphism.

Factoring by $(K^\times)^q$, we obtain an $\mathbb{F}_q[G]$-homomorphism $\mathcal{S} \xrightarrow{f} \mathcal{H}[q]$ with

$$\ker f = \mathcal{U}_q (K^\times)^q / (K^\times)^q = \mathcal{U}_q / \mathcal{U}^q.$$

Since the ring $\mathbb{F}_q[G]$ is semi-simple (see Remark 11.1), the Selmer group $\mathcal{S}$ is $\mathbb{F}_q[G]$-isomorphic to $\ker f \oplus f(\mathcal{S})$ (see Remark C.8). It follows that $\mathcal{S}$ is isomorphic to a submodule of $\mathcal{U}_q / \mathcal{U}^q \oplus \mathcal{H}[q]$. Finally, Theorem E.1 implies the $\mathbb{F}_q[G]$-isomorphism $\mathcal{H}[q] \cong \mathcal{H}/\mathcal{H}^q$. This proves the proposition.   □

Thus, Theorem 11.5 will be proved if we produce a nonzero $\Theta \in \mathbb{F}_q[G]^+$, of weight 0, annihilating both $\mathcal{U}_q / \mathcal{U}^q$ and $\mathcal{H}/\mathcal{H}^q$. To find such an annihilator, we should investigate the group $\mathcal{U}_q / \mathcal{U}^q$. First of all, we determine the Galois module structure of $\mathcal{U}/\mathcal{U}^q$.

11.3 Units as Galois Module

In this section p is, as usual, an odd prime number and K is the pth cyclotomic field. We denote by $\mathcal{O} = \mathcal{O}_K$ its ring of integers, by $\mathcal{U}$ its unit group, and by $G = \mathrm{Gal}(K/\mathbb{Q})$ its Galois group.

The principal goal of this section is to study the Galois module structure of the group $\mathcal{U}/\mathcal{U}^q$, where q is another odd prime number. However, we begin with the unit group itself, more precisely, with its *free part*, that is, the quotient $\tilde{\mathcal{U}} = \mathcal{U}/\Omega$, where $\Omega = \Omega_K$ stands for the torsion subgroup of $K^\times$.

Since both $\mathcal{U}$ and Ω are G-modules, so is $\tilde{\mathcal{U}}$. We want to calculate its annihilator. In other words, we want to determine all $\Theta \in \mathbb{Z}[G]$ such that η^Θ is a root of unity for any $\eta \in \mathcal{U}$.

Obviously, the norm element $\mathcal{N} = \sum_{\sigma \in G} \sigma$ belongs to the annihilator. Also, for any unit η the quotient $\eta/\bar\eta$ is a root of unity; see Lemma 4.11. Hence $1 - \iota$ (where ι is the complex conjugation) is in the annihilator as well. Thus, the annihilator contains the ideal $(\mathcal{N}, 1 - \iota)$. In fact, it is slightly bigger.

Theorem 11.7. *The annihilator of* $\tilde{\mathcal{U}}$ *is the ideal* $\mathcal{I}$ *of* $\mathbb{Z}[G]$ *consisting of all* Θ *such that* $2\Theta \in (\mathcal{N}, 1 - \iota)$.

For the proof we need a lemma, which is, essentially, due to Dirichlet. Recall that $G = \{\sigma_1, \ldots, \sigma_{p-1}\}$, where σ_k is defined by $\zeta \mapsto \zeta^k$. Also, let $r = (p-3)/2$ be the rank of the unit group.

---

[2]Recall that the *q-torsion subgroup* of a multiplicatively written abelian group A is $A[q] = \{x \in A : x^q = 1\}$.

Lemma 11.8. *Let* $a_1, \ldots, a_r \in \mathbb{Z}$ *be such that* $\Theta = a_1\sigma_1 + \cdots + a_r\sigma_r$ *annihilates* $\tilde{\mathcal{U}}$. *Then* $a_1 = \cdots = a_r = 0$.

Proof. Consider the map $\lambda : \mathcal{U} \to \mathbb{R}^r$, which associates to each $\eta \in \mathcal{U}$ the vector $(\log|\eta^{\sigma_1}|, \ldots, \log|\eta^{\sigma_r}|)$. The kernel of λ is Ω and the image $\lambda(\mathcal{U})$ is a lattice in $\mathbb{R}^r$. If Θ annihilates $\tilde{\mathcal{U}}$ then

$$a_1 \log|\eta^{\sigma_1}| + \cdots + a_r \log|\eta^{\sigma_r}| = 0$$

for any $\eta \in \mathcal{U}$. If $a_1, \ldots, a_r$ were not all zero then the image $\lambda(\mathcal{U})$ would have belonged to a proper subspace of $\mathbb{R}^r$. This, however, is impossible because $\lambda(\mathcal{U})$ is a lattice. The lemma is proved.   □

Proof of Theorem 11.7. As we have already seen, $(\mathcal{N}, 1 - \iota) \subseteq \mathrm{ann}(\tilde{\mathcal{U}})$. Moreover, if $m\Theta \in \mathrm{ann}(\tilde{\mathcal{U}})$ for a nonzero integer m, then $\Theta \in \mathrm{ann}(\tilde{\mathcal{U}})$ (if $\eta^{m\Theta}$ is a root of unity, then so is η^{Θ}). Thus, $\mathcal{I} \subset \mathrm{ann}(\tilde{\mathcal{U}})$.

To complete the proof, we have to show that $\mathrm{ann}(\tilde{\mathcal{U}}) \subseteq \mathcal{I}$. Put

$$\mathcal{N}' = \sigma_1 + \cdots + \sigma_{(p-1)/2},$$

Since $\sigma_{p-k} = \iota\sigma_k$, we have $\sigma_k - \sigma_{p-k} = \sigma_k(1 - \iota)$. This implies that

$$2\mathcal{N}' = \mathcal{N} - \mathcal{N}'(1 - \iota) \in (\mathcal{N}, 1 - \iota),$$

which means that $\mathcal{N}' \in \mathcal{I}$.

Now let Θ annihilate $\tilde{\mathcal{U}}$. Again using the relation $\sigma_{p-k} = \iota\sigma_k$, we may write $\Theta = \Theta' + \iota\Theta''$, where both Θ' and Θ'' are linear combinations of $\sigma_1, \ldots, \sigma_m$ with $m = (p - 1)/2$. Say,

$$\Theta' = a_1'\sigma_1 + \cdots + a_m'\sigma_m, \qquad \Theta'' = a_1''\sigma_1 + \cdots + a_m''\sigma_m.$$

Now

$$\Theta + (1 - \iota)\Theta'' - \left(a_m' + a_m''\right)\mathcal{N}'$$

is a linear combination of $\sigma_1, \ldots, \sigma_{m-1}$, annihilating $\tilde{\mathcal{U}}$. By Lemma 11.8, it must vanish. Thus,

$$\Theta = -(1 - \iota)\Theta'' + \left(a_m' + a_m''\right)\mathcal{N}',$$

which implies that $\Theta \in \mathcal{I}$. The theorem is proved.   □

Now we fix an odd prime number q and study the $\mathbb{F}_q[G]$-module structure of the quotient group $\mathcal{U}/\mathcal{U}^q$.

Theorem 11.9. *Assume that q does not divide $p - 1$. Then $\mathcal{U}/\mathcal{U}^q$ is a cyclic $\mathbb{F}_q[G]$-module. Its annihilator is $(\mathcal{N}) \oplus (1 - \iota)$.*

Proof. The kernel of the natural G-epimorphism

$$\mathcal{U} \to \tilde{\mathcal{U}} \to \tilde{\mathcal{U}}/\tilde{\mathcal{U}}^q$$

is $\Omega\mathcal{U}^q$. Since all elements of Ω are qth powers, the kernel is $\mathcal{U}^q$, which means that $\mathcal{U}/\mathcal{U}^q$ and $\tilde{\mathcal{U}}/\tilde{\mathcal{U}}^q$ are isomorphic as G-modules. Hence they are isomorphic as $\mathbb{F}_q[G]$-modules as well. Therefore it suffices to show that $\tilde{\mathcal{U}}/\tilde{\mathcal{U}}^q$ is a cyclic $\mathbb{F}_q[G]$-module and determine its annihilator.

For the annihilator we shall apply Theorem C.5. Since q does not divide $p-1$, the ring $\mathbb{F}_q[G]$ is semi-simple, and, in particular, it has no nilpotents. Hence $q\mathbb{Z}[G]$ is a *radical ideal* of the ring $\mathbb{Z}[G]$ (as defined in Appendix C.2). Moreover, since any quotient of $\mathbb{F}_q[G]$ is semi-simple as well, any ideal of $\mathbb{Z}[G]$ containing $q\mathbb{Z}[G]$ is a radical ideal. In particular, so is the ideal $q\mathbb{Z}[G] + \operatorname{ann}_{\mathbb{Z}[G]}(\tilde{\mathcal{U}})$. Now Theorem C.5 implies that the $\mathbb{F}_q[G]$-annihilator of $\tilde{\mathcal{U}}/\tilde{\mathcal{U}}^q$ is the image of $\operatorname{ann}_{\mathbb{Z}[G]}(\tilde{\mathcal{U}})$. Hence, by Theorem 11.7, the ideal $\operatorname{ann}_{F_q(G)}(\tilde{\mathcal{U}}/\tilde{\mathcal{U}}^q)$ consists of the elements $\Theta \in \mathbb{F}_q[G]$ such that $2\Theta \in (\mathcal{N}, 1-\iota)$, where, from now on, $\mathcal{N}$ stands for the norm element of the ring $\mathbb{F}_q[G]$ (and not of $\mathbb{Z}[G]$). But 2 is invertible in $\mathbb{F}_q[G]$, which implies that

$$\operatorname{ann}_{F_q(G)}(\tilde{\mathcal{U}}/\tilde{\mathcal{U}}^q) = (\mathcal{N}, 1-\iota).$$

Further, we have $\mathrm{w}(\mathcal{N}) = p-1 \neq 0$ in $\mathbb{F}_q$ (here we again use the assumption $p \not\equiv 1 \bmod q$). But, by Proposition D.2, the principal ideal $(\mathcal{N})$ is equal to $\mathbb{F}_q\mathcal{N}$, and, in particular, the nonzero elements of $(\mathcal{N})$ have nonzero weight. It follows that the ideals $(\mathcal{N})$ and $(1-\iota)$ have nonzero intersection, and we have

$$\operatorname{ann}_{F_q(G)}(\tilde{\mathcal{U}}/\tilde{\mathcal{U}}^q) = (\mathcal{N}) \oplus (1-\iota).$$

In the sequel we denote $\operatorname{ann}_{\mathbb{F}_q(G)}(\tilde{\mathcal{U}}/\tilde{\mathcal{U}}^q)$ by $\mathcal{I}$.

It remains to show that $\tilde{\mathcal{U}}/\tilde{\mathcal{U}}^q$ is a cyclic $\mathbb{F}_q[G]$-module. By Corollary C.16, it suffices to prove that $|\tilde{\mathcal{U}}/\tilde{\mathcal{U}}^q| = |F_q[G]/\mathcal{I}|$. Since $\tilde{\mathcal{U}}$ is a free abelian group of rank $(p-3)/2$, we have $|\tilde{\mathcal{U}}/\tilde{\mathcal{U}}^q| = q^{(p-3)/2}$.

Now let us determine the cardinality of $F_q[G]/\mathcal{I}$. First of all, let us find the dimension of $\mathcal{I}$ as an $\mathbb{F}_q$-vector space. Since $(\mathcal{N}) = \mathbb{F}_q\mathcal{N}$, the $\mathbb{F}_q$-dimension of the principal ideal $(\mathcal{N})$ is 1. Further, the principal ideal $(1-\iota)$ is exactly the minus part $\mathbb{F}_q[G]^-$, and its $\mathbb{F}_q$-dimension is[3] $(p-1)/2$. It follows that $\dim_{\mathbb{F}_q}\mathcal{I} = (p+1)/2$. Thus, $|\mathcal{I}| = q^{(p+1)/2}$ and, consequently,

$$|F_q[G]/\mathcal{I}| = q^{(p-3)/2} = |\tilde{\mathcal{U}}/\tilde{\mathcal{U}}^q|.$$

The theorem is proved.   □

---

[3]Alternatively, one could determine both dimensions using Proposition D.13.

Next, we decompose $\mathcal{U}/\mathcal{U}^q$ into a direct sum of smaller modules, using the theory of cyclotomic units (see Chap. 10). But before this, we establish an important property of these units.

11.4 q-Primary Cyclotomic Units

For the proof of Theorem 11.5, it is crucial that not all cyclotomic units are q-primary (see Sect. 11.1 for the definition of q-primary numbers). More precisely, we have the following statement.

Theorem 11.10 (Mihăilescu). *Let p and q be odd prime numbers, and assume that $p > q$. Then not all cyclotomic units of the pth cyclotomic field are q-primary.*

Proof. To begin with, introduce the polynomial

$$f(T) = \frac{(1 + T)^q - 1 - T^q}{q} \in \mathbb{Z}[T]. \tag{11.1}$$

It is a nonzero monic polynomial of degree $q - 1$.

Now assume that all cyclotomic units of the pth cyclotomic field K are q-primary. In particular, so is $1 + \zeta^q = (1 - \zeta^{2q})/(1 - \zeta^q)$. Thus, there exists $\beta \in \mathcal{O}_K$ such that $1 + \zeta^q \equiv \beta^q \bmod q^2$. Then $(1 + \zeta)^q \equiv 1 + \zeta^q \equiv \beta^q \bmod q$. Lemma 6.7 implies that $(1 + \zeta)^q \equiv \beta^q \bmod \mathfrak{q}^2$ for any prime ideal $\mathfrak{q} \mid q$. Since q is unramified in K, this yields the congruence $(1 + \zeta)^q \equiv \beta^q \bmod q^2$

Thus, $(1 + \zeta)^q \equiv 1 + \zeta^q \bmod q^2$. This can be rewritten as $f(\zeta) \equiv 0 \bmod q$, where $f(T)$ is the polynomial defined in (11.1).

Applying the Galois conjugation, we obtain $f(\zeta^\sigma) \equiv 0 \bmod q$ for any $\sigma \in G$. If now $\mathfrak{q}$ is a prime ideal above q, then we have the $p - 1$ congruences

$$f(\zeta^\sigma) \equiv 0 \bmod \mathfrak{q} \qquad (\sigma \in G). \tag{11.2}$$

Since $\zeta^\sigma \not\equiv \zeta^\tau \bmod \mathfrak{q}$ for distinct $\sigma, \tau \in G$, congruences (11.2) imply that

$$p - 1 \leq \deg f = q - 1,$$

which contradicts our assumption $p > q$. The theorem is proved.   □

It is an interesting question whether the assumption $p > q$ is relevant here. To the best of our knowledge, the answer is unknown.

11.5 Proof of Theorem 11.5

The preparation is over. We arrived at the culmination point of this chapter and, probably, of the whole book.

As usual, in this section p and q are distinct odd prime numbers, K is the pth cyclotomic field, and G its Galois group, $\mathcal{U}$ its group of units, and $\mathcal{U}_q$ the subgroup of q-primary units. We shall assume throughout this section that q does not divide $p - 1$. In particular, the ring $\mathbb{F}_q[G]$ is semi-simple. In this section we denote this ring by R.

First of all, we deliver the promised decomposition of $\mathcal{U}/\mathcal{U}^q$ into a sum of smaller R-modules. We denote by $\mathcal{C}$ the group of cyclotomic units and by $\mathcal{C}_q$ the subgroup of q-primary cyclotomic units.

Proposition 11.11. *The group $\mathcal{U}/\mathcal{U}^q$ is R-isomorphic to*

$$\mathcal{U}/\mathcal{C}\mathcal{U}^q \oplus \mathcal{C}_q/(\mathcal{C} \cap \mathcal{U}^q) \oplus \mathcal{C}/\mathcal{C}_q.$$

The group $\mathcal{U}_q/\mathcal{U}^q$ is R-isomorphic to a submodule of $\mathcal{U}/\mathcal{C}\mathcal{U}^q \oplus \mathcal{C}_q/(\mathcal{C} \cap \mathcal{U}^q)$.

Proof. Since R is semi-simple, we have the R-isomorphisms (see Remark C.8)

$$\mathcal{U}/\mathcal{U}^q \cong \mathcal{U}/\mathcal{C}\mathcal{U}^q \oplus \mathcal{C}\mathcal{U}^q/\mathcal{U}^q \cong \mathcal{U}/\mathcal{C}\mathcal{U}^q \oplus \mathcal{C}\mathcal{U}^q/\mathcal{C}_q\mathcal{U}^q \oplus \mathcal{C}_q\mathcal{U}^q/\mathcal{U}^q.$$

Further,

$$\mathcal{C}_q\mathcal{U}^q/\mathcal{U}^q \cong \mathcal{C}_q/(\mathcal{C}_q \cap \mathcal{U}^q) = \mathcal{C}_q/(\mathcal{C} \cap \mathcal{U}^q),$$
$$\mathcal{C}\mathcal{U}^q/\mathcal{C}_q\mathcal{U}^q = \mathcal{C}\mathcal{C}_q\mathcal{U}^q/\mathcal{C}_q\mathcal{U}^q \cong \mathcal{C}/(\mathcal{C} \cap \mathcal{C}_q\mathcal{U}^q) = \mathcal{C}/\mathcal{C}_q,$$

which proves the first statement.

Similarly, $\mathcal{U}_q/\mathcal{U}^q$ is R-isomorphic to $\mathcal{U}_q/\mathcal{C}_q\mathcal{U}^q \oplus \mathcal{C}_q\mathcal{U}^q/\mathcal{U}^q$. As we have seen above, the second term here is $\mathcal{C}_q/(\mathcal{C} \cap \mathcal{U}^q)$. Further, the kernel of the natural homomorphism

$$\mathcal{U}_q \hookrightarrow \mathcal{U} \to \mathcal{U}/\mathcal{C}\mathcal{U}^q$$

is $\mathcal{C}_q\mathcal{U}^q$, which means that $\mathcal{U}_q/\mathcal{C}_q\mathcal{U}^q$ is a submodule of $\mathcal{U}/\mathcal{C}\mathcal{U}^q$. This proves the second statement. $\qquad\square$

As we have seen in Sect. 11.2, Theorem 11.5 will be proved if we find a nonzero $\Theta \in R^+$, of weight 0, annihilating both $\mathcal{U}_q/\mathcal{U}^q$ and $\mathcal{H}/\mathcal{H}^q$. Proposition 11.11, together with Theorem 11.10, solves half of this problem.

Corollary 11.12. *Assume that $p > q$. Then there exists a nonzero $\Theta \in R^+$, of weight 0, annihilating both $\mathcal{U}/\mathcal{C}\mathcal{U}^q$ and $\mathcal{C}_q/(\mathcal{C} \cap \mathcal{U}^q)$; in particular, it annihilates $\mathcal{U}_q/\mathcal{U}^q$.*

Proof. In Proposition 11.11 we established the R-isomorphism

$$\mathcal{U}/\mathcal{U}^q \cong M \oplus \mathcal{C}/\mathcal{C}_q,$$

where $M = \mathcal{U}/\mathcal{C}\mathcal{U}^q \oplus \mathcal{C}_q/(\mathcal{C} \cap \mathcal{U}^q)$. From Theorem 11.10 we know that the group $\mathcal{C}/\mathcal{C}_q$ is nontrivial, which means that M is isomorphic to a *proper submodule* of $\mathcal{U}/\mathcal{U}^q$.

Recall now (see Theorem 11.9) that $\mathcal{U}/\mathcal{U}^q$ is a cyclic R-module with annihilator $\mathcal{I} = (\mathcal{N}) \oplus (1 - \iota)$. Since $\mathcal{U}/\mathcal{U}^q$ is cyclic, and M is its proper submodule, the annihilator of M is *strictly greater* than $\mathcal{I}$. It follows that the annihilator of M has a nonzero intersection with the direct complement $\mathcal{I}^\perp$; see Proposition C.14. Using this proposition, we find that

$$\mathcal{I}^\perp = (\mathcal{N})^\perp (1 - \iota)^\perp = R^{\mathrm{aug}} R^+,$$

where R^{aug} is the augmentation ideal. Thus, M is annihilated by a nonzero element of $R^{\mathrm{aug}} R^+$. This proves the corollary.   □

Still, this is insufficient: we have to annihilate $\mathcal{H}/\mathcal{H}^q$ as well. This is accomplished by the fundamental theorem due to the Brazilian mathematician Thaine [132].

Theorem 11.13 (Thaine).   *Assume that q does not divide $p - 1$. Then any $\Theta \in R^+$, annihilating $\mathcal{U}/\mathcal{C}\mathcal{U}^q$, annihilates $\mathcal{H}/\mathcal{H}^q$ as well.*

Using this theorem, we complete the proof of Theorem 11.5 at once: any Θ from Corollary 11.12 annihilates both $\mathcal{U}_q/\mathcal{U}^q$ and $\mathcal{H}/\mathcal{H}^q$; hence it annihilates the Selmer group $\mathcal{C}$ by Proposition 11.6.

We congratulate the reader who arrived to this point. Now you know how Catalan's problem was finally solved!

The theorem of Thaine will be proved in Chap. 12.

Chapter 12
The Theorem of Thaine

In this chapter we prove the theorem of Thaine, which was used in Sect. 11.5.

12.1 Introduction

In this section we employ our traditional notation: unless the contrary is stated explicitly, p and q are distinct odd prime numbers, $K = \mathbb{Q}(\zeta_p)$ is the p th cyclotomic field, $\mathcal{O} = \mathcal{O}_K$ is its ring of integers, and $G = \mathrm{Gal}(K/\mathbb{Q})$ is the Galois group. We also denote by $\mathcal{U}$ the group of Dirichlet units of the field K, by $\mathcal{C}$ the group of cyclotomic units, and by $\mathcal{H}$ the class group of K.

Stickelberger's theorem provided a nontrivial annihilator for the class group of the cyclotomic field K. However, as it has already been mentioned, Stickelberger's theorem is, essentially, a "relative" or "minus" result. Indeed, the plus-part of Stickelberger's ideal is $\mathcal{N}\mathbb{Z}$ (see Proposition 7.23), and $\mathcal{N}$ is an obvious annihilator of the class group; this means that Stickelberger's theorem does not tell us anything interesting about the plus-part R^+ of the group ring $R = \mathbb{Z}[G]$.

For a multiplicatively written abelian group A denote by $[A]_q$ the group A/A^q. Thaine [132] proved the following theorem.

Theorem 12.1 (Thaine). *Let q be a prime number not dividing $p - 1$. Then any $\Theta \in R^+$, annihilating $[\mathcal{U}/\mathcal{C}]_q$, annihilates $[\mathcal{H}]_q$ as well.*

Since R^+ trivially annihilates the relative class group $\mathcal{H}^-$, Theorem 12.1 is equivalent to the following formally weaker statement: any $\Theta \in R^+$, annihilating $[\mathcal{U}/\mathcal{C}]_q$, annihilates $[\mathcal{H}^+]_q$.

In fact, Thaine proved even more: under the same assumption $q \nmid (p - 1)$ any $\Theta \in R^+$, annihilating the q-Sylow subgroup of $\mathcal{U}/\mathcal{C}$, annihilates the q-Sylow subgroup of $\mathcal{H}^+$ as well. We do not prove this more precise statement.

© Springer International Publishing Switzerland 2014
Y.F. Bilu et al., *The Problem of Catalan*, DOI 10.1007/978-3-319-10094-4_12

As we have already mentioned in Remark 10.6, while groups $\mathcal{U}/\mathcal{C}$ and $\mathcal{H}^+$ have the same cardinality, it is widely believed that they need not be isomorphic. To the best of our knowledge, no example of non-isomorphism of these groups is currently known. What is known for sure is that they may differ as Galois modules: see [136, Remark after Theorem 8.2]. Thaine's theorem tells us that still, the G-modules $\mathcal{U}/\mathcal{C}$ and $\mathcal{H}^+$ cannot be "too independent."

The structure of the proof of Thaine's theorem is quite analogous to that of Stickelberger's theorem, though the details are different and more involved. To prove Stickelberger's theorem, we showed that, for any prime ideal $\mathfrak{l}$ of degree 1 (which means exactly that the prime ℓ below it is 1 mod p) and for any Θ from Stickelberger's ideal, the ideal $\mathfrak{l}^\Theta$ is principal. In this argument Gauss sums played a key role. Next, we used the Chebotarev density theorem and the Class Field Theory to show that every ideal class contains a prime ideal of degree 1.

For Thaine's theorem, we argue similarly. We fix a "suitable" (this will be made precise) prime number ℓ and show that for any prime ideal $\mathfrak{l}$ of the cyclotomic field K, lying above ℓ, the ideal $\mathfrak{l}^\Theta$ is a qth power of another ideal as soon as $\Theta \in R$ annihilates $[\mathcal{U}/\mathcal{C}]_q$. For the proof, we shall use a certain "substitute" for the Gauss sums, which will be obtained from the so-called Hilbert's "Theorem 90." Next, we use the Chebotarev density theorem and the Class Field Theory to show that every ideal class from $\mathcal{H}^+$ has a prime ideal above a "suitable" ℓ.

Now let us be more specific. Call a prime number ℓ *suitable* if it satisfies the following condition: if a unit $\eta \in \mathcal{U}$ is a qth power modulo[1] ℓ, then $\eta \in \mathcal{U}^q$. Alternatively, ℓ is suitable if the natural map $[\mathcal{U}]_q \to [(\mathcal{O}/\ell\mathcal{O})^\times]_q$ is injective.

Theorem 12.1 is an immediate consequence of the following two statements.

Theorem 12.2. *Assume that $q \nmid (p-1)$. Let ℓ be a suitable prime number satisfying $\ell \equiv 1$ mod pq, and let $\mathfrak{l}$ be a prime ideal of K above ℓ. Assume that $\Theta \in R$ annihilates $[\mathcal{U}/\mathcal{C}]_q$. Then $\mathfrak{l}^\Theta$ is equivalent to a qth power of an ideal of K.*

Theorem 12.3. *Assume that $q \nmid (p-1)$. Then any ideal class of the field K contains a prime ideal such that the underlying prime number is suitable and is 1 mod pq.*

These two theorems will be proved in the subsequent sections.

12.2 Preparations

In this section we collect miscellaneous facts, to be used in the proof of Theorems 12.2 and 12.3.

---

[1]That is, $\eta \equiv \alpha^q$ mod ℓ for some $\alpha \in \mathcal{O}_K$.

12.2.1 A Property of Cyclotomic Units

We begin with the following simple lemma.

Lemma 12.4. *Let p and ℓ be odd prime numbers. Put $K = \mathbb{Q}(\zeta_p)$ and $L = \mathbb{Q}(\zeta_p, \zeta_\ell)$. Assume that $\ell \equiv 1 \bmod p$. Then $\mathcal{N}_{L/K}(\zeta_\ell - \zeta_p) = 1$.*

Proof. Proposition 4.17 (with K as L and ℓ as p) implies that $[L : K] = \ell - 1$ and that the full system of conjugates of ζ_ℓ over K is $\zeta_\ell, \dots, \zeta_\ell^{\ell-1}$. Hence

$$\mathcal{N}_{L/K}(\zeta_\ell - \zeta_p) = \prod_{k=1}^{\ell-1} \left(\zeta_\ell^k - \zeta_p\right) = \frac{\zeta_p^\ell - 1}{\zeta_p - 1}.$$

Since $\ell \equiv 1 \bmod p$, we have $\zeta_p^\ell = \zeta_p$. Whence the result. □

(More generally, for any k not divisible by p, we have $\mathcal{N}_{L/K}\left(\zeta_\ell - \zeta_p^k\right) = 1$.)

Lemma 12.4 looks quite innocent, but it has a consequence which is absolutely crucial for the proof of Thaine's theorem.

Corollary 12.5. *Let η be a cyclotomic unit of the field K. Then there exists a unit ε of the field L such that $\mathcal{N}_{L/K}\varepsilon = 1$ and $\varepsilon \equiv \eta \bmod (\zeta_\ell - 1)$.*

Proof. For $\eta = \frac{1-\zeta_p^k}{1-\zeta_p^m}$ take $\varepsilon = \frac{\zeta_\ell - \zeta_p^k}{\zeta_\ell - \zeta_p^m}$. For an arbitrary cyclotomic unit take the suitable product. □

12.2.2 The "Theorem 90"

Another crucial lemma is the already mentioned Hilbert's "Theorem 90." If L/K is a finite Galois extension of fields, then for any $\alpha \in L^\times$ and any $\tau \in \mathrm{Gal}(L/K)$, the element $\varepsilon = \alpha/\alpha^\tau$ satisfies $\mathcal{N}_{L/K}(\varepsilon) = 1$. The "Theorem 90" tells us that for the *cyclic* extensions[2] the converse is true.

Lemma 12.6 (Hilbert's "Theorem 90"). *Let L/K be a finite cyclic extension of fields, and let τ be a generator of its Galois group. Then for any $\varepsilon \in L$ with $\mathcal{N}_{L/K}(\varepsilon) = 1$ there exists $\alpha \in K^\times$ such that $\varepsilon = \alpha/\alpha^\tau$.*

Proof. We follow Lang [58, Sect. 6.8]. Let m be the order of τ. Then the automorphisms $\mathrm{id}, \tau, \dots, \tau^{m-1}$ are pairwise distinct. Proposition D.5 implies that they are linearly independent over L. In particular, the linear combination

$$\mathrm{id} + \varepsilon\tau + \varepsilon^{1+\tau}\tau^2 + \cdots + \varepsilon^{1+\tau+\cdots+\tau^{m-2}}\tau^{m-1}$$

---

[2] A finite Galois extension of fields is called *cyclic* if its Galois group is cyclic.

defines a map which is not identically 0 on L. Hence there exists $\beta \in L$ such that

$$\alpha = \beta + \varepsilon \beta^\tau + \varepsilon^{1+\tau} \beta^{\tau^2} + \cdots + \varepsilon^{1+\tau+\cdots+\tau^{m-2}} \beta^{\tau^{m-1}}$$

is nonzero.

We have

$$\varepsilon \alpha^\tau = \varepsilon \beta^\tau + \varepsilon^{1+\tau} \beta^{\tau^2} + \cdots + \varepsilon^{1+\tau+\cdots+\tau^{m-2}} \beta^{\tau^{m-1}} + \varepsilon^{1+\tau+\cdots+\tau^{m-1}} \beta^{\tau^m}. \qquad (12.1)$$

Since m is the order of τ, we have $\beta^{\tau^m} = \beta$. Also, since τ generates the Galois group,

$$\varepsilon^{1+\tau+\cdots+\tau^{m-1}} = \mathcal{N}_{L/K}(\varepsilon) = 1.$$

It follows that the last term in (12.1) is β, and the right-hand side is, consequently, equal to α. This proves the lemma.   □

12.2.3   Reduction mod ℓ

Next, we establish some very simple properties of the residues modulo a prime number.

In this section K is a number field, $\mathcal{O}$ its ring of integers, and ℓ an odd prime number, unramified in K. We denote by Λ the residue ring $\mathcal{O}/\ell\mathcal{O}$. Since ℓ is unramified, we have

$$\Lambda = \prod_{\mathfrak{l}|\ell} \mathcal{O}/\mathfrak{l},$$

where the product is over all prime ideals above ℓ. In particular, Λ is a semi-simple ring.

If K is a Galois extension of $\mathbb{Q}$, then the Galois group acts on Λ. It turns out that the multiplicative group $\Lambda^\times$ is a cyclic Galois module.

Proposition 12.7. *Assume that K is a Galois extension of $\mathbb{Q}$. Then there exists $\alpha \in \mathcal{O}$ such that the multiplicative group $\Lambda^\times$ is generated by (the Λ-images of) the Galois conjugates of α. In other words, α generates $\Lambda^\times$ as a Galois module.*

Proof. Fix a prime ideal $\mathfrak{l}$ above ℓ. The multiplicative group $(\mathcal{O}/\mathfrak{l})^\times$ is cyclic. By the Chinese Remainder Theorem, there exists $\alpha \in \mathcal{O}$ such that the image of α in $\mathcal{O}/\mathfrak{l}$ generates $(\mathcal{O}/\mathfrak{l})^\times$ and the image of α in $\mathcal{O}/\mathfrak{l}'$ is 1 for any prime ideal $\mathfrak{l}' \mid \ell$ distinct from the fixed $\mathfrak{l}$. Then (the image of) α generates the subgroup $(\mathcal{O}/\mathfrak{l})^\times$ of $\Lambda^\times$. More generally, for any $\sigma \in G$ the element α^σ generates the subgroup $(\mathcal{O}/\mathfrak{l}^\sigma)^\times$.

Since the group $G = \mathrm{Gal}(K/\mathbb{Q})$ acts transitively on the set of prime ideals above ℓ, the set $\{\alpha^\sigma : \sigma \in G\}$ generates $\Lambda^\times$ as an abelian group. Hence α generates $\Lambda^\times$ as a G-module. □

It follows that, given another prime number q, the group $[\Lambda^\times]_q = \Lambda^\times/(\Lambda^\times)^q$ is a cyclic G-module, or, equivalently, a cyclic $\mathbb{F}_q[G]$-module (generated by the image of α). Under some additional assumptions it becomes a *free* cyclic module.

Proposition 12.8. *In the setup of Proposition 12.7, assume that ℓ completely splits in K. Also, let q be a prime number dividing $\ell - 1$. Then $[\Lambda^\times]_q$ is a free cyclic $\mathbb{F}_q[G]$-module, where G is the Galois group of $K/\mathbb{Q}$.*

The proposition applies, in particular, when K is the p th cyclotomic field and $\ell \equiv 1 \bmod p\, q$.

Proof. Since we already know that $[\Lambda^\times]_q$ is a cyclic $\mathbb{F}_q[G]$-module, it suffices to show that $\left|[\Lambda^\times]_q\right| = \left|\mathbb{F}_q[G]\right|$. Fix a prime ideal $\mathfrak{l} \mid \ell$. Since ℓ splits completely, we have

$$\Lambda = \prod_{\sigma \in G} \mathcal{O}/\mathfrak{l}^\sigma, \qquad \Lambda^\times = \prod_{\sigma \in G} (\mathcal{O}/\mathfrak{l}^\sigma)^\times,$$

and each $(\mathcal{O}/\mathfrak{l}^\sigma)^\times$ is isomorphic to $(\mathbb{Z}/\ell\mathbb{Z})^\times$. Since $q \mid (\ell - 1)$, every $\left[(\mathcal{O}/\mathfrak{l}^\sigma)^\times\right]_q$ is the q-element cyclic group. Hence

$$\left|[\Lambda^\times]_q\right| = \left|\prod_{\sigma \in G} \left[(\mathcal{O}/\mathfrak{l}^\sigma)^\times\right]_q\right| = q^{|G|} = \left|\mathbb{F}_q[G]\right|,$$

as wanted. □

Remark 12.9. A careful reader could have noticed that we defined the notion of cyclic module only for commutative rings. Hence the notion of cyclic G-module is defined in this book only for a commutative group G, so, formally, one should assume in Propositions 12.7 and 12.8 that K is an abelian extension of $\mathbb{Q}$ (which is, of course, sufficient for our purposes). However, both propositions are valid, with obvious definitions, for any Galois extensions, the word "G-module" being replaced by the "right G-module."

12.2.4 Galois-Invariant Prime Ideals

We shall also need a simple property of Galois extensions of number fields. If L/K is a Galois extension, then every element of L, invariant under the action of the Galois group, belongs to K (sometimes this is called the "main theorem of the Galois theory"). This property does not extend to ideals. For instance, if $\mathfrak{P}$ is a prime ideal of L totally ramified over K, then $\mathfrak{P}$ is invariant under the Galois action, but is not an ideal of K. Still, a slightly corrected "ideal analogue" of the "main theorem" does hold; the proof is a simple exercise, left to the reader.

Proposition 12.10. *Let L/K be a Galois extension of number fields, and let $\mathfrak{A}$ be an ideal of L invariant under the action of $\mathrm{Gal}(L/K)$. Then $\mathfrak{A} = \mathfrak{a}\mathfrak{b}$, where $\mathfrak{a}$ is an ideal of K and $\mathfrak{b}$ is a product of prime ideals of L ramified over K.*

12.2.5 Decomposition of Prime Numbers in Cyclotomic Fields

Finally, we recall the decomposition of the prime number ℓ in the fields $K = \mathbb{Q}(\zeta_p)$ and $L = \mathbb{Q}(\zeta_p, \zeta_\ell)$, under the assumption $\ell \equiv 1 \bmod p$. By this assumption, ℓ splits completely in K. Hence, if we fix a K-ideal $\mathfrak{l}$ above ℓ, then

$$(\ell) = \prod_{\sigma \in G} \mathfrak{l}^\sigma,$$

where $G = \mathrm{Gal}(K/\mathbb{Q})$.

Further, by Proposition 4.17, the prime ideal $\mathfrak{l}$ totally ramifies in L, the ramification index being $\ell - 1$, and the same is true for every $\mathfrak{l}^\sigma$. If $\mathfrak{L}$ is the prime ideal of L above $\mathfrak{l}$, then $\mathfrak{l} = \mathfrak{L}^{\ell-1}$. Further, if we extend σ to L by putting $\sigma(\zeta_\ell) = \zeta_\ell$ (see Proposition 4.26), then $\mathfrak{L}^\sigma$ is the prime ideal above $\mathfrak{l}^\sigma$, and $\mathfrak{l}^\sigma = (\mathfrak{L}^\sigma)^{\ell-1}$. Thus, in the field L the prime number ℓ factorizes as

$$(\ell) = \left(\prod_{\sigma \in G} \mathfrak{L}^\sigma \right)^{\ell-1}.$$

12.3 Proof of Theorem 12.2

We use the setup of Sect. 12.1. Let $\Theta \in \mathbb{Z}[G]$ annihilate $[\mathcal{U}/\mathcal{C}]_q$. We have to show that $\mathfrak{l}^\Theta$ is equivalent to a qth power of an ideal of K.

The proof consists of several steps. First of all, we use Lemma 12.4 (more precisely Corollary 12.5), and Hilbert's "Theorem 90", to produce a nontrivial annihilator Ψ for the class of $\mathfrak{l}$ in $[\mathcal{H}]_q$. Next, we apply Kummer's argument, as in the proof of Theorem 7.11, to derive a congruence involving the coefficients Ψ, similar to congruence (7.8). Next, this congruence is rewritten as an identity in the multiplicative group $[(\mathcal{O}/\ell\mathcal{O})^\times]_q$, involving both Ψ and Θ. This latter relation would allow us to conclude that Θ annihilates the class of $\mathfrak{l}$ as well.

12.3.1 A Nontrivial Annihilator

The assumption "Θ annihilates $[\mathcal{U}/\mathcal{C}]_q$" means that for any unit $\upsilon \in \mathcal{U}$ there exists a cyclotomic unit $\eta \in \mathcal{C}$ such that $\upsilon^\Theta \in \eta\mathcal{U}^q$. Let us specify $\upsilon \in \mathcal{U}$ to be such a unit that its image in $[\mathcal{U}]_q$ generates $[\mathcal{U}]_q$ as a G-module. (This is possible because $[\mathcal{U}]_q$ is a cyclic G-module; see Theorem 11.9.)

Let η be a cyclotomic unit as in the previous paragraph: $\upsilon^\Theta \in \eta\mathcal{U}^q$. Corollary 12.5 implies that there is a unit ε of the field $L = K(\zeta_\ell)$ such that $\mathcal{N}_{L/K}(\varepsilon) = 1$ and

$$\varepsilon \equiv \eta \bmod (\zeta_\ell - 1). \tag{12.2}$$

The Galois group of the extension L/K is $(\mathbb{Z}/\ell\mathbb{Z})^\times$; in particular, it is cyclic. Hence we may apply Hilbert's "Theorem 90" (Lemma 12.6): fix a generator τ of $\mathrm{Gal}(L/K)$ and find $\alpha \in L$ such that $\varepsilon = \alpha^\tau/\alpha$ (we apply the "Theorem 90" to τ^{-1} rather than to τ itself). As we shall see, the number α will play in the proof the same role as the Gauss sum played in the proof of Stickelberger's theorem. In particular, the relation $\alpha^\tau = \varepsilon\alpha$ is analogous to the relation $g(\chi)^{\tau_b} = \bar{\chi}(b)g(\chi)$ (see Lemma 7.13), which was crucial in Sect. 7.4.

The proof of Stickelberger's theorem relied on the prime factorization of the Gauss sum (Theorem 7.11). To prove Thaine's theorem, we shall factorize α. Since $\alpha^\tau = \varepsilon\alpha$, the principal ideal (α) is τ-invariant. Hence (α) is invariant with respect to $\mathrm{Gal}(L/K)$ (recall that τ generates this group). The only prime ideals of L ramified over K are the ideals above ℓ; that is, they are the ideals $\mathfrak{L}^\sigma$, defined at the end of Sect. 12.2. Hence, according to Proposition 12.10,

$$(\alpha) = \mathfrak{a}\prod_{\sigma \in G}(\mathfrak{L}^\sigma)^{s_\sigma},$$

where $\mathfrak{a}$ is an ideal of K and all s_σ are integers. Taking the norm, we obtain

$$\left(\mathcal{N}_{L/K}(\alpha)\right) = \mathfrak{a}^{\ell-1}\prod_{\sigma \in G}(\mathfrak{l}^\sigma)^{s_\sigma} = \mathfrak{a}^{\ell-1}\mathfrak{l}^\Psi,$$

where $\Psi = \sum_{\sigma \in G} s_\sigma \sigma$. Since, by the assumption, $q \mid (\ell - 1)$, we have proved that ideal $\mathfrak{l}^\Psi$ is equivalent to a qth power of an ideal of K.

12.3.2 Kummer's Argument

To identify the coefficients s_σ, we use Kummer's argument, in a similar fashion as we did in the proof of Theorem 7.11.

Let $b \in \mathbb{Z}$ be such that $\zeta_\ell^\tau = \zeta_\ell^b$. Since τ generates $\mathrm{Gal}(L/K)$, the image of b in $\mathbb{Z}/\ell\mathbb{Z}$ generates the multiplicative group $(\mathbb{Z}/\ell\mathbb{Z})^\times$. This observation is irrelevant now, but will be important at the final stage of the proof.

The algebraic number

$$\beta = \frac{(1-\zeta_\ell)^{s_\sigma}}{\alpha}$$

is an $\mathfrak{L}^\sigma$-adic unit, because $\mathrm{Ord}_{\mathfrak{L}^\sigma}(1-\zeta_\ell) = 1$ by Lemma 7.12. Using this lemma with $\mathfrak{L}$ replaced by $\mathfrak{L}^\sigma$, and the relation $\alpha^\tau = \varepsilon\alpha$, we obtain

$$\beta^\tau = \frac{(1-\zeta_\ell^b)^{s_\sigma}}{\varepsilon\alpha} = \frac{\beta}{\varepsilon}\left(\frac{1-\zeta_\ell^b}{1-\zeta_\ell}\right)^{s_\sigma} \equiv \beta\frac{b^{s_\sigma}}{\varepsilon} \bmod \mathfrak{L}^\sigma.$$

On the other hand, Proposition 4.19 implies that $\beta^\tau \equiv \beta \bmod \mathfrak{L}^\sigma$. We obtain the congruence $b^{s_\sigma} \equiv \varepsilon \bmod \mathfrak{L}^\sigma$. Combining this with (12.2), we obtain $b^{s_\sigma} \equiv \eta \bmod \mathfrak{L}^\sigma$.

In this latter congruence both sides belong to the field K. Hence the ideal $\mathfrak{L}^\sigma$ can be replaced by the underlying ideal $\mathfrak{l}^\sigma$. We obtain, for every $\sigma \in G$, the congruence

$$b^{s_\sigma} \equiv \eta \bmod \mathfrak{l}^\sigma. \tag{12.3}$$

The field L and the unit ε played their role, and we do not need them anymore. From now on, we work entirely in the field K.

12.3.3   Projecting All This to $[(\mathcal{O}/\ell\,\mathcal{O})^\times]_q$

Now we project congruences (12.3) to the residue ring $\Lambda = \mathcal{O}/\ell\mathcal{O}$ and yet further, to the quotient multiplicative group $[\Lambda^\times]_q = \Lambda^\times/(\Lambda^\times)^q$.

Recall that ℓ splits in K as $(\ell) = \prod_{\sigma \in G}\mathfrak{l}^\sigma$, where $\mathfrak{l}$ is a fixed prime ideal above ℓ (see end of Sect. 12.2). It follows that $\Lambda = \prod_{\sigma \in G}\mathcal{O}/\mathfrak{l}^\sigma$. By the Chinese Remainder Theorem, there exists $\alpha \in \mathcal{O}$ such that

$$\alpha \equiv b \bmod \mathfrak{l}, \qquad \alpha \equiv 1 \bmod \mathfrak{l}^\sigma \quad (\sigma \neq 1).$$

Then $\alpha^\psi \equiv b^{s_\sigma} \bmod \mathfrak{l}^\sigma$. Hence (12.3) can be rewritten as $\eta \equiv \alpha^\psi \bmod \ell$, which can be viewed as an identity in Λ. Combining this with the initial condition $\upsilon^\Theta \in \eta\mathcal{U}^q$, we obtain the congruence

$$\upsilon^\Theta \equiv \alpha^\psi\gamma^q \bmod \ell$$

with some $\gamma \in \mathcal{U}$.

Since both sides of the latter congruence are invertible mod ℓ, we can project this to the quotient multiplicative group $[\Lambda^\times]_q$. We obtain

$$\bar{\upsilon}^\Theta = \bar{\alpha}^\Psi, \tag{12.4}$$

where $\bar{\alpha}$ and $\bar{\upsilon}$ are the images of α and υ in $[\Lambda^\times]_q$.

12.3.4 Conclusion

We are almost done. In this final stage of the proof it will be more convenient to work mod q. Thus, let $\bar{\Psi}$ and $\bar{\Theta}$ are the images of Ψ and Θ in $\mathbb{F}_q[G]$. We know that $\bar{\Psi}$ annihilates the class of $\mathfrak{l}$ in $[\mathcal{H}]_q$ and want to prove the same for $\bar{\Theta}$.

Proposition 12.8 implies that $[\Lambda^\times]_q$ is a free cyclic $\mathbb{F}_q[G]$-module. (Here we use the condition $\ell \equiv 1 \bmod p\,q$.) Furthermore, as we have seen in Sect. 12.3.2, the image of b generates the multiplicative group $(\mathbb{Z}/\ell\mathbb{Z})^\times$. Hence, arguing as in the proof of Proposition 12.7, we conclude that α generates $\Lambda^\times$ as a G-module. It follows that $\bar{\alpha}$ generates $[\Lambda^\times]_q$ as an $\mathbb{F}_q[G]$-module. Therefore there is an $\mathbb{F}_q[G]$-isomorphism $[\Lambda^\times]_q \cong \mathbb{F}_q[G]$ such that $\bar{\alpha} \in [\Lambda^\times]_q$ corresponds to $1 \in \mathbb{F}_q[G]$. Let $\Upsilon \in \mathbb{F}_q[G]$ correspond to $\bar{\upsilon} \in [\Lambda^\times]_q$ under this isomorphism. Then (12.4) can be rewritten as

$$\bar{\Psi} = \bar{\Theta}\Upsilon. \tag{12.5}$$

If Υ were an invertible element of $\mathbb{F}_q[G]$, we could have written $\bar{\Theta} = \Upsilon^{-1}\bar{\Psi}$, completing the proof. Unfortunately, we cannot directly argue this way because Υ has no reasons to be invertible (and it is not, in fact). However, a more delicate argument of this sort would work, as we shall see in a while.

Since ℓ is a *suitable* prime, the natural map $[\mathcal{U}]_q \to [\Lambda^\times]_q$ is injective. Recall also that υ was chosen as a generator of the cyclic module $[\mathcal{U}]_q$. It follows that Υ generates a submodule of $\mathbb{F}_q[G]$ isomorphic to $[\mathcal{U}]_q$. According to Theorem 11.9, the annihilator of this submodule is $(\mathcal{N}) \oplus (1 - \iota)$, where $\mathcal{N}$, as usual, is the norm element. Since $\mathbb{F}_q[G]$ is a semi-simple ring, this implies that

$$1 \in (\Upsilon) \oplus (\mathcal{N}) \oplus (1 - \iota).$$

(This can be expressed as "the element $\bar{\Psi}$ is invertible modulo the ideal $(\mathcal{N}) \oplus (1 - \iota)$".)

Multiplying by $\bar{\Theta}$ and using (12.5), we obtain

$$\bar{\Theta} \in (\bar{\Psi}) + (\mathcal{N}) + ((1 - \iota)\bar{\Theta}).$$

Now recall that $\Theta \in \mathbb{Z}[G]^+$ (it is only here where we use this assumption), which implies that $(1 - \iota)\bar{\Theta} = 0$, and we finally obtain $\bar{\Theta} \in (\bar{\Psi}) + (\mathcal{N})$. Since both $\bar{\Psi}$ and $\mathcal{N}$ annihilate the class of $\mathfrak{l}$ in $[\mathcal{H}]_q$, so does $\bar{\Theta}$. The theorem is proved.

12.4   Reduction of a Multiplicative Group Modulo a Prime Ideal

Let q be an odd prime number. In this section we prove the following theorem.

Theorem 12.11. *Let K be a number field and $\beta \in K^\times$ not a qth power in K. Assume that q is unramified in K. Then every ideal class of K contains infinitely many prime ideals $\mathfrak{l}$, of degree 1 (over $\mathbb{Q}$), such that the underlying prime is $1 \bmod q$ and such that β is not a qth power modulo $\mathfrak{l}$.*

This theorem is not formally needed here. However, its proof is very similar to the proof of Theorem 12.3, but technically simpler. Therefore it can provide a good motivation for the proof of Theorem 12.3, which otherwise looks somewhat messy.

In this and the subsequent sections we systematically use the notions of q-radical extension and Kummer's pairing; see Appendix F.2 for the definitions. Let K be a field (of characteristic distinct from q) containing the group μ_q of qth roots of unity, $L = K(\sqrt[q]{B})$ a q-radical extension of K with Galois group $\Gamma = \mathrm{Gal}(L/K)$, and $B \times \Gamma \xrightarrow{f} \mu_q$ Kummer's pairing. We say that $\beta \in B$ and $\gamma \in \Gamma$ are *Kummer-orthogonal* if $f(\beta, \gamma) = 1$. We say that β is Kummer-orthogonal to a subset $S \subseteq \Gamma$ if it is Kummer-orthogonal to every element of S.

Our starting point is the following general observation.

Proposition 12.12. *Let K be a number field containing the qth roots of unity, let B be a subgroup of $K^\times$, and let L and Γ be as above. Then for any prime ideal $\mathfrak{l}$ of K, unramified in L, an element $\beta \in B$ is a qth power $\bmod\, \mathfrak{l}$ if and only if β is Kummer-orthogonal to $\varphi_{\mathfrak{l}} \in \Gamma$, the Frobenius element of $\mathfrak{l}$.*

Proof. It is a direct consequence of the definitions of the Frobenius element and Kummer's pairing. □

To warm up, we first establish two "weaker" versions of Theorem 12.11. Everywhere below K is a number field and $\beta \in K^\times$ is not a qth power in K. A prime ideal $\mathfrak{l}$ of K such that β is not a qth power modulo $\mathfrak{l}$ will be called *suitable*.

We start from the "extra-light" version, where we do not require $\mathfrak{l}$ to belong to the given ideal class and assume in addition that $\zeta_q \in K$.

Proposition 12.13. *Assume that K contains the qth roots of unity. Then there exist infinitely many suitable prime ideals of K, of degree 1 (over $\mathbb{Q}$), such that the underlying prime is $1 \bmod q$.*

Proof. We apply Proposition 12.12 with $B = \langle \beta \rangle$. Since β is not a qth power, the Galois group Γ is cyclic of order q, and the only element of Γ, Kummer-orthogonal to β, is 1. Thus, if $\mathfrak{l}$ is such that $\varphi_{\mathfrak{l}} \neq 1$, then β is not a qth power modulo $\mathfrak{l}$. By the Chebotarev density theorem, there exist infinitely many prime ideals $\mathfrak{l}$ with this property and of degree 1. Finally, since $\mathfrak{l}$ is of degree 1 and $\zeta_q \in K$, the underlying prime is $1 \bmod q$. $\square$

Next, the "simply light" version: we drop the assumption $\zeta_q \in K$, but still do not require our prime ideals to belong to the given class.

Proposition 12.14. *There exist infinitely many suitable prime ideals of K, of degree 1, such that the underlying prime is $1 \bmod q$.*

Proof. Put $L = K(\zeta_q, \beta^{1/q})$, so that we have a tower of fields $K \subseteq K(\zeta_q) \subseteq L$, the subgroup $\mathrm{Gal}(L/K(\zeta_q))$ of the group $\mathrm{Gal}(L/K)$ being cyclic. By the Chebotarev density theorem, there are infinitely many prime ideals $\mathfrak{l}$ of K of degree 1 such that the Artin symbol $\left[\frac{\mathfrak{l}}{L/K}\right]$ is contained in $\mathrm{Gal}(L/K(\zeta_q))$ and generates this group.

Fix one such $\mathfrak{l}$. Since $\left[\frac{\mathfrak{l}}{L/K}\right] \subset \mathrm{Gal}(L/K(\zeta_q))$, the ideal $\mathfrak{l}$ totally splits in $K(\zeta_q)$. Let $\mathfrak{l} = \mathfrak{L}_1 \cdots \mathfrak{L}_s$ be the decomposition of $\mathfrak{l}$ in $K(\zeta_q)$. Then

$$\left[\frac{\mathfrak{l}}{L/K}\right] = \{\varphi_{\mathfrak{L}_1}, \ldots, \varphi_{\mathfrak{L}_s}\},$$

where $\varphi_{\mathfrak{L}_i}$ is the Frobenius of $\mathfrak{L}_i$ in $\mathrm{Gal}(L/K(\zeta_q))$.

Now assume that β is a qth power $\bmod \mathfrak{l}$. Then it is a qth power modulo every $\mathfrak{L}_i$. It follows that β is Kummer-orthogonal to the set $\left[\frac{\mathfrak{l}}{L/K}\right]$ and, consequently, to the group Γ, generated by this set. Thus, β is a qth power in $K(\zeta_q)$. Corollary F.2 now implies that it is a qth power in K, contradicting the assumption. Hence $\mathfrak{l}$ is suitable.

Finally, since $\mathfrak{l}$ is of degree 1 and totally splits in $K(\zeta_q)$, every $\mathfrak{L}_i$ is of degree 1, whence the underlying prime is $1 \bmod q$. The proposition is proved. $\square$

Notice that neither Proposition 12.13 nor even Proposition 12.14 is formally weaker than Theorem 12.11, because in them we do not assume that q is unramified in K. This additional assumption is needed only in the "full" version of Theorem 12.11, to ensure that the prime ideals $\mathfrak{l}$ can be selected in a given ideal class.

Proof of Theorem 12.11. As in Proposition 12.14 we set $L = K(\zeta_q, \beta^{1/q})$, and we let E be the Hilbert Class Field of K (see Appendix A.11). We view both L and E as subfields of a fixed algebraic closure of K, so that the intersection $L \cap E$ and the composite LE are well defined.

We claim that $L \cap E = K$. Indeed, the field $K' = L \cap E$ is both abelian and unramified over K. Since q is unramified in K, the qth root of unity ζ_q is not in K. Theorem F.7 implies that $K' \subseteq K(\zeta_q)$. But every prime ideal of K above q totally ramifies in $K(\zeta_q)$ (Proposition 4.17). Since K' is unramified over K, we must have $K' = K$. Thus, $L \cap E = K$ and, consequently,

$$\mathrm{Gal}(LE/K) = \mathrm{Gal}(E/K) \times \mathrm{Gal}(L/K).$$

The Galois group $\text{Gal}(E/K)$ can be identified, via the Artin map, with the class group $\mathcal{H} = \mathcal{H}_K$. Hence $\text{Gal}(LE/K)$ may be identified with the product $\mathcal{H} \times \text{Gal}(L/K)$. If $\mathfrak{l}$ is a prime ideal of K, unramified in L, then

$$\left[\frac{\mathfrak{l}}{LE/K} \right] = \{\text{cl}(\mathfrak{l})\} \times \left[\frac{\mathfrak{l}}{L/K} \right],$$

where $\text{cl}(\mathfrak{l})$ is the ideal class of $\mathfrak{l}$ in K.

Applying the Chebotarev density theorem to the extension LE/K, we find, for every ideal class $C \in \mathcal{H}$, infinitely many prime ideals $\mathfrak{l}$ of degree 1, such that $\text{cl}(\mathfrak{l}) = C$ and $\left[\frac{\mathfrak{l}}{L/K} \right]$ is as in the proof of Proposition 12.14 (that is, is contained in $\text{Gal}(L/K(\zeta_q))$ and generates this group). Arguing as in the proof of Proposition 12.14, we show that any such $\mathfrak{l}$ is suitable, and its underlying prime is $1 \bmod q$. The theorem is proved. $\square$

12.5 Reduction of a Multiplicative Group Modulo a Prime Number and Proof of Theorem 12.3

Informally, the results of Sect. 12.4 can be stated as follows: *if β is not a qth power then the group $[\langle \beta \rangle]_q$ faithfully reduces modulo certain prime ideals*. Next, instead of a cyclic subgroup $\langle \beta \rangle$ we consider an arbitrary finitely generated subgroup $B \le K^{\times}$ and ask for conditions on when the reduction of the group $[B]_q$ is faithful. The analogue of the condition "β is not a qth power in K" is

$$B \cap (K^{\times})^q = B^q. \tag{12.6}$$

Of course, even when (12.6) is satisfied, we cannot expect that $[B]_q$ faithfully reduces modulo prime ideals, because the group $[B]_q$ is usually not cyclic, and the multiplicative group modulo a prime ideal is cyclic. However, under certain Galois conditions upon B, the reduction modulo many *rational* prime numbers is faithful.

As in the previous section, let q be an odd prime number. In this section prove the following theorem.

Theorem 12.15. *Let K be a finite abelian extension of $\mathbb{Q}$ with Galois group G and let B be a G-invariant finitely generated subgroup of $K^{\times}$, satisfying (12.6). Assume that q is unramified in K and does not divide the degree $[K : \mathbb{Q}]$. Assume also that $[B]_q$ is a cyclic G-module. Then, for any ideal class C of K, there exist infinitely many prime numbers ℓ with the following properties:*

- *ℓ totally splits in K;*
- *$\ell \equiv 1 \bmod q$;*
- *the reduction homomorphism $[B]_q \to [(\mathcal{O}_K/\ell\mathcal{O}_K)^{\times}]_q$ is injective;*
- *the class C contains a prime ideal above ℓ.*

Theorem 12.3 is a particular case of Theorem 12.15.

Proof of Theorem 12.3 (assuming Theorem 12.15). If p is an odd prime number distinct from q and such that $q \nmid (p-1)$, then q is unramified in the field $K = \mathbb{Q}(\zeta_p)$ and does not divide the degree $[K : \mathbb{Q}]$. Recall also that $[\mathcal{U}]_q$ (where $\mathcal{U}$ is the group of Dirichlet units of K) is a cyclic Galois module (Theorem 11.9). Applying Theorem 12.15 with $B = \mathcal{U}$, we find, for every ideal class C of K, infinitely many prime numbers ℓ with the properties listed therein. Since ℓ totally splits in K, we have $\ell \equiv 1 \bmod p$ as well. This proves Theorem 12.3. $\qquad\square$

As in Sect. 12.4, we first prove a "light" version, without requiring that there is a prime ideal above ℓ in the given class.

In the sequel K is a finite abelian extension of $\mathbb{Q}$ with Galois group G and B is a G-invariant finitely generated subgroup of $K^{\times}$, satisfying (12.6). Call a prime number ℓ *suitable* if the reduction homomorphism $[B]_q \to [(\mathcal{O}/\ell\mathcal{O})^{\times}]_q$ is injective.

Proposition 12.16. *Assume that q does not divide $[K : \mathbb{Q}]$ and that*

$$K \cap \mathbb{Q}(\zeta_q) = \mathbb{Q}. \tag{12.7}$$

Assume also that $[B]_q$ is a cyclic G-module. Then there exist infinitely many suitable prime numbers, congruent to $1 \bmod q$ and which totally split in K.

Proof. The proof is similar to that of Proposition 12.14. We put $L = K(\sqrt[q]{B})$, where

$$\sqrt[q]{B} = \left\{ \beta \in \bar{K}^{\times} : \beta^q \in B \right\},$$

in particular $\zeta_q \in \sqrt[q]{B}$. Then we have a tower of fields $K \subseteq K(\zeta_q) \subseteq L$, and the subgroup $\Gamma = \mathrm{Gal}(L/K(\zeta_q))$ of the group $\mathrm{Gal}(L/K)$ is isomorphic, by Proposition F.6, to $B/B \cap (K^{\times})^q$, which is $[B]_q$ because of (12.6). Moreover, the two groups are isomorphic as G-modules; see Appendix F.4 (here the assumption (12.7) is used). Since $[B]_q$ is a cyclic G-module by the assumption, so is Γ.

By the Chebotarev density theorem, there is infinitely many prime ideals $\mathfrak{l}$ of K of degree 1 such that the Artin symbol $\left[\frac{\mathfrak{l}}{L/K} \right]$ is contained in Γ and contains a generator of Γ as a G-module. Fix one such $\mathfrak{l}$, and denote by ℓ its underlying rational prime. Let $\beta \in B$ be a qth power $\bmod \ell$. Then it is a qth power $\bmod \mathfrak{l}^{\sigma}$ for any $\sigma \in G$. It follows that β is Kummer-orthogonal to the set $\left[\frac{\mathfrak{l}^{\sigma}}{L/K} \right]$ for any $\sigma \in G$. A straightforward argument shows that $\left[\frac{\mathfrak{l}^{\sigma}}{L/K} \right] = \left[\frac{\mathfrak{l}}{L/K} \right]^{\sigma}$. Hence β is Kummer-orthogonal to the set

$$\bigcup_{\sigma \in G} \left[\frac{\mathfrak{l}}{L/K} \right]^{\sigma}. \tag{12.8}$$

Since the set $\left[\frac{\mathfrak{l}}{L/K}\right]$ generates Γ as G-module, the set (12.8) generates it as an abelian group. Hence β is Kummer-orthogonal to Γ, which means that it is a qth power in $K(\zeta_q)$ and even in K, by Corollary F.2.

We have just proved that the kernel of the reduction homomorphism $B \to [(\mathcal{O}_K/\ell\mathcal{O}_K)^\times]_q$ is $B \cap (K^\times)^q$, which is B^q by (12.6). Hence ℓ is suitable.

Finally, since $\mathfrak{l}$ is of degree 1 and totally splits in $K(\zeta_q)$, the prime ℓ totally splits in $K(\zeta_q)$, whence $\ell \equiv 1 \bmod q$. The proposition is proved. $\qquad\square$

Notice that the condition "$[B]_q$ is a cyclic G-module" is not only sufficient, but also necessary. Indeed, since $q \nmid [K : \mathbb{Q}] = |G|$, the ring $\mathbb{F}_q[G]$ is semi-simple; hence it is a principal ideal ring. On the other hand, the group $(\mathcal{O}_K/\ell\mathcal{O}_K)^\times$ is a cyclic $\mathbb{F}_q[G]$-module (Proposition 12.7), and so are its submodules. In particular, so should be $[B]_q$.

Notice also that Proposition 12.16 is not formally weaker than Theorem 12.15, because the hypothesis "q is unramified in K" is replaced by a formally weaker hypothesis (12.7).

Now we are well prepared to prove Theorem 12.15.

Proof of Theorem 12.15. It combines the proofs of Theorem 12.11 and Proposition 12.16. We set $L = K(\zeta_q, \sqrt[q]{B})$. Since q is unramified in K, we have $[K(\zeta_q) : K] = q - 1$, which implies (12.7). Hence, as in the proof of Proposition 12.16, the group $\Gamma = \mathrm{Gal}(L/K(\zeta_q))$ is G-isomorphic to $[B]_q$ and, consequently, is a cyclic G-module.

Denote by E be the Hilbert Class Field of K. As in the proof of Theorem 12.11, we show that $L \cap E = K$ (here we again use the hypothesis that q is unramified in K) and, consequently, $\quad\bullet$

$$\mathrm{Gal}(LE/K) = \mathrm{Gal}(E/K) \times \mathrm{Gal}(L/K).$$

We again identify $\mathrm{Gal}(E/K)$ with the class group $\mathcal{H} = \mathcal{H}_K$ and $\mathrm{Gal}(LE/K)$ with the direct product $\mathcal{H} \times \mathrm{Gal}(L/K)$.

Applying the Chebotarev density theorem to the extension LE/K, we find infinitely many prime ideals $\mathfrak{l}$ of degree 1, such that $\mathrm{cl}(\mathfrak{l}) = C$ and $\left[\frac{\mathfrak{l}}{L/K}\right]$ is as in the proof of Proposition 12.16 (that is, is contained in $\mathrm{Gal}(L/K(\zeta_q))$ and generates this group as a G-module). Arguing as in the proof of Proposition 12.16, we show that any such $\mathfrak{l}$ is suitable, and its underlying prime is $1 \bmod q$. $\qquad\square$

This proves the Theorem of Thaine.

Chapter 13
Baker's Method and Tijdeman's Argument

This chapter is somewhat isolated and can be read (almost) independently of the others. We only assume some (very modest) knowledge of the algebraic number theory (Sects. A.1–A.8 of Appendix A) and basics about the heights (Appendix B). In this chapter we discuss the application of *Baker's method* to Diophantine equations of Catalan type. We give a brief introduction to this method, show how it applies to classical Diophantine equations, and reproduce the beautiful argument of Tijdeman, who proved that Catalan's equation has only finitely many solutions. Moreover, the solutions are bounded by an absolute effective constant (that is, a constant that can, in principle, be explicitly determined), which reduces the problem to a finite computation. Before the work of Mihăilescu this was the top achievement on Catalan's problem.

We also consider the more general *equation of Pillai* and show that it has finitely many solutions when one of the four variables is fixed.

To make this chapter self-contained and independent of the rest of the book (except Appendices A and B), we sometimes re-prove statements already proved elsewhere in the previous chapters.

Convention. As it is routinely done in this book, we fix an embedding $\bar{\mathbb{Q}} \hookrightarrow \mathbb{C}$ and view all the algebraic numbers occurring in this section as complex numbers.

13.1 Introduction: Thue, Gelfond, and Baker

In this section we briefly recall the prehistory and history of Baker's theory.

13.1.1 The Theorem of Thue

It is classically known that equation $ax^2 + bxy + cy^2 = 1$ with $a, b, c \in \mathbb{Q}^\times$ may have infinitely many solutions in $x, y \in \mathbb{Z}$. In 1909 the Norwegian mathematician Thue [133] proved the following theorem.

© Springer International Publishing Switzerland 2014
Y.F. Bilu et al., *The Problem of Catalan*, DOI 10.1007/978-3-319-10094-4_13

Theorem 13.1 (Thue). *Let*

$$f(x, y) = a_n y^n + a_{n-1} y^{n-1} x + \cdots + a_0 x^n \in \mathbb{Z}[x, y]$$

be a $\mathbb{Q}$-irreducible homogeneous polynomial of degree $n \geq 3$, and $A \in \mathbb{Z}$ a nonzero integer. Then the equation

$$f(x, y) = A \tag{13.1}$$

has only finitely many solutions in $x, y \in \mathbb{Z}$.

Thue proved his theorem by reduction to Diophantine approximation. Decompose the polynomial $f(x, y)$ into linear factors over $\mathbb{C}$:

$$f(x, y) = a_n (y - \theta_1 x) \cdots (y - \theta_n x) \tag{13.2}$$

Then, if (x, y) is a solution of our equation, we have

$$|\theta_1 - y/x| \cdots |\theta_n - y/x| = |A| |x|^{-n}.$$

If now θ is the nearest to y/x among the roots $\theta_1, \ldots, \theta_n$, then, by the triangle inequality,

$$|\theta_i - y/x| \geq \frac{1}{2} |\theta - \theta_i|$$

for $\theta_i \neq \theta$, which implies the inequality

$$|\theta - y/x| \leq c |x|^{-n}, \tag{13.3}$$

where c is a constant depending on f and A.

Now, to prove Thue's theorem, it suffices to show that the latter inequality is impossible for large x and y. Liouville's inequality (see Remark B.4) implies that $|\theta - y/x| \geq c_1 |x|^{-n}$ with another constant c_1, but this is, obviously, insufficient.

Thue proved that for any (real) algebraic number of degree $n > 1$ and any $\varepsilon > 0$ the inequality

$$|\theta - y/x| \geq |x|^{-n/2 - 1 - \varepsilon} \tag{13.4}$$

holds for sufficiently large integers x and y. For $n = 2$ this is weaker than Liouville's inequality, but for $n \geq 3$ it is stronger than it, and for large x it contradicts (13.3). This proves the finiteness.

Thue's work had a profound impact on the subsequent development of the Diophantine analysis. Many basic results in this discipline generalize his work or rely on his ideas. Among many of the improvements of Thue's inequality (13.4) mention the celebrated theorem of Roth [119], asserting that in the same setup the stronger inequality $|\theta - y/x| \geq |x|^{-2 - \varepsilon}$ takes place.

Unfortunately, Thue's proof of (13.4) has a serious defect, inherited by all of the subsequent generalizations. What he actually proves is the existence of a big number $X = X(\theta)$ with the following property: if (13.4) has a solution (x_0, y_0) with $|x_0| > X$, then any other solution of (13.4) is bounded in terms of this (x_0, y_0).

Of course, this implies finiteness for the number of solutions of (13.4), but this argument gives no method to bound the size of these solutions, because a priori we are not given a solution (x_0, y_0) with $|x_0| > X$. One says that Thue's argument is *ineffective*, and the same is true for its subsequent generalizations.

13.1.2   Logarithmic Forms

Now let us turn to a seemingly unrelated subject. Let $\gamma_1, \ldots, \gamma_n$ be nonzero complex algebraic numbers. Given $\mathbf{b} = (b_1, \ldots, b_n) \in \mathbb{Z}^n$, put

$$\gamma = \gamma(\mathbf{b}) = \gamma_1^{b_1} \cdots \gamma_n^{b_n}.$$

Assume that $\gamma \neq 1$ and ask the following question: how close can γ be to 1? The simplest lower estimate for $|\gamma - 1|$ follows again from Liouville's inequality: using Propositions B.3 and B.2, it is easy to show that

$$|\gamma - 1| \geq e^{-c\|\mathbf{b}\|}, \tag{13.5}$$

where $\|\mathbf{b}\| = |b_1| + \cdots + |b_n|$ is the ℓ_1-norm of the vector $\mathbf{b}$ and c is a positive constant depending on $\gamma_1, \ldots, \gamma_n$ (which can be easily made explicit).

Bounding from below $|\gamma - 1|$ is, essentially, equivalent to bounding from below the quantity $\Lambda = \Lambda(\mathbf{b}) = \log \gamma = b_1 \log \gamma_1 + \cdots + b_n \log \gamma_n$, with a suitable choice of the complex logarithms $\log \gamma_1, \ldots, \log \gamma_n$. This is why the theory of Gelfond–Baker is also called the *theory of logarithmic forms*.

Estimate (13.5) is too weak to have any interesting consequences. We call it *trivial*. In his book [37] (see also [38]) Gelfond showed that any *nontrivial* estimate, that is, an estimate of the shape $|\gamma - 1| \geq e^{-o(\|\mathbf{b}\|)}$, would imply the theorem of Thue. In particular, any *explicit* nontrivial lower bound for logarithmic forms would imply an *effective* proof for the theorem of Thue.

Gelfond himself obtained a nontrivial lower bound for a logarithmic form in $n = 2$ variables, as a by-product of his work in transcendence theory. In 1934 Gelfond [34, 35] and, independently, Schneider [123] solved *Hilbert's seventh problem*: if γ and β are complex algebraic numbers, with $\gamma \neq 0, 1$ and $\beta \notin \mathbb{Q}$ then γ^β is transcendent. Later, Gelfond [36] gave a quantitative version of this theorem, by estimating from below the difference $\beta \log \gamma_1 - \log \gamma_2$, where γ_1, γ_2 and β are algebraic numbers, $\gamma_i \neq 0, 1$ and $\beta \notin \mathbb{Q}$.

Gelfond's crucial observation was that the latter estimate remains true even when $\beta \in \mathbb{Q}$, of course, provided $\beta \log \gamma_1 - \log \gamma_2 \neq 0$. Writing the rational β as $-b_1/b_2$, one obtains a nontrivial lower estimate for the logarithmic form $\Lambda = b_1 \log \gamma_1 + b_2 \log \gamma_2$. Gelfond's original estimate [36] was of the shape

$|\Lambda| \geq e^{-c(\log \|\mathbf{b}\|)^{4+\varepsilon}}$, where the constant c depends on the numbers γ_i and on $\varepsilon > 0$. Later it was improved by many authors. The currently best known estimate for a binary logarithmic form is due to Laurent et al.; see [63–65], where further bibliography can be found. The result of [65] was extensively used for the numerical solution of various cases of Catalan's equation (see [89–91]).

Gelfond also observed in [37] that Diophantine approximation results of Thue type imply a nontrivial lower bound for linear forms in $n \geq 3$ logarithms. However, this bound inherits the non-effectiveness of Thue's method and cannot be used for the effective analysis of Diophantine equations.

In 1966 Baker [4] finally obtained a nontrivial effective lower bound for linear forms in $n \geq 3$ logarithms. In its qualitative form, Baker's result can be stated as follows.

Theorem 13.2 (Baker). *Let $\gamma_1, \ldots, \gamma_n$ be nonzero complex algebraic numbers. For every $\varepsilon > 0$ there exists a positive real number $B = B(\gamma_1, \ldots, \gamma_n, \varepsilon)$, which can be explicitly expressed in terms of $\gamma_1, \ldots, \gamma_n$ and ε, such that for $\mathbf{b} \in \mathbb{Z}^n$ with $0 < |\gamma(\mathbf{b}) - 1| \leq e^{-\varepsilon \|\mathbf{b}\|}$ we have $\|\mathbf{b}\| \leq B$.*

Informally speaking, either $\gamma(\mathbf{b}) = 1$ or $|\gamma(\mathbf{b}) - 1| \geq e^{-o(\|\mathbf{b}\|)}$.

This result belongs to the top arithmetical achievements of the twentieth century. Baker derived from his bound effective proofs of several Diophantine finiteness results, including the theorem of Thue [5]. In Sect. 13.4 we show some of the Diophantine applications of Baker's effective theorems.

In 1970 Baker was awarded a Fields medal for his work in Diophantine analysis and transcendence.

For applications one often needs a more explicit version of Theorem 13.2, with an explicit form for the exponent $o(\|\mathbf{b}\|)$. Baker [4] proved for

$$\Lambda = \Lambda(\mathbf{b}) = b_1 \log \gamma_1 + \cdots + b_n \log \gamma_n$$

the estimate $|\Lambda| \geq e^{-c(\log \|\mathbf{b}\|)^{n+1+\varepsilon}}$, where the constant c depends on the numbers γ_i and on $\varepsilon > 0$. Since then, this was refined many times by many authors. The modern estimate [7, 79, 135] is as follows.

Theorem 13.3 (Baker, Wüstholz, Waldschmidt, Matveev, …). *In the above setup, we have either $\Lambda = 0$ or*

$$|\Lambda| \geq e^{-c(n)d^{n+2}\mathrm{h}'(\gamma_1)\cdots\mathrm{h}'(\gamma_n)\log\|\mathbf{b}\|}. \tag{13.6}$$

Here $d = [\mathbb{Q}(\gamma_1, \ldots, \gamma_n) : \mathbb{Q}]$ and $\mathrm{h}'(\gamma_i) = \max\{\mathrm{h}(\gamma_i), d^{-1}\}$, where $\mathrm{h}(\cdot)$ is the height function (see Appendix B). Also, $c(n)$ is a constant depending on n.

The articles quoted above provide various explicit expressions for $c(n)$.

Using the Dirichlet approximation theorem (like Theorem VI in [17, Chap. 1]), one can find a nonzero vector $\mathbf{b} = (b_1, \ldots, b_n) \in \mathbb{Z}^n$, such that $|\Lambda(\mathbf{b})| \leq \|\mathbf{b}\|^{-n/2}$ (and even $|\Lambda(\mathbf{b})| \leq \|\mathbf{b}\|^{-n}$ if all the logarithms are real). Hence $\log \|\mathbf{b}\|$ cannot be replaced in (13.6) by a smaller function of $\|\mathbf{b}\|$. On the other hand, one can, probably, improve on the dependence in the algebraic numbers γ_i; for instance, the product $\mathrm{h}'(\gamma_1)\cdots\mathrm{h}'(\gamma_n)$ can, probably, be replaced by the sum $\mathrm{h}'(\gamma_1)+\cdots+\mathrm{h}'(\gamma_n)$, which would have many important consequences. However, such an improvement would require substantially new ideas and seems to be beyond our present knowledge.

The present effort is concentrated on the refinement of the constant $c(n)$. The best result in this direction is due to Matveev [79], who proved that one may take $c(n) = c^n$ with an absolute constant c.

Unfortunately, within the frames of this book, we cannot even give an idea of the proof of Theorem 13.2, let alone Theorem 13.3. For this we address the reader to Waldschmidt's and Nesterenko's contributions in the Cetraro volume [78]. See also the Baker Festschrift volume [139] for the history of the subject and the present state of the art.

13.2 Heights in Finitely Generated Groups

Many arguments in the subsequent sections rely on a simple property, which links $\|\mathbf{b}\| = |b_1| + \cdots + |b_n|$ and the height of the multiplicative combination

$$\gamma = \gamma(\mathbf{b}) = \gamma_1^{b_1}\cdots\gamma_n^{b_n}.$$

Basic properties of heights (see Appendix B[1]) imply that $\mathrm{h}(\gamma) \le c\|\mathbf{b}\|$, where the constant c depends on the numbers γ_i; for instance,

$$c = \max\{\mathrm{h}(\gamma_1),\ldots,\mathrm{h}(\gamma_n)\}$$

would do. It turns out that, when the numbers γ_i are multiplicatively independent, the opposite inequality holds as well.

Theorem 13.4. *Let $\gamma_1,\ldots,\gamma_n$ be multiplicatively independent nonzero algebraic numbers. Then there exists a positive constant c (depending on the numbers γ_i) such that for any $\mathbf{b} \in \mathbb{Z}^n$ we have $\|\mathbf{b}\| \le c\,\mathrm{h}(\gamma(\mathbf{b}))$.*

Proof. The proof goes back to Dirichlet. Let K be the number field generated by the numbers γ_i, and let S be the set of all valuations $v \in M_K$ such that $|\gamma_i|_v \ne 1$ for some i. We put $s = |S|$ and consider the map $K^\times \xrightarrow{\mathcal{L}} \mathbb{R}^s$ defined by $\alpha \mapsto (\log|\alpha|_v)_{v\in S}$.

Let Γ be the multiplicative group generated by $\gamma_1,\ldots,\gamma_n$. By the assumption, Γ is a free abelian group of rank n. For $\gamma \in \Gamma$ we have

$$\|\mathcal{L}(\gamma)\| = 2d\,\mathrm{h}(\gamma), \tag{13.7}$$

---

[1]Attention: notation $\|\cdot\|$ has a different meaning in Appendix B.

where $d = [K : \mathbb{Q}]$ and $\| \cdot \|$ is the ℓ_1-norm on $\mathbb{R}^s$. Indeed, by the very definition of S, we have $|\gamma|_v = 1$ for $\gamma \in \Gamma$ and $v \notin S$. It follows that

$$d\,\mathrm{h}(\gamma) = \sum_{v \in S} \log\max\{1, |\gamma|_v\}.$$

Hence, for $\gamma \in \Gamma$, we have

$$\|\mathcal{L}(\gamma)\| = \sum_{v \in S} \left|\log |\gamma|_v\right| = \sum_{v \in S} \left(\log\max\{1, |\gamma|_v\} + \log\max\{1, |\gamma^{-1}|_v\}\right)$$

$$= d\left(\mathrm{h}(\gamma) + \mathrm{h}(\gamma^{-1})\right) = 2d\,\mathrm{h}(\gamma),$$

as wanted.

Kronecker's theorems [Proposition B.2, items (2) and (3)] imply that the map $\mathcal{L}$ is injective on Γ and that $\mathcal{L}(\Gamma)$ is a discrete subgroup of $\mathbb{R}^s$. Indeed, if $\mathcal{L}(\gamma) = 0$, then $\mathrm{h}(\gamma) = 0$, which implies that γ is a root of unity by Kronecker's first theorem. Since Γ is a free abelian group, this is possible only if $\gamma = 1$. This proves that $\mathcal{L}|_\Gamma$ is injective.

Similarly, Kronecker's second theorem implies that there is $\varepsilon > 0$ such that $\mathrm{h}(\gamma) \geq \varepsilon$ if $\gamma \in \Gamma$ and $\gamma \neq 1$. Hence for $\lambda \in \mathcal{L}(\Gamma)$ and $\lambda \neq 0$ we have $\|\lambda\| \geq 2d\varepsilon$, which proves that $\mathcal{L}(\Gamma)$ is a discrete subgroup.

We have proved that $\mathcal{L}(\Gamma)$ is a discrete subgroup of $\mathbb{R}^s$ of rank n. Hence it generates a vector space V of dimension n. Now we are ready to finish the proof. The isomorphism $\mathbb{Z}^n \to \Gamma$, defined by $\mathbf{b} \mapsto \gamma(\mathbf{b})$, continues to an isomorphism $\mathbb{Z}^n \xrightarrow{\psi} \mathcal{L}(\Gamma)$, and the latter extends by linearity to an $\mathbb{R}$-isomorphism $\mathbb{R}^n \xrightarrow{\psi} V$. Since ψ is a non-singular linear map, there exists $\kappa > 0$ such that $\|\psi(\mathbf{b})\| \geq \kappa\|\mathbf{b}\|$ for $\mathbf{b} \in \mathbb{R}^n$. In particular, $\|\mathbf{b}\| \leq \kappa^{-1}\|\mathcal{L}(\gamma(\mathbf{b}))\|$ for $\mathbf{b} \in \mathbb{Z}^n$. In view of (13.7) this proves the theorem with $c = 2d\kappa^{-1}$. $\qquad\square$

With the help of Theorem 13.4, Baker's Theorem 13.2 can be restated as follows.

Theorem 13.5. *Let Γ be a finitely generated multiplicative group of complex algebraic numbers. Then for any $\varepsilon > 0$ there exists a positive real $h_0 = h_0(\varepsilon, \Gamma)$ such that for any $\gamma \in \Gamma$ with $|\gamma - 1| \leq e^{-\varepsilon\mathrm{h}(\gamma)}$ we have $\mathrm{h}(\gamma) \leq h_0$.*

Proof. Since Γ is finitely generated, its torsion subgroup is finite and cyclic (because a finite group of roots of unity is cyclic), of order, say, m. It follows that Γ has a basis of the form $\xi, \gamma_1, \ldots, \gamma_n$, where ξ is a root of unity and $\gamma_1, \ldots, \gamma_n$ are multiplicatively independent. Then every element of γ has a unique presentation as $\xi^a \gamma(\mathbf{b})$, where $0 \leq a < m$ and $\mathbf{b} \in \mathbb{Z}^n$. Since multiplication by a root of unity does not affect the height of an algebraic number, we have

$$c_1\|\mathbf{b}\| \leq \mathrm{h}(\gamma(\mathbf{b})) = \mathrm{h}(\xi^a \gamma(\mathbf{b})) \leq c_2\|\mathbf{b}\|,$$

with some positive c_1 and c_2 depending on $\gamma_1, \ldots, \gamma_n$. Here the inequality on the right follows from the basic properties of heights (as indicated in the beginning of this section), and the inequality on the left is Theorem 13.4.

Now if $\gamma = 1$ then the result is obvious, and if $0 < |\gamma - 1| \le e^{-\varepsilon h(\gamma)}$ then we complete the proof applying Theorem 13.2. $\qquad\square$

13.3 Almost nth Powers

This is another preparatory section. We recall one simple algebraic principle that is widely used in the Diophantine analysis from the time of Diophantus.

In its simplest form, the principle tells the following: if the product xy of coprime positive integers x and y is a square in $\mathbb{Z}$, then each of x and y is itself a square. Of course, this is well known and obvious.

More generally, if xy is "almost a square" (that is, a square times a given integer c), and if x, y are "almost coprime" (that is, $\gcd(x, y)$ divides a given integer d), then both x and y are "almost squares." This means that there exists a finite set $M \subset \mathbb{Z}$ (depending on c and d) such that $x = aX^2$ and $b = bY^2$, where $a, b \in M$ and $X, Y \in \mathbb{Z}$.

Of course, the same statement is true not only for squares but for nth powers with an arbitrary n. Arguments of this sort appeared in the book several times, in particular, in Chap. 2, in the proof of Cassels' relations, and in the "most important" Lemma 6.1.

In the following proposition we extend this property to number fields. It would also be convenient not to restrict to (algebraic) integers, but to allow our x and y to have a bounded denominator. In other words, we select them in a given fractional ideal of a number field.

Recall the definition of divisibility of fractional ideals. Given fractional ideals $\mathfrak{a}$ and $\mathfrak{b}$ of some number field, we say that $\mathfrak{a} \mid \mathfrak{b}$ if there exists an integral ideal $\mathfrak{q}$ such that $\mathfrak{b} = \mathfrak{a}\mathfrak{q}$. Equivalently, $\mathfrak{a} \mid \mathfrak{b}$ if $\mathfrak{b} \subseteq \mathfrak{a}$.

Lemma 13.6. *Let K be a number field and $\mathfrak{c}$ a nonzero fractional ideal of K. Let also $\mathfrak{d}$ be a nonzero integral ideal and n a positive integer:*

1. *There exists a finite set M of fractional ideals of K, depending on $\mathfrak{c}$ and $\mathfrak{d}$, but independent of n, such that the following holds. If $x, y \in \mathfrak{c}$ are such that $xy \in (K^\times)^n$ and[2] $\gcd(x, y) \mid \mathfrak{d}$, then there exist $\mathfrak{a}, \mathfrak{b} \in M$ such that (x) is $\mathfrak{a}$ times an nth power of an integral ideal and (y) is $\mathfrak{b}$ times an nth power of an integral ideal.*
2. *There exists a finite set $M_n \subset K^\times$, depending on $\mathfrak{c}$, $\mathfrak{d}$ and n such that the following holds. If $x, y \in \mathfrak{c}$ are such that $xy \in (K^\times)^n$ and $\gcd(x, y) \mid \mathfrak{d}$, then there exist $\alpha, \beta \in M_n$ and $X, Y \in \mathcal{O}_K$ such that $x = \alpha X^n$ and $y = \beta Y^n$.*

---

[2] By $\gcd(x, y)$ we mean the fractional ideal generated by x and y.

Proof. If $\mathfrak{p}$ is a prime ideal of K and x, y are as in the hypothesis, then

$$\mathrm{Ord}_\mathfrak{p}\mathfrak{c} \leq \min\{\mathrm{Ord}_\mathfrak{p}x, \mathrm{Ord}_\mathfrak{p}y\} \leq \mathrm{Ord}_\mathfrak{p}\mathfrak{d},$$

$$\mathrm{Ord}_\mathfrak{p}x + \mathrm{Ord}_\mathfrak{p}y \equiv 0 \bmod n.$$

It follows that there exists a finite set of integers $A_\mathfrak{p}$ (not depending on n) such that the smallest of the numbers $\mathrm{Ord}_\mathfrak{p}x$ or $\mathrm{Ord}_\mathfrak{p}y$ belongs to $A_\mathfrak{p}$, and the other one is congruent modulo n to some element of $A_\mathfrak{p}$. Moreover, for all but finitely many $\mathfrak{p}$, we have

$$\mathrm{Ord}_\mathfrak{p}\mathfrak{c} = \mathrm{Ord}_\mathfrak{p}\mathfrak{d} = 0,$$

and for all such $\mathfrak{p}$ we have $A_\mathfrak{p} = \{0\}$. Hence the set M, consisting of the ideals $\mathfrak{a}$ satisfying $\mathrm{Ord}_\mathfrak{p}\mathfrak{a} \in A_\mathfrak{p}$ for every $\mathfrak{p}$, is finite and, obviously, has the required property. This proves Part 1.

Now fix an integral representative in every ideal class of K, and fix a generator for every *principal* ideal of the form $\mathfrak{a}\mathfrak{h}^{-n}$, where $\mathfrak{a}$ runs over M and $\mathfrak{h}$ runs over the set of the fixed representatives. The set M_n' of the fixed generators has the following property: for any x, y as in the hypothesis, there exist $\alpha, \beta \in M_n'$ such that each of $(x\alpha^{-1})$ and $(y\beta^{-1})$ is an nth power of a *principal* integral ideal: $(x) = (\alpha X^n)$ and $(y) = (\beta Y^n)$ for some $X, Y \in \mathcal{O}_K$.

To complete the proof, we invoke the Dirichlet unit theorem. It implies that the group $\mathcal{U}_K/\mathcal{U}_K^n$ is finite. We fix a representative for every class of $\mathcal{U}_K$ modulo $\mathcal{U}_K^n$, and we define M_n to be the set of all $\alpha\eta$, where $\alpha \in M_n'$ and η runs over the set of the fixed representatives. Obviously, the set M_n is as wanted. The lemma is proved. □

We shall use this lemma in the following context: if a is a simple root of a polynomial f and $f(x)$ is an nth power for some integer x, then $x - a$ is an "almost" nth power for this x. Precisely speaking, we have the following consequence.

Corollary 13.7. *Let K be a number field, $f(x) \in K[x]$ a nonzero polynomial, $a \in K$ a simple root of f, and n a positive integer:*

1. *There exists a finite set M of fractional ideals of K, depending on f, but independent of n, such that the following holds. If $x \in \mathcal{O}_K$ is such that $f(x) \in (K^\times)^n$ then there exists $\mathfrak{a} \in M$ such that $(x - a)$ is $\mathfrak{a}$ times an nth power of an integral ideal.*

2. *There exists a finite set $M_n \subset K^\times$, depending on f and n such that the following holds. If $x \in \mathcal{O}_K$ is such that $f(x) \in (K^\times)^n$ then there exist $\alpha \in M_n$ and $X \in \mathcal{O}_K$ such that $x - a = \alpha X^n$.*

Proof. Define polynomials $g(x), h(x) \in K[x]$ from $f(x) = (x - a)g(x)$, and $g(x) - g(a) = (x - a)h(x)$. Recall that $g(a) \neq 0$ by the assumption.

Clearly, there exists a fractional ideal $\mathfrak{c}$, depending only on f such that both $x - a$ and $g(x)$ are in $\mathfrak{c}$ for any $x \in \mathcal{O}_K$. Also, since $g(a) \neq 0$, the relation

$$g(x) - (x - a)h(x) = g(a)$$

implies that there exists a nonzero integral ideal $\mathfrak{d}$ such that for any $x \in \mathcal{O}_K$ we have $\gcd(x - a, g(x)) \mid \mathfrak{d}$. Now applying Lemma 13.6 with $x - a$ as x and with $g(x)$ as y, we obtain the result. $\square$

Remark 13.8. More generally, if a is a root of order r, then, defining

$$n' = n/\gcd(n, r),$$

one shows that $(x - a)$ is $\mathfrak{a}$ times an n'th power of an integral ideal, where again $\mathfrak{a}$ is selected in a finite set depending only on f. Indeed, writing $f(x) = (x - a)^r g(x)$ and arguing as above, one proves that the principal ideal $((x - a)^r)$ is "almost" an nth power, whence $(x - a)$ is "almost" an n'th power. Similarly, $x - a = \alpha X^{n'}$, where $X \in \mathcal{O}_K$ and α is selected in a finite set, depending on f and n.

Part 1 of Lemma 13.6 holds even when $n = 0$. Moreover, in this case, the gcd assumption is obsolete, and we have the following statement.

Lemma 13.9. *Let K be a number field, $\mathfrak{c}$ a nonzero fractional ideal of K, and a a nonzero element of K. Then there exists a finite set M (depending on $\mathfrak{c}$ and a) of principal ideals of K such that the following holds. If $x, y \in \mathfrak{c}$ are such that $xy = a$ then the principal ideals (x) and (y) belong to M.*

The proof is very simple and left as an exercise.

A fargoing generalization of Lemma 13.6 is known as the *Chevalley-Weil Theorem* (see [59, Sect. 2.8]).

13.4   Effective Analysis of Classical Diophantine Equations

In this section we use Baker's inequality (13.6) to give effective proofs of the theorem of Thue and of the theorem of Siegel on the "superelliptic equations."

In this and the subsequent sections the following conventions will apply, unless the contrary is stated explicitly:

- By a *solution* of the equations we deal with we always mean a solution (x, y) in $\mathbb{Z}^2$ or in $\mathbb{Z} \times \mathbb{Q}$.
- By *constants* we mean positive real numbers that may depend on the equation (and can be explicitly expressed in terms of the equation) but which are independent of the solution we consider. This convention extends to the constants implied by the $O(\cdot)$ notation. Even in the course of the same argument the same letter (say, c) may denote different constants, when this does not lead to a confusion. All these constants are *effective*, that is, explicitly computable in terms of the equation.

13.4.1   The Thue Equation

First of all, we prove an effective version of Theorem 13.1.

Theorem 13.10 (Thue, Baker).   *Let $f(x, y) \in \mathbb{Z}[x, y]$ be a $\mathbb{Q}$-irreducible homogeneous polynomial of degree $n \geq 3$, and $A \in \mathbb{Z}$ a nonzero integer. Then the equation*

$$f(x, y) = A$$

has only finitely many solutions in $x, y \in \mathbb{Z}$, and the size of these solutions can be bounded by an explicitly computable constant (depending on f and A).

This effective version of Thue's theorem was proved by Baker [5].

Proof. As in Sect. 13.1.1, we write

$$f(x, y) = a_n(y - \theta_1 x) \cdots (y - \theta_n x)$$

and show that, given a solution (x, y) with $x \neq 0$, there exists a root θ among $\theta_1, \ldots, \theta_n$ such that y/x is "very close" to θ; precisely speaking,

$$|\theta - y/x| \leq c|x|^{-n},$$

with some constant c. It is not relevant for the proof, but can be remarked that the root θ may be assumed to be real; for any non-real theta there is an obvious lower bound for the difference $|\theta - y/x|$: we have $|\theta - y/x| \geq |\operatorname{Im} \theta|$.

In the sequel we shall assume (as we may, without loss of generality) that $\theta = \theta_1$. Now consider the quotient

$$\frac{y - \theta_2 x}{y - \theta_3 x} = \frac{y/x - \theta_2}{y/x - \theta_3}$$

(here we use the assumption $n \geq 3$). It is "very close" to $(\theta_1 - \theta_2)/(\theta_1 - \theta_3)$. In other words, the quantity

$$\gamma = \frac{\theta_1 - \theta_3}{\theta_1 - \theta_2} \cdot \frac{y - \theta_2 x}{y - \theta_3 x}$$

is "very close" to 1:

$$|\gamma - 1| \leq c|x|^{-n}, \tag{13.8}$$

with another constant c.

We want now to express the height of γ in terms of our solution (x, y). Writing

$$\gamma = \frac{\theta_1 - \theta_3}{\theta_1 - \theta_2} \cdot \frac{y/x - \theta_2}{y/x - \theta_3},$$

and using the properties of heights from Appendix B (most significantly, Part 9 of Proposition B.2), we find that

$$h(\gamma) = h(y/x) + O(1) = \max\{\log|x/d|, \log|y/d|\} + O(1),$$

where $d = \gcd(x, y)$. Since d divides $f(x, y) = A$, we obtain

$$h(\gamma) = \max\{\log|x|, \log|y|\} + O(1). \tag{13.9}$$

Finally, since y/x is "close" to θ_1, we have $\log|y| = \log|x| + O(1)$. We obtain

$$h(\gamma) = \log|x| + O(1). \tag{13.10}$$

Now let K be the number field generated by the roots $\theta_1, \theta_2, \theta_3$. Lemma 13.9 implies that there is a finite set $M \subset K^\times$ such that for any solution (x, y), each of the numbers $y - \theta_i x$ can be presented as product of an element of M and a Dirichlet unit. Hence every $y - \theta_i x$ belongs to the multiplicative group Γ', generated by the set M and the units of K. It follows that, for any solution (x, y), the quantity γ defined above belongs to the group Γ, generated by Γ' and the quotient $(\theta_1 - \theta_2)/(\theta_1 - \theta_3)$.

Now it is the time to apply Theorem 13.5. Fix $\varepsilon > 0$, to be specified later. Theorem 13.5 combined with estimate (13.10) implies that either $|x|$ is bounded or

$$|\gamma - 1| \geq e^{-\varepsilon \log|x| + O(1)}.$$

When ε is chosen to be strictly smaller than n (say, $\varepsilon = n/2$ would do), this contradicts (13.8) for large $|x|$. The theorem is proved.   □

13.4.2   The Superelliptic Equation

Another classical Diophantine equation studied by Baker is the superelliptic equation $y^n = f(x)$.

Theorem 13.11 (Siegel, Baker). *Let $f(x) \in \mathbb{Q}[x]$ be a polynomial with at least 2 simple roots (over $\mathbb{C}$). Then for any integer $n \geq 3$ the equation $y^n = f(x)$ has only finitely many solutions in $x \in \mathbb{Z}$ and $y \in \mathbb{Q}$, and the size of these solutions can be bounded by an explicitly computable constant (depending on f and n). If f has at least three simple roots, then the same is true for the solutions of $y^2 = f(x)$.*

A noneffective version of this theorem is due to Siegel [127], and Baker [6] proved an effective version.

To illustrate the idea of the proof, assume that $f(x)$ belongs to $\mathbb{Z}[x]$ and has two simple roots in $\mathbb{Z}$; call them a and b. Corollary 13.7 implies that for any solution

(x, y) of our equation we have $x - a = \alpha u^n$, where α is selected in some finite set and $u \in \mathbb{Z}$. Similarly, $x - b = \beta v^n$, where β belongs to a finite set and $v \in \mathbb{Z}$.

Thus, we reduced our equation to finitely many Thue equations of the form

$$\alpha u^n - \beta v^n = b - a, \tag{13.11}$$

which proves the theorem in the special case. In the general case, one reduces the superelliptic equation to equations of the same shape as (13.11), but now the coefficients α and β belong to some number field K, and one looks for solutions in the ring of integers $\mathcal{O}_K$. One can analyze this "generalized Thue equation" in, basically, the same way as one does with the classical Thue equation over $\mathbb{Q}$. However, it is more practical (and more instructive) to reduce the superelliptic equation directly to Baker's inequality, without passing through the Thue equation.

To understand how this should be done, let us return to the case when $f(x)$ has distinct simple roots $a, b \in \mathbb{Z}$. As we have seen, in this case, the original equation reduces to several Thue equations of the form (13.11). Let us analyze one such equation using the method of Sect. 13.4.1. We write

$$\alpha u^n - \beta v^n = \alpha(u - \theta_1 v) \cdots (u - \theta_n v),$$

where

$$\theta_k = \zeta^{k-1} \sqrt[n]{\beta/\alpha} \qquad (k = 1, \ldots, n)$$

with $\zeta = \zeta_n$ a primitive nth root of unity. If (u, v) is a "big solution," then the quotient u/v is "very close" to one of the roots $\theta_1, \ldots, \theta_n$, say, to θ_1.

Now as in the proof of Theorem 13.10, we introduce the quantity γ:

$$\gamma = \frac{\theta_1 - \theta_3}{\theta_1 - \theta_2} \cdot \frac{u - \theta_2 v}{u - \theta_3 v}.$$

Expressing γ in terms of the variables x and y of the original equation $y^n = f(x)$, we find

$$\gamma = \frac{1 - \zeta^2}{1 - \zeta} \cdot \frac{\sqrt[n]{x - a} - \zeta \sqrt[n]{x - b}}{\sqrt[n]{x - a} - \zeta^2 \sqrt[n]{x - b}} \tag{13.12}$$

for some determination of the roots $\sqrt[n]{x - a}$ and $\sqrt[n]{x - b}$.

Notice that γ can be defined, using (13.12), without any additional assumption on the roots a and b. This suggests the following strategy for proving Theorem 13.11: define γ as in (13.12) (one should only be careful here with the correct determination of the roots), and then show that γ belongs to some finitely generated group (independent of x) and is close to 1, so that one can apply Baker's inequality.

We can proceed now with the formal proof.

Proof of Theorem 13.11. Let $a \in \bar{\mathbb{Q}}$ be a simple root of f. If $(x, y) \in \mathbb{Z} \times \mathbb{Q}$ is a solution of $y^n = f(x)$, then Corollary 13.7 implies $x - a = \alpha u^n$, where α is selected in some finite set A and $u \in \mathbb{Q}(a)$. It follows that for any solution (x, y) all the roots $\sqrt[n]{x - a}$ belong to the number field $\mathbb{Q}(a, \sqrt[n]{A})$, generated over $\mathbb{Q}(a)$ by the nth roots of the elements of A and by the nth roots of unity.

Similarly, if b is another simple root of f then there exists a finite set $B \subset \mathbb{Q}(b)$ such that for any solution (x, y) all the roots $\sqrt[n]{x - b}$ belong to the number field $\mathbb{Q}(b, \sqrt[n]{B})$.

Denote by K the number field $\mathbb{Q}(a, b, \sqrt[n]{A}, \sqrt[n]{B})$ and fix a primitive nth root of unity $\zeta = \zeta_n$. Lemma 13.9 and the identity

$$\prod_{k=0}^{n-1} \left(\sqrt[n]{x - a} - \zeta^k \sqrt[n]{x - b} \right) = b - a$$

imply the following: there exists a finite set of principal ideals of the field K, not depending on the solution (x, y), such that for $k \in \{0, \ldots, n - 1\}$ the principal ideals $\left(\sqrt[n]{x - a} - \zeta^k \sqrt[n]{x - b} \right)$ belong to this set. It follows that there exists a finitely generated subgroup Γ' of $K^\times$ (not depending on the solution) such that

$$\sqrt[n]{x - a} - \zeta^k \sqrt[n]{x - b} \in \Gamma' \qquad (k = 0, \ldots, n - 1). \tag{13.13}$$

Now it is the time to define the roots $\sqrt[n]{x - a}$ and $\sqrt[n]{x - b}$ as complex numbers. We may restrict ourselves to solutions (x, y) with $x > 0$; for solutions with $x < 0$ just consider the equation $y^n = f(-x)$. Further, we may assume that $x > \max\{|a|, |b|\}$ and define the roots $\sqrt[n]{1 - a/x}$ and $\sqrt[n]{1 - b/x}$ as the sums of the corresponding binomial series:

$$\sqrt[n]{1 - \frac{a}{x}} = \sum_{k=0}^{\infty} \binom{1/n}{k} \left(\frac{-a}{x} \right)^k,$$

and similarly for $\sqrt[n]{1 - b/x}$. With this definition we have

$$\sqrt[n]{1 - \frac{a}{x}} = 1 + O\left(\frac{1}{x} \right), \qquad \sqrt[n]{1 - \frac{b}{x}} = 1 + O\left(\frac{1}{x} \right). \tag{13.14}$$

Next, we define $\sqrt[n]{x - a} = \sqrt[n]{x} \sqrt[n]{1 - a/x}$, where $\sqrt[n]{x}$ is the only positive nth root of x, and similarly for $\sqrt[n]{x - b}$.

Now assume that $n \geq 3$. Relations (13.14) imply that the quantity

$$\gamma = \gamma(x) = \frac{1 - \zeta^2}{1 - \zeta} \cdot \frac{\sqrt[n]{x - a} - \zeta \sqrt[n]{x - b}}{\sqrt[n]{x - a} - \zeta^2 \sqrt[n]{x - b}}$$

satisfies

$$|\gamma - 1| \leq Cx^{-1} \tag{13.15}$$

with some constant C. Writing

$$\rho = \sqrt[n]{\frac{x-b}{x-a}}, \qquad \gamma = \frac{1-\zeta^2}{1-\zeta} \cdot \frac{1-\zeta\rho}{1-\zeta^2\rho}$$

and using Proposition B.2 (Part 9 and others), we find

$$h(\rho^n) = h(x) + O(1), \qquad h(\gamma) = h(\rho) + O(1),$$

whence

$$h(\gamma) = \frac{1}{n}h(x) + O(1) = \frac{1}{n}\log x + O(1). \tag{13.16}$$

Now we complete the proof using Theorem 13.5. Clearly, $\gamma \in \Gamma$, the group generated by Γ' and $(1 - \zeta^2)/(1 - \zeta)$. Theorem 13.5 applied with suitably chosen ε and combined with estimate (13.16) implies that either x is bounded or

$$|\gamma - 1| \geq e^{-(1/2)\log x + O(1)},$$

contradicting (13.15) for large x. The case $n \geq 3$ is proved.

If $n = 2$ and f has three simple roots a, b, c, the proof is very similar. This time we put

$$\gamma = \frac{\sqrt{x-a} + \sqrt{x-b}}{\sqrt{x-a} + \sqrt{x-c}}.$$

We again have

$$|\gamma - 1| \leq Cx^{-1}. \tag{13.17}$$

As for the height, we cannot now have an asymptotical estimate like (13.16), but we can estimate the height in terms of x from above and from below, which would suffice.

The upper estimate is a straightforward application of Proposition B.2:

$$h(\gamma) \leq 2\log x + O(1). \tag{13.18}$$

To estimate the height from below, write

$$\rho = \sqrt{\frac{x-b}{x-a}}, \qquad \gamma = \frac{1+\rho}{1 + \sqrt{\lambda\rho^2 + \mu}},$$

where

$$\lambda = \frac{a-c}{a-b}, \qquad \mu = \frac{c-b}{a-b}.$$

Resolving this in ρ, we find

$$\rho = \gamma - 1 \pm \gamma \sqrt{(\lambda - \mu)\gamma^2 - 2\lambda\gamma + \lambda + \mu}.$$

Using repeatedly various parts of Proposition B.2, we obtain the estimates $h(\rho) \leq (5/2)h(\gamma) + O(1)$ and $\log x = h(x) = 2h(\rho) + O(1)$, resulting in

$$\log x \leq 5h(\gamma) + O(1). \tag{13.19}$$

Applying Theorem 13.5 with suitably chosen ε and combining it with (13.18) and (13.19), we deduce that either x is bounded or

$$|\gamma - 1| \geq e^{-(1/2)\log x + O(1)},$$

contradicting (13.17) for large x. The theorem is proved in the case $n = 2$ as well.

$\square$

13.5   The Theorem of Schinzel and Tijdeman and the Equation of Pillai

Schinzel and Tijdeman [122] considered the superelliptic equation with the "varying" exponent n and proved the following remarkable theorem.

Theorem 13.12 (Schinzel, Tijdeman). *Let $f(x) \in \mathbb{Q}[x]$ be a polynomial with at least two distinct roots (over $\mathbb{C}$). Then there exists a positive effectively computable constant $n_0 = n_0(f)$ such that for any integer $n \geq n_0(f)$ the equation $y^n = f(x)$ has no solutions in $x \in \mathbb{Z}$ and $y \in \mathbb{Q}$ with $y \neq 0, \pm 1$.*

The proof of this theorem again relies on Baker's inequality, but Theorem 13.5 is no longer sufficient. In fact, now we have to deal not with elements of a finitely generated group Γ, but of the group $(K^\times)^n \Gamma$, generated by Γ and all nth powers in the field K. For elements of this group we have the following Baker-type statement.

Theorem 13.13. *Let K be a number field, Γ a finitely generated subgroup of $K^\times$, and $n \geq 2$ an integer. Then for any $\gamma \in (K^\times)^n \Gamma$ we have either $\gamma = 1$ or*

$$|\gamma - 1| \geq e^{-c_1 \frac{\log n}{n} h(\gamma) - c_2}, \tag{13.20}$$

where the positive constants c_1 and c_2 depend on K and on Γ, but not on n.

Notice that this theorem is of interest only for large n; for small n an apparently better estimate $|\gamma - 1| \geq e^{-d(h(\gamma) + \log 2)}$, where $d = [K : \mathbb{Q}]$, follows from "Liouville's inequality" (Proposition B.3).

Proof. Fix a system $\gamma_1, \ldots, \gamma_r$ of generators of Γ. Then any $\gamma \in (K^\times)^n \Gamma$ can be written as $\gamma = \gamma_0^n \gamma_1^{b_1} \cdots \gamma_r^{b_r}$ with $\gamma_0 \in K^\times$ and $0 \leq b_i < n$. For the height of γ_0 we have the estimate

$$h(\gamma_0) \leq \frac{1}{n}\big(h(\gamma) + b_1 h(\gamma_1) + \cdots + b_r h(\gamma_r)\big) \leq \frac{h(\gamma)}{n} + O(1),$$

the implicit constant depending on Γ. Hence, $h'(\gamma_0) \leq h(\gamma)/n + O(1)$. Applying Theorem 13.3, we obtain $\gamma = 1$ or

$$|\gamma - 1| \geq e^{-c(K, \Gamma) h'(\gamma_0) \log n} \geq e^{-c_1 \frac{\log n}{n} h(\gamma) - c_2},$$

as wanted.   □

Remark 13.14. It is useful to express the constants c_1 and c_2 explicitly in terms of a system $\gamma_1, \ldots, \gamma_r$ of generators of Γ. The proof implies that one may take

$$c_1 = c(r, d) h'(\gamma_1) \cdots h'(\gamma_r), \qquad c_2 = c_1 \left(h(\gamma_1) + \cdots + h(\gamma_r) + 1\right),$$

where, as above, $d = [K : \mathbb{Q}]$. This will be used in Sect. 13.6.

Proof of Theorem 13.12. To simplify the notation, we shall give a detailed proof in the special case when the polynomial f has two distinct *simple* roots (which is sufficient for the subsequent applications). At the end, we indicate the few changes to be made to adapt the proof to the general case.

Notice first of all that n is bounded in terms of x. Indeed, if $f(x) \neq 0, \pm 1$ then the set of n such that $f(x)$ is a pure nth power is finite. And if $f(x) \in \{0, \pm 1\}$ then from $y^n = f(x)$ we find that either $n = 0$ or $y \in \{0, \pm 1\}$, which is excluded.

Thus, in the sequel, we may assume $|x|$ as big as we please. In particular, we may assume that $x \neq 0$.

Now let a and b be two distinct simple roots of f, and put $K = \mathbb{Q}(a, b)$. All constants in the proof, either implied by the $O(\cdot)$-notation or denoted by $c, c_1, c_2 \ldots$, may depend on f but not on n.

Corollary 13.7 implies that the principal ideal $(x - a)$ is $\mathfrak{a}\mathfrak{u}^n$, where $\mathfrak{u}$ is an integral ideal of the field K and $\mathfrak{a}$ belongs to a finite set of (fractional) ideals, which depends on f but is independent of x and n.

Now let $h = h_K$ be the class number[3] of the field K. Then the ideals $\mathfrak{a}^h$ and $\mathfrak{u}^h$ are principal. It follows that $(x - a)^h = \alpha \eta \mathfrak{u}^n$, where α belongs to a finite set (independent of x and n), u is an algebraic integer, and η is a unit of the field K. Hence $(x - a)^h$ belongs to $(K^\times)^n \Gamma$, where Γ is a finitely generated multiplicative

---

[3] We hope that no confusion occurs between the height function $h(\cdot)$ and the class number h.

group, which does not depend on x and n. Similarly, $(x - b)^h \in (K^\times)^n \Gamma'$ with some finitely generated Γ', and we may assume that $\Gamma' = \Gamma$. Hence,

$$\gamma = \left(\frac{x - a}{x - b}\right)^h \in (K^\times)^n \Gamma,$$

and Theorem 13.13 applies here. To use it, we need to express the height of γ in terms of x. Using items (8) and (9) of Proposition B.2 and recalling that $h(x) = \log |x|$ (because $x \neq 0$), we obtain $h(\gamma) = h \log |x| + O(1)$. Substituting this to (13.20), we obtain that either $\gamma = 1$ or

$$|\gamma - 1| \geq e^{-c_1 h \frac{\log n}{n} \log |x| - c_2}. \tag{13.21}$$

On the other hand, $\gamma = 1 + O(|x|^{-1})$ and $\gamma \neq 1$ when $|x| \geq 2 \max\{|a|, |b|\}$. Combining this with (13.21), we obtain that either $n \leq c_3$ or $|x| \leq c_4$. Since n is bounded in terms of x, this completes the proof in the case when the roots a and b are simple.

Now assume that a and b are distinct roots of $f(x)$ of multiplicities r and s, respectively. Using Remark 13.8 instead of Corollary 13.7, we obtain that $(x - a)^h \in (K^\times)^{n/\gcd(n,r)} \Gamma$ and $(x - b)^h \in (K^\times)^{n/\gcd(n,s)} \Gamma$ with some finitely generated group Γ. Hence $\gamma \in (K^\times)^{n'} \Gamma$, where

$$n' = \frac{n}{\text{lcm}(\gcd(n, r), \gcd(n, s))}.$$

Now, arguing as above, we bound n', which implies a bound on n. The theorem is proved.   □

Remark 13.15. Alternatively, one can define

$$\gamma = \left(\frac{y^n}{d(x - a)^m}\right)^h, \qquad h = h_{\mathbb{Q}(a)},$$

where m is the degree and d is the leading coefficient of the polynomial f. Since $(x - a)^h$ is an "almost" nth power in $\mathbb{Q}(a)$, so is γ, and since $y^n = f(x) = dx^m + O(|x|^{m-1})$, we have $\gamma = 1 + O(|x|^{-1})$. Finally, since a is not the single root of f, we have $\gamma \neq 1$ for big x.

This approach might be useful in some cases, because it permits to deal with the field $\mathbb{Q}(a)$ instead of $\mathbb{Q}(a, b)$. We shall see how to profit from this in Sect. 13.6.

Combining Theorems 13.11 and 13.12, we obtain the following.

Theorem 13.16. *Let $f(x) \in \mathbb{Q}[x]$ be a polynomial with at least two simple roots. Then the equation $y^n = f(x)$ has only finitely many solutions in $x \in \mathbb{Z}$, $y \in \mathbb{Q}$, and $n \in \mathbb{Z}$ with $n \geq 3$ and $y \neq 0 \pm 1$; moreover, the size of these solutions can be explicitly bounded. If f has at least three simple roots, then the assumption $n \geq 3$ may be replaced by $n \geq 2$.*

We apply this to the *Diophantine equation of Pillai*. The Indian mathematician Pillai studied in [114, 115] the Diophantine equation $x^m - y^n = c$ with a fixed nonzero integer c. He conjectured that this equation has only finitely many nontrivial solutions in positive integers x, y, m, n with $m, n > 1$. This conjecture is proved now only for $c = \pm 1$.

One may state a more general conjecture.

Conjecture 13.17. Let a, b, c be nonzero integers. Then the equation

$$ax^m + by^n = c \tag{13.22}$$

has finitely many solutions in integers x, y and positive integers m, n such that

$$x, y \neq 0, \pm 1, \quad m, n > 1, \quad (m, n) \neq (2, 2). \tag{13.23}$$

(One has to exclude the case $m = n = 2$ because equation $ax^2 + by^2 = c$ may have infinitely many solutions.)

This conjecture is usually called *generalized Pillai's conjecture*, and Eq. (13.22) is called *generalized Pillai's equation*. Solutions satisfying (13.23) will be called *nontrivial*.

No instance of Pillai's conjecture with $|abc| > 1$ is currently proved.

Theorem 13.16 implies that this equation may have only finitely many nontrivial solutions with one of the four variables fixed.

Theorem 13.18. *Let a, b, c be nonzero integers. Fix integers $x_0 \neq 0, \pm 1$ and $m_0 > 1$. Then Eq. (13.22) has at most finitely many nontrivial solutions (x, y, m, n) with $x = x_0$ and at most finitely many nontrivial solutions with $m = m_0$.*

Proof. When $m = m_0$ is fixed, we write our equation as $y^n = f(x)$ with $f(x) = -(ax^{m_0} - c)/b$ and apply Theorem 13.16.

Now fix $x = x_0$ and write $m = 3\mu + \rho$, where $\rho \in \{0, 1, 2\}$. Then (13.22) can be rewritten as $y^n = f(X)$ with $f(X) = -(AX^3 - c)/b$, where $A = ax_0^\rho$ and $X = x_0^\mu$. For every ρ this equation has finitely many solutions by Theorem 13.16. Whence the result. $\square$

13.6 Tijdeman's Argument

We conclude this chapter with the remarkable result of Tijdeman [134], who proved in 1976 that Catalan's problem reduces to a finite calculation.

Theorem 13.19 (Tijdeman). *There exists an absolute effective constant C such that any nontrivial solution (x, y, p, q) of Catalan's equation $x^p - y^q = 1$ satisfies*

$$\max\{|x|, |y|, p, q\} \leq C.$$

By a *nontrivial solution* we mean here a solution in integers $x, y \neq 0$ and primes p, q.

Before the work of Mihăilescu this was the best result on Catalan's problem. See Sect. 1.4 (p. 5) and Sect. 1.6 (p. 7) of the Historical Account, where Tijdeman's work is put into the historical context.

Notice first of all that if one of the exponents p, q is 2 then the result follows from Theorem 13.18. Hence, we may assume that p, q are odd prime numbers. This allows us to use the symmetry: if (x, y, p, q) is a solution, then so is $(-y, -x, q, p)$.

Also, again by Theorem 13.18, we may assume that

$$|x|, |y| \geq 10. \tag{13.24}$$

13.6.1 Preparations

To prove Tijdeman's theorem, we write Catalan's equation as

$$y^q = x^p - 1, \tag{13.25}$$

and we may apply Theorem 13.12 to bound q in terms of p. We want to make this explicit. If we imitate the proof of Theorem 13.12 in this special case, we will have to deal with the quantity

$$\left(\frac{x-a}{x-b}\right)^{h_p},$$

where h_p is the class number of the cyclotomic field $K_p = \mathbb{Q}(\zeta_p)$ and a, b are two distinct pth roots of unity (in particular, one of them can be 1). This quantity must be an "almost" qth power in K_p.

It is more efficient, however, to do as suggested in Remark 13.15 and take

$$\gamma = \frac{y^q}{(x-1)^p}. \tag{13.26}$$

This would allow us to work over the field $\mathbb{Q}$ rather that the pth cyclotomic field. Another important advantage is that we have the following very precise version of Corollary 13.7. We denote by $\Phi_p(x)$ the pth cyclotomic polynomial:

$$\Phi_p(x) = \frac{x^p - 1}{x - 1} = x^{p-1} + \cdots + x + 1.$$

Proposition 13.20. *Let (x, y, p, q) be a solution of Catalan's equation (that is, x, y are nonzero integers and p, q are (distinct) odd prime numbers, satisfying $x^p - y^q = 1$). Then either*

$$\gcd(x - 1, \Phi_p(x)) = 1,$$

in which case $x - 1, \Phi_p(x) \in (\mathbb{Q}^\times)^q$, *or*

$$\gcd(x - 1, \Phi_p(x)) = p \,,$$

in which case $x - 1 \in p^{-1}(\mathbb{Q}^\times)^q$ *and* $\Phi_p(x) \in p\,(\mathbb{Q}^\times)^q$. *In addition, in the second case,* $\mathrm{Ord}_p \Phi_p(x) = 1$. *Symmetrically, we have either* $y + 1 \in (\mathbb{Q}^\times)^p$ *or* $y + 1 \in q^{-1}(\mathbb{Q}^\times)^p$.

Proof. We may quote here certain statements from Chap. 2 or 3, but we give a quick independent proof for the reader's convenience. It will be more practical to deal with the polynomial $g(t) = \Phi_p(t + 1)$. Since $g(0) = \Phi_p(1) = p$, we have clearly $\gcd(t, g(t)) \in \{1, p\}$ for $t \in \mathbb{Z}$. Moreover, writing

$$g(t) = \frac{(t + 1)^p - 1}{t} = p + \sum_{k=2}^{p-1} \binom{p}{k} t^{k-1} + t^{p-1}$$

we see that $g(t) \equiv t^{p-1} \bmod p$, which means that, for $t \in \mathbb{Z}$, either both t and $g(t)$ are not divisible by p (in which case they are coprime) or both are divisible by p (in which case their gcd is p). In this second case we have $g(t) \equiv p \bmod p^2$, in particular $\mathrm{Ord}_p g(t) = 1$.

Now let (x, y, p, q) be a solution of Catalan's equation. Setting $t = x - 1$, we rewrite the equation as $t g(t) = y^q$. Thus, if $\gcd(t, g(t)) = 1$ then both t and $g(t)$ are pure qth powers, and if $\gcd(t, g(t)) = p$ then one of the numbers t and $g(t)$ is $p u^q$ and the other is $p^{q-1} v^q$, where $u, v \in \mathbb{Z}$ and $p \nmid u$. Since $\mathrm{Ord}_p g(t) = 1$, it cannot be $p^{q-1} v^q$. Thus, $g(t) = p u^q \in p\,(\mathbb{Q}^\times)^q$ and $t = p^{q-1} v^q \in p^{-1}(\mathbb{Q}^\times)^q$. This completes the proof.   □

Remark 13.21. According to Cassels' Theorem 3.3 from Chap. 3, we actually have $\gcd(x - 1, \Phi_p(x)) = p$. This result (which is quite involved; see Sect. 3.3) is crucial for the work of Mihăilescu, but we do not need it to prove Theorem 13.19. Therefore we prefer to ignore this additional knowledge at this time.

Corollary 13.22. *We have* $|x| \geq 2^q - 1$ *and* $|y| \geq 2^p - 1$.

Proof. If $x - 1 \in (\mathbb{Q}^\times)^q$ then either $x - 1 \in \{0, \pm 1\}$, which is impossible by the assumption (13.24), or $|x - 1| \geq 2^q$, which implies that $|x| \geq 2^q - 1$. If $x - 1 \in p^{-1}(\mathbb{Q}^\times)^q$ then $|x - 1| \geq p^{q-1}$, and a quick inspection shows that $p^{q-1} > 2^q$ for any distinct odd primes p and q. This proves that $|x| \geq 2^q - 1$, and $|y| \geq 2^p - 1$ follows by symmetry.   □

Again, much sharper results are available (see Sect. 3.4), but the estimates from Corollary 13.22 are sufficient here.

We also need some estimates concerning γ.

Proposition 13.23. *The rational number* γ *defined in (13.26) is distinct from* ± 1. *If* $p < q$ *then* γ *satisfies*

$$\gamma = 1 + O(p/x), \qquad \mathrm{h}(\gamma) = (p - 1) \log |x| + O(\log p), \qquad (13.27)$$

the implied constants being absolute.

Proof. Since

$$\gamma = \frac{x^p - 1}{(x - 1)^p},$$ (13.28)

we clearly have $\gamma \neq \pm 1$. Further, since $|x| \geq 2^q - 1 \geq 2^p - 1$ by Corollary 13.22, we have $\gamma = 1 + O(p/x)$.

To estimate the height of γ, we write further

$$\gamma = \frac{x^{p-1} + \cdots + x + 1}{(x - 1)^{p-1}}.$$

Proposition 13.20 implies that the greatest common divisor of the denominator and the numerator is 1 or p. Hence, using Proposition B.2(5), we obtain

$$\mathrm{h}(\gamma) = \log \max\left\{|x^{p-1} + \cdots + x + 1|, |(x - 1)^{p-1}|\right\} + O(\log p).$$ (13.29)

Again using $|x| \geq 2^p - 1$, we find that the right-hand side of (13.29) is $(p - 1) \log |x| + O(\log p)$. This completes the proof. $\qquad\qquad\square$

13.6.2 Proof of Theorem 13.19

We now start the proof of Theorem 13.19. By Theorem 13.18 it suffices to bound one of the exponents p or q. We will do even better, bounding both of them.

We assume that $q > p$. The proof consists of two steps: first we bound q in terms of p, and after we bound p in terms of q; the combination of both bounds would imply an absolute bound for p and q. Unless the contrary is stated explicitly, all constants denoted by c or implied by the $O(\cdot)$ notation are absolute.

Bounding q in Terms of p

Our first objective will be the inequality

$$q \leq cp(\log p)^2$$ (13.30)

with some absolute constant c. The exact form of the inequality is of little importance; we only need to know that q cannot be "too large" compared with p.

As indicated above, this will be a reenactment of the proof of Theorem 13.12, but with γ defined as in (13.26), and with Proposition 13.20 replacing Corollary 13.7.

Proposition 13.20 implies that $\gamma \in \Gamma_p(\mathbb{Q}^\times)^q$, where Γ_p is the subgroup of $\mathbb{Q}^\times$ generated by p. Since $\gamma \neq 1$ (Proposition 13.23), Theorem 13.13 implies that

$$|\gamma - 1| \geq e^{-K_1 \frac{\log q}{q} \mathrm{h}(\gamma) - K_2},$$

where quantities K_1 and K_2 depend on Γ_p, that is, on p. Moreover, as follows from Remark 13.14, one has $K_1 = O(\log p)$ and $K_2 = O\left((\log p)^2\right)$ (recall that the constants implied by the $O(\cdot)$-notation are absolute). Using also the height estimate from (13.27), we obtain

$$|\gamma - 1| \ge e^{-c\,\frac{(p-1)\log p \log q}{q}\,\log|x| - c(\log p)^2} \tag{13.31}$$

with some absolute constant c.

On the other hand, the first estimate from (13.27) can be rewritten as

$$0 < |\gamma - 1| \le e^{-\log|x| + O(\log p)}.$$

Comparing this with (13.31), we obtain

$$\log|x| \le c\,\frac{p \log p \log q}{q}\,\log|x| + c(\log p)^2$$

with a different c. Using the inequality $|x| \ge 2^q - 1$, we deduce from this that

$$q \le cp \log p \log q + c\,\frac{q}{\log(2^q - 1)}\,(\log p)^2,$$

which implies that $q \le cp(\log p)^2$ with yet another constant c. This proves estimate (13.30).

Bounding p in Terms of q

Thus, we have

$$p \le q \le cp(\log p)^2. \tag{13.32}$$

Now we want to bound p in terms of q. We may write the equation as $x^p = y^q + 1$ and try the quantity

$$\frac{x^p}{(y + 1)^q},$$

but this would imply something like $p \le cq(\log q)^2$, quite uninteresting.

Tijdeman's bright idea was to use

$$\beta = \frac{(x - 1)^p}{(y + 1)^q}.$$

The crucial point is that β is an "almost pqth power"; precisely, Proposition 13.20 implies that $\beta \in \Gamma_{p,q}(\mathbb{Q}^\times)^{pq}$, where $\Gamma_{p,q}$ is the multiplicative group generated by p and q. As we will see in a while, applying Theorem 13.13 to β would result in a very sharp upper bound for p in terms of q.

In what follows we treat β very similarly to our treatment of γ above. First of all, we have for β analogues of estimates (13.27):

$$1 \neq \beta = 1 + O(q/y), \qquad h(\beta) = q \log|y| + O(q/y). \qquad (13.33)$$

It is clear that $\beta \neq 1$: observe, for instance, that the denominator and the numerator in the definition of β are coprime, because $x - 1$ divides y^q. To prove that $\beta = 1 + O(q/y)$, write

$$\beta = \frac{(x-1)^p}{x^p - 1} \left(\frac{y}{y+1} \right)^q.$$

Using Corollary 13.22 and (13.32), we find

$$\beta = (1 + O(p/x))(1 + O(q/y)).$$

Finally, since $q > p$, we have $|y| < |x|$, which implies that $\beta = 1 + O(q/y)$.

In a similar fashion one estimates the height of β. Since the denominator and the numerator in the definition of β are coprime, we have

$$h(\beta) = \max\left\{ p \log|x-1|, q \log|y+1| \right\},$$

and, using, as above, Corollary 13.22 and (13.32), we find

$$h(\beta) = \max\left\{ p \log|x|, q \log|y| \right\} + O(q/y).$$

Finally, since

$$p \log|x| = \log|y^q + 1| = q \log|y| + O(y^{-q}),$$

we obtain $h(\beta) = q \log|y| + O(q/y)$.

Now we are ready to apply Theorem 13.13, which, together with Remark 13.14, gives the lower bound

$$|\beta - 1| \geq e^{-K_1 \frac{\log(pq)}{pq} h(\beta) - K_2}, \qquad K_1 = O\left((\log q)^2\right), \qquad K_2 = O\left((\log q)^3\right).$$

Combining this with the height estimate from (13.33), we obtain

$$|\beta - 1| \geq e^{-c \frac{(\log q)^3}{p} \log|y| - c(\log q)^3} \qquad (13.34)$$

with some absolute constant c.

Rewriting the first estimate from (13.33) as

$$0 < |\beta - 1| \le e^{-\log|y| + O(\log q)}$$

and comparing this with (13.34), we obtain

$$\log|y| \le c \frac{(\log q)^3}{p} \log|y| + c(\log q)^3$$

with a different c; in other words,

$$p \le c(\log q)^3 + c \frac{p}{\log|y|}(\log q)^3. \tag{13.35}$$

By Corollary 13.22 we have $|y| \ge 2^p - 1$. Hence the right-hand side of (13.35) is $O\left((\log q)^3\right)$. Thus,

$$p \le c(\log q)^3$$

with yet another c.

This is the promised bound for p in terms of q. Combining it with (13.32) we bound both p and q by an absolute constant, completing the proof of Theorem 13.19.   □

Appendix A
Number Fields

This is a brief synopsis of the algebraic number theory used in the present book. It cannot serve as introduction to algebraic number theory: its only purpose is to refresh the terminology and recall the basic facts. We give almost no proofs and almost no references, assuming that every reader will find the proofs of all (or, at least, most) of the statements from this appendix in his favorite algebraic number theory textbook(s).

A.1 Embeddings, Integral Bases, and Discriminant

An *algebraic number field* or, shorter, a *number field* is a finite extension of the field $\mathbb{Q}$ of rational numbers. Let K be a number field of degree $n = [K:\mathbb{Q}]$. Then there exist exactly n distinct embeddings $\sigma_1, \ldots, \sigma_n : K \hookrightarrow \mathbb{C}$. An embedding σ is called *real* if $\sigma(K) \subset \mathbb{R}$ and *complex* otherwise. Complex embeddings enter the list $\sigma_1, \ldots, \sigma_n$ in pairs: if σ is a complex embedding, then its complex conjugate $\bar{\sigma}$ is another complex embedding of K. In particular, the number of complex embeddings is always even.

A field with only real embeddings is called *totally real*, and a field with no real embeddings is called *totally imaginary*.

Denote by $\mathcal{O}_K$ the ring of algebraic integers from the number field K. The additive group of $\mathcal{O}_K$ is a free abelian group of rank n. Let $w_1, \ldots, w_n$ be a $\mathbb{Z}$-basis of $\mathcal{O}_K$ (such a basis is usually called an *integral basis of K*). The quantity

$$\mathcal{D}_K = \left(\det \left[\sigma_i(w_j) \right]_{1 \leq i, j \leq n} \right)^2$$

is independent of the choice of the integral basis $w_1, \ldots, w_n$ and is called the *discriminant of K*. It is a nonzero rational integer.

Recall also *Kronecker's theorem*: if $\alpha \in \mathcal{O}_K$ satisfies $|\sigma(\alpha)| \leq 1$ for any embedding σ of K, then either $\alpha = 0$ or α is a root of unity. [It is proved in Appendix B; see Proposition B.2(2).]

A.2   Units, Regulator

The invertible elements of the ring $\mathcal{O}_K$ are called the *units* (or *Dirichlet units*) of K. The multiplicative group $\mathcal{U}_K = \mathcal{O}_K^\times$ of units is a finitely generated abelian group. More precisely, let t_1 and $2t_2$ be the number of real and complex embeddings of K, respectively. Let Ω_K be the group of roots of unity from K (it is a finite cyclic group). The fundamental *Dirichlet unit theorem* asserts that $\mathcal{U}_K/\Omega_K$ **is a free abelian group of rank** $r = t_1 + t_2 - 1$. (The inequality $r \leq t_1 + t_2 - 1$ is relatively easy to establish; the nontrivial part is $r \geq t_1 + t_2 - 1$.)

Let $\sigma_1, \ldots, \sigma_{r+1}$ be a selection of embeddings of K, containing all real embeddings and one from each pair of complex embeddings ($r + 1 = t_1 + t_2$), and let $\eta_1, \ldots, \eta_r$ be a basis of the infinite part of $\mathcal{U}_K$ (usually called a *system of fundamental units*). Put $e_i = 1$ if σ_i is real and $e_i = 2$ if σ_i is complex, and consider the map $\mathcal{U}_K \to \mathbb{R}^r$ which associates to every $\eta \in \mathcal{U}_K$ the vector

$$(e_1 \log |\sigma_1(\eta)|, \ldots, e_r \log |\sigma_r(\eta)|).$$

The kernel of this map is Ω_K, and its image is a lattice in $\mathbb{R}^r$. The fundamental volume of this lattice, that is, the quantity

$$\left| \det \left[e_i \log |\sigma_i(\eta_j)| \right]_{1 \leq i,j \leq r} \right|$$

(which is independent of the choice of the basic units $\eta_1, \ldots, \eta_r$ and of our selection of the embeddings $\sigma_1, \ldots, \sigma_{r+1}$), is called the *regulator of K* and is denoted by $\mathcal{R}_K$.

A.3   Ideals, Factorization

The ring $\mathcal{O}_K$ is a Dedekind ring; that is, it is Noetherian and integrally closed, and all its nonzero prime ideals are maximal. Hence it has the unique factorization property for ideals: **every nonzero ideal of $\mathcal{O}_K$ has a unique (up to the order) presentation as a product of prime ideals**. In other words, the nonzero ideals of $\mathcal{O}_K$ form a free multiplicative semigroup, the (nonzero) prime ideals being its free generators.

A *fractional ideal* is a finitely generated $\mathcal{O}_K$-submodule of K. (Equivalently, an $\mathcal{O}_K$-module $\mathfrak{a} \subset K$ is a fractional ideal if there exists a nonzero $\alpha \in K$ such that $\alpha\mathfrak{a} \subset \mathcal{O}_K$.) For every nonzero fractional ideal $\mathfrak{a}$ there exists a unique fractional ideal $\mathfrak{a}^{-1}$ such that $\mathfrak{a}\mathfrak{a}^{-1} = (1)$. It follows that nonzero fractional ideals form a free abelian group, with prime ideals as free generators.

One often abuses the language, calling the fractional ideals as the *ideals of K*. Ideals of $\mathcal{O}_K$ are then referred to as *integral ideals of K*, and nonzero prime ideals of $\mathcal{O}_K$ are called the *prime ideals of K* (or just the *primes of K*). We follow this tradition in the present book.

If $\mathfrak{a}$ is an ideal of a number field K, and L a finite extension of K, then $\mathfrak{a}\mathcal{O}_L$ is an ideal of L, which is usually denoted by $\mathfrak{a}$ as well. (In case of confusion, one can specify which field is in mind, but this is seldom needed.) For instance, a prime ideal $\mathfrak{p}$ of K often becomes composite in L, and we can speak about the prime decomposition of $\mathfrak{p}$ in L.

A.4 Norm of an Ideal

Let $\mathfrak{a}$ be a nonzero ideal of $\mathcal{O}_K$. The *norm* of $\mathfrak{a}$, denoted by $\mathcal{N}\mathfrak{a}$, is, by definition, the cardinality of the residue ring $\mathcal{O}_K/\mathfrak{a}$. The norm of the zero ideal is, by definition, 0.

The **norm is multiplicative**: $\mathcal{N}(\mathfrak{ab}) = \mathcal{N}\mathfrak{a}\mathcal{N}\mathfrak{b}$. By multiplicativity, the norm function extends to all fractional ideals, defining a homomorphism from the group of (nonzero) fractional ideals to $\mathbb{Q}^{\times}$. This function is compatible with the usual norm map $K \to \mathbb{Q}$: the norm of the principal ideal (α) is equal to the absolute value of the norm of α.

If K is a Galois extension of $\mathbb{Q}$ with Galois group G, then for any ideal $\mathfrak{a}$ we have $\prod_{\sigma \in G} \mathfrak{a}^{\sigma} = (\mathcal{N}\mathfrak{a})$. (We write the Galois action by G exponentially.)

The norm defined above is called sometimes the *absolute norm*. More generally, let K be a number field, L a finite extension of K, and $\mathfrak{a}$ an ideal of L. Then one can define the L/K-norm (called also *relative norm*) $\mathcal{N}_{L/K}\mathfrak{a}$, which will be an ideal of K. The L/K-norm is multiplicative, and one also has the "transitivity relation": if $K \subset L \subset M$ is a tower of number fields and $\mathfrak{a}$ is an ideal of M, then

$$\mathcal{N}_{L/K}\left(\mathcal{N}_{M/L}\mathfrak{a}\right) = \mathcal{N}_{M/K}\mathfrak{a}. \tag{A.1}$$

If L is a Galois extension of K with Galois group G and $\mathfrak{a}$ an ideal of L, then $\prod_{\sigma \in G} \mathfrak{a}^{\sigma} = \mathcal{N}_{L/K}\mathfrak{a}$.

If $\mathfrak{a}$ is an ideal of the number field K, then one can define the absolute norm $\mathcal{N}\mathfrak{a}$, which is a nonnegative rational number, and the $K/\mathbb{Q}$-norm $\mathcal{N}_{K/\mathbb{Q}}\mathfrak{a}$, which is an ideal of $\mathbb{Q}$. The two definitions agree in the sense that $\mathcal{N}_{K/\mathbb{Q}}\mathfrak{a}$ is the ideal generated by $\mathcal{N}\mathfrak{a}$.

A.5 Ideal Classes, the Class Group

The multiplicative group of fractional ideals has a subgroup consisting of principal ideals (it is isomorphic to $K^{\times}/\mathcal{O}_K^{\times}$). The corresponding quotient group is called the *group of classes of ideals* or, shorter, the *class group* of K. The class group is denoted by $\mathcal{H}_K$; its elements are called *classes of ideals*, or *ideal classes*, or simply *classes*.

It is fundamental and nontrivial that **the class group is finite**. Its cardinality is called the class number of K; it is denoted by h_K. Together with the discriminant $\mathcal{D}_K$ and regulator $\mathcal{R}_K$, the class number belongs to the most important numerical invariants of the field K.

A.6   Prime Ideals, Ramification

Any prime ideal of K contains exactly one prime number, called the *underlying prime* of this prime ideal. Let $\mathfrak{p}$ be a prime ideal and p its underlying prime. Then $\mathcal{O}_K/\mathfrak{p}$ is the finite field $\mathbb{F}_{p^f}$, where the integer f satisfies $f \leq n = [K:\mathbb{Q}]$ and is called the *residue field degree* of $\mathfrak{p}$ (over $\mathbb{Q}$).

Further, the positive integer e such that $\mathfrak{p}^e$ divides p, but $\mathfrak{p}^{e+1}$ does not, is called the *ramification index* of $\mathfrak{p}$ (over $\mathbb{Q}$). The ideal $\mathfrak{p}$ is *ramified* (over $\mathbb{Q}$) if $e > 1$ and *unramified* otherwise. The product ef is called the *local degree*, or simply the *degree* of $\mathfrak{p}$ (over $\mathbb{Q}$).

Conversely, let p be a prime number. Then only finitely many (actually, no more than n) prime ideals of K lie above p. Let $\mathfrak{p}_1, \ldots, \mathfrak{p}_s$ be these prime ideals, $e_1, \ldots, e_s$ their ramification indices, and $f_1, \ldots, f_s$ their residue field degrees, respectively. Then we have the factorization $(p) = \mathfrak{p}_1^{e_1} \cdots \mathfrak{p}_s^{e_s}$. Comparing the absolute norms of both parts, we obtain the basic identity $e_1 f_1 + \cdots + e_s f_s = n$ ("global degree is the sum of local degrees").

We say that *p splits completely in K* if p decomposes in K into a product of n distinct prime ideals; equivalently, p splits completely if all prime ideals of K above p are of local degree 1.

We say that the prime p is *ramified* in K if at least one of the ideals $\mathfrak{p}_1, \ldots, \mathfrak{p}_s$ is ramified over $\mathbb{Q}$, that is, if at least one of the numbers $e_1, \ldots, e_s$ is greater than 1. Otherwise, the prime p is *unramified* in K. The prime p is *totally ramified* in K if $s = 1, e_1 = n, f_1 = 1$, that is, if $(p) = \mathfrak{p}^n$ for some prime ideal $\mathfrak{p}$. In this case, p is also totally ramified in any subfield of K (distinct from $\mathbb{Q}$).

One proves the existence of an integral ideal $\mathfrak{d}_K$, called the *different* of K, with the following properties: $\mathcal{N}(\mathfrak{d}_K) = |\mathcal{D}_K|$, and a prime ideal $\mathfrak{p}$ is ramified if and only if $\mathfrak{p} \mid \mathfrak{d}_K$. Consequently, **a prime number p is ramified in K if and only if it divides the discriminant $\mathcal{D}_K$**. In particular, there are only finitely many ramified primes.

Minkowski's theorem asserts that $|\mathcal{D}_K| > 1$ for $K \neq \mathbb{Q}$. Equivalently, **if $K \neq \mathbb{Q}$ then at least one prime number is ramified in K**.

More generally, let K be a number field, L a finite extension of K, and $\mathfrak{P}$ a prime ideal of L. Then $\mathfrak{p} = \mathfrak{P} \cap \mathcal{O}_K$ is a prime ideal of K, called the *underlying prime ideal of $\mathfrak{P}$ in K*. One can define now the ramification index and the residue field degree of $\mathfrak{P}$ over K, the relative different $\mathfrak{d}_{L/K}$, the relative discriminant $\mathcal{D}_{L/K} = \mathcal{N}_{L/K}(\mathfrak{d}_{L/K})$, and so on. We shall use the following properties of the different:

- Differents are multiplicative in towers; that is, if $K \subseteq L \subseteq M$ is a tower of number fields, then $\mathfrak{d}_{M/K} = \mathfrak{d}_{M/L}\mathfrak{d}_{L/K}$.
- If $\alpha \in \mathcal{O}_L$ is such that $L = K(\alpha)$ and $f(x) \in \mathcal{O}_K[x]$ is the minimal polynomial of α over K then $\mathfrak{d}_{L/K}$ divides $f'(\alpha)$.

A.7 Galois Extensions

Let K be a finite Galois extension of $\mathbb{Q}$ with the Galois group $G = \mathrm{Gal}(K/\mathbb{Q})$. Let p be a prime number and $\mathfrak{p}$ a prime ideal of K above p. Then for any $\sigma \in G$ the prime ideal $\mathfrak{p}^\sigma$ again lies above p. Thus, G acts in the obvious way on the set $\{\mathfrak{p}_1, \ldots, \mathfrak{p}_s\}$ of the prime ideals of K above p. It is important that this action is transitive, which implies that the ramification indices $e_1, \ldots, e_s$ are all equal, and the same is true for the residue field degrees $f_1, \ldots, f_s$. One calls the number $e = e_1 = \cdots = e_s$ the *ramification index of p in K*; similarly, $f = f_1 = \cdots = f_s$ is called the *residue field degree of p in K*.

Again, let $\mathfrak{p}$ be a prime ideal of K and p the underlying prime number. The *decomposition group* of $\mathfrak{p}$ is the subgroup of G stabilizing $\mathfrak{p}$:

$$G_\mathfrak{p} = \{\sigma \in G : \mathfrak{p}^\sigma = \mathfrak{p}\}.$$

The *inertia group* is the subgroup of $G_\mathfrak{p}$ defined by

$$I_\mathfrak{p} = \{\sigma \in G_\mathfrak{p} : \alpha^\sigma \equiv \alpha \bmod \mathfrak{p} \text{ for all } \alpha \in \mathcal{O}_K\}.$$

We have $|G_\mathfrak{p}| = ef$ and $|I_\mathfrak{p}| = e$. The inertia group is a normal subgroup of the decomposition group, the quotient $G_\mathfrak{p}/I_\mathfrak{p}$ being canonically isomorphic to the Galois group of the residue field $\mathcal{O}_K/\mathfrak{p}$ over $\mathbb{Z}/p\mathbb{Z}$. In particular, it is a cyclic group of f elements.

If $\mathfrak{p}$ is unramified then $I_\mathfrak{p}$ is trivial and $G_\mathfrak{p}$ is itself a cyclic group of f elements. Moreover, it has a canonical generator $\varphi = \varphi_\mathfrak{p} \in G_\mathfrak{p}$, called the *Frobenius element*. The latter is defined as the unique element of $G_\mathfrak{p}$ such that $\alpha^\varphi \equiv \alpha^p \bmod \mathfrak{p}$ for all $\alpha \in \mathcal{O}_K$. Thus, to every unramified prime ideal $\mathfrak{p}$ of K, we associate its Frobenius element $\varphi_\mathfrak{p}$.

One has similar terminology and statements in the relative case, for the Galois extensions L/K. We shall be very brief. Let $\mathfrak{P}$ be a prime ideal of L and $\mathfrak{p}$ the underlying prime ideal of K. The subgroup $G_{\mathfrak{P}/\mathfrak{p}}$ of $G = \mathrm{Gal}(L/K)$ preserving $\mathfrak{P}$ is called the *decomposition group of $\mathfrak{P}$ over K*. If $\mathfrak{P}$ is unramified over K then $G_{\mathfrak{P}/\mathfrak{p}}$ is a cyclic group generated by the *Frobenius element* $\varphi = \varphi_{\mathfrak{P}/\mathfrak{p}}$, which is the unique element of $G_{\mathfrak{P}/\mathfrak{p}}$ satisfying $\alpha^\varphi \equiv \alpha^{N\mathfrak{p}} \bmod \mathfrak{P}$ for all $\alpha \in \mathcal{O}_L$.

A.8   Valuations

Let K be a field (not just a number field). A *valuation* v on K is a real-valued function $x \mapsto |x|_v$ on K with the following properties:

1. We have $|\alpha|_v \geq 0$ for all $\alpha \in K$; also, $|\alpha|_v = 0$ if and only if $\alpha = 0$.
2. For any $\alpha, \beta \in K$ we have $|\alpha\beta|_v = |\alpha|_v |\beta|_v$.
3. There exists a constant $c_v > 0$ such that for any $\alpha, \beta \in K$ we have

$$|\alpha + \beta|_v \leq c_v \max\{|\alpha|_v, |\beta|_v\} . \tag{A.2}$$

The valuation, defined by $|0|_v = 0$ and $|\alpha|_v = 1$ for any $\alpha \neq 0$, is called *trivial*. In the sequel, trivial valuations are excluded from consideration; thus, from now on, when we say *valuation*, we mean *nontrivial valuation*.

If v is a valuation and a a positive real number then the function $x \mapsto |x|_v^a$ is again a valuation (with the constant c_v^a instead of c_v). Two valuations v and v' are called *equivalent* if there exists $a > 0$ such that $|x|_{v'} = |x|_v^a$ for any $x \in K$.

A valuation v is called *non-Archimedean* if (A.2) holds with $c_v = 1$ and *Archimedean* otherwise. If v is (non-)Archimedean, then all equivalent valuations are (non-)Archimedean as well.

One usually denotes by M_K the set of all classes of equivalent valuations of the field K (these classes are called *places*). It is practical, however, to choose a representative in each class ("normalize the valuations") and to denote by M_K the set of chosen representatives.

In general, there is no "canonical" way to normalize the valuations of a number field, and, except the case $K = \mathbb{Q}$, there is even no commonly accepted normalization: different normalizations are used in different sources. Below we describe the normalizations adopted in this book.

First of all, recall the commonly used normalization for $K = \mathbb{Q}$. We have

$$M_{\mathbb{Q}} = \{\text{the prime numbers}\} \cup \{\infty\},$$

where the symbol ∞ corresponds to the usual absolute value: $|\alpha|_\infty = |\alpha|$ for $\alpha \in \mathbb{Q}$, and each prime number p corresponds to the standard p-adic valuation, defined as follows. Let α be a nonzero rational number. We define the p-adic order $\mathrm{Ord}_p \alpha$ as the unique integer m such that $\alpha = p^m a/b$ with $a, b \in \mathbb{Z}$ satisfying $\gcd(p, ab) = 1$. Next, we define $|\alpha|_p = p^{-\mathrm{Ord}_p(\alpha)}$. By definition, one puts $\mathrm{Ord}_p 0 = +\infty$ and $|0|_p = 0$. With this normalization, for any nonzero $\alpha \in \mathbb{Q}$, one has the *product formula* $\prod_{v \in M_{\mathbb{Q}}} |\alpha|_v = 1$.

Now let K be an arbitrary number field. Then

$$M_K = \{\text{the prime ideals of } K\} \cup \{\text{the real embeddings of } K\} \cup$$

$$\{\text{the pairs of complex conjugate embeddings of } K\}.$$

If $v \in M_K$ is a real embedding σ, then, for $\alpha \in K$, we define $|\alpha|_v = |\sigma(\alpha)|$. If v is a pair of complex conjugate embeddings $\sigma, \bar{\sigma}$, then we define $|\alpha|_v = |\sigma(\alpha)|^2$. Finally, if v is a prime ideal $\mathfrak{p}$ then $|\alpha|_v = (\mathcal{N}\mathfrak{p})^{-\mathrm{Ord}_\mathfrak{p}(\alpha)}$ (the $\mathfrak{p}$-*adic valuation*). Here $\mathrm{Ord}_\mathfrak{p}(\alpha)$ is the $\mathfrak{p}$-adic order of α, defined by $(\alpha) = \mathfrak{p}^{\mathrm{Ord}_\mathfrak{p}(\alpha)}\mathfrak{a}\mathfrak{b}^{-1}$, where $\mathfrak{a}, \mathfrak{b}$ are integral ideals of K coprime with $\mathfrak{p}$ (with the convention $\mathrm{Ord}_\mathfrak{p}(0) = +\infty$). With this normalization we again have the product formula

$$\prod_{v \in M_K} |\alpha|_v = 1 \qquad (\alpha \in K^\times).$$

The valuations (and places) corresponding to the embeddings of K are called *infinite*; they are Archimedean. The $\mathfrak{p}$-adic valuations are called *finite*; they are non-Archimedean.

Let K be a number field and L a finite extension of K. We say that $w \in M_L$ *lies above* $v \in M_K$ and v *lies below* w (notation: $w \mid v$) if the restriction of w to K is equivalent to v. There are only finitely many $w \in M_L$ above a given $v \in M_K$. With our normalization, for any $\alpha \in L$, we have the identity

$$\left|\mathcal{N}_{L/K}\alpha\right|_v = \prod_{w \mid v} |\alpha|_w.$$

In particular, if $\alpha \in K$ then

$$\prod_{w \mid v} |\alpha|_w = |\alpha|_v^{[L:K]}. \tag{A.3}$$

Now let $\tau : K \to K'$ be an isomorphism of number fields. Then τ induces a natural map $\tau_* : M_K \to M_{K'}$. Indeed, for a $v \in M_K$, we define a valuation $\tau_* v$ of K' by $|\beta|_{\tau_* v} = |\tau^{-1}(\beta)|_v$. A priori, $\tau_* v$ is merely equivalent to a valuation from $M_{K'}$; however, it is easy to show that it is "correctly" normalized and thereby belongs to $M_{K'}$.

Indeed, if v is $\mathfrak{p}$-adic, then $\tau_* v$ is $\mathfrak{p}'$-adic, where $\mathfrak{p}' = \tau(\mathfrak{p})$. It remains to notice that prime ideals $\mathfrak{p}$ of K and $\mathfrak{p}'$ of K' have the same norm. If v is infinite and corresponds to the embedding σ of K, then $\tau_* v$ corresponds to the embedding $\sigma \circ \tau$ of K', and if σ is real (respectively, complex), then $\sigma \circ \tau$ is real (respectively, complex) as well.

Recall in conclusion the following simple property of non-Archimedean valuations, which is widely exploited in the present book. It will be more convenient to use the notion of $\mathfrak{p}$-adic order rather than valuation.

Lemma A.1. *Let K be a number field, $\alpha_0, \ldots, \alpha_m \in K$, and $\mathfrak{p}$ a prime ideal of K. Assume that $\alpha_0 \neq 0$ and that $\mathrm{Ord}_\mathfrak{p}(\alpha_0) < \mathrm{Ord}_\mathfrak{p}(\alpha_k)$ for $k = 1, \ldots, m$. Then*

$$\mathrm{Ord}_\mathfrak{p}(\alpha_0 + \cdots + \alpha_m) = \mathrm{Ord}_\mathfrak{p}(\alpha_0).$$

In particular, $\alpha_0 + \cdots + \alpha_m \neq 0$.

A.9   Dedekind ζ-Function

The *Riemann ζ-function* is defined, for $s > 1$, by

$$\zeta(s) = \sum_{n=1}^{\infty} n^{-s}. \tag{A.4}$$

The arithmetical significance of this function comes from Euler's observation that it decomposes into the following infinite product:

$$\zeta(s) = \prod_p (1 - p^{-s})^{-1}, \tag{A.5}$$

where p runs over all prime numbers (the *Euler product formula*).

Since the series $\sum_{n=1}^{\infty} n^{-1}$ diverges to infinity, we have $\zeta(s) \to +\infty$ as s tends to 1 from the right. Using the obvious inequality

$$(n+1)^{-s} \le \int_n^{n+1} x^{-s} dx \le n^{-s},$$

we obtain

$$\int_1^{+\infty} x^{-s} dx \le \zeta(s) \le 1 + \int_1^{+\infty} x^{-s} dx.$$

Since $\int_1^{+\infty} x^{-s} dx = (s-1)^{-1}$, we find the exact asymptotic behavior of $\zeta(s)$ as s approaches 1 from the right:

$$\zeta(s) = (s-1)^{-1} + O(1).$$

In particular,

$$\lim_{s \downarrow 1}(s-1)\zeta(s) = 1. \tag{A.6}$$

The Dedekind ζ-function is a proper generalization of the Riemann ζ-function for the needs of algebraic number theory. Given a number field K, we define the Dedekind ζ-function $\zeta_K(s)$ by

$$\zeta_K(s) = \sum_{\mathfrak{a}} \mathcal{N}(\mathfrak{a})^{-s},$$

the sum being over all nonzero integral ideals of K. Obviously, $\zeta_{\mathbb{Q}}(s) = \zeta(s)$. Again, the series converges for $s > 1$, and we have the *Euler product formula*

$$\zeta_K(s) = \prod_{\mathfrak{p}} (1 - \mathcal{N}(\mathfrak{p})^{-s})^{-1}, \tag{A.7}$$

the product being over the prime ideals of K. One also has an analogue for the "residue formula" (A.6):

$$\lim_{s \downarrow 1}(s-1)\zeta_K(s) = \frac{2^{t_1}(2\pi)^{t_2}\mathcal{R}_K h_K}{\omega\sqrt{|\mathcal{D}_K|}}, \tag{A.8}$$

where t_1 and $2t_2$ are the numbers of real and of complex embeddings and ω is the number of roots of unity in K. The proof again relies on approximating certain sums by integrals, but it is much more involved.

The series (A.4) converges not only for real $s > 1$ but also for complex s with $\operatorname{Re} s > 1$, and the sum is analytic on the half-plane $\operatorname{Re} s > 1$. Riemann showed that it satisfies a functional equation, which allows one to continue $\zeta(s)$ to a meromorphic function on $\mathbb{C}$ with a single pole at $s = 1$. The same is true for the Dedekind ζ-function $\zeta_K(s)$. The complex analytic theory of the ζ-function is deep and important, but we do not use it in this book.

A.10   Chebotarev Density Theorem

One uses the ζ-function to establish various "density theorems" about prime numbers and prime ideals. The most important of them is the Chebotarev density theorem.

First of all, we define the notion of *Dirichlet density*. Combining the Euler product formula (A.5) and the residue formula (A.6), we obtain

$$\sum_p -\log(1 - p^{-s}) = \log\frac{1}{s-1} + O(1) \tag{A.9}$$

for $s > 1$. Using that $-\log(1 - z) = z + O(|z|^2)$ for $|z| \le 1/2$, we may write the left-hand side of (A.9) as

$$\sum_p p^{-s} + O\left(\sum_p p^{-2s}\right).$$

The $O(\cdot)$-term is obviously bounded for $s > 1$, and we obtain

$$\sum_p p^{-s} = \log\frac{1}{s-1} + O(1). \tag{A.10}$$

Now let A be a set of prime numbers. We say that the set A is *regular* if there exists a real number a such that

$$\sum_{p \in A} p^{-s} = a \log \frac{1}{s-1} + O(1)$$

for $s > 1$. The number a is called the *Dirichlet density* of the set A.

Now we are ready to state the theorem of Chebotarev. Let K be a finite Galois extension of $\mathbb{Q}$ and let G be its Galois group. Recall that to every prime ideal $\mathfrak{p}$ of K, unramified over $\mathbb{Q}$, we associated its Frobenius element $\varphi_{\mathfrak{p}}$ (see Sect. A.7). Now, in the opposite direction, to every prime number p, unramified in K, we associate the subset of G, consisting of the Frobenius elements of all prime ideals above p. This subset is a full conjugacy class[1] of G; it is called the *Artin symbol* of p (with respect to K) and is denoted by $\left[\frac{p}{K}\right]$.

Chebotarev proved that conversely, every element from G serves as the Frobenius element for infinitely many prime ideals of K, and, consequently, each conjugacy class serves as the Artin symbol for infinitely many prime numbers. In particular, the *Artin map*

$$\{\text{prime numbers unramified in } K\} \to \{\text{conjugacy classes of } G\}$$

$$p \mapsto \left[\frac{p}{K}\right]$$

is surjective. In fact, Chebotarev proved much more than this: given a conjugacy class S of G, **the set of prime numbers with Artin symbol S is regular and has the "correct" Dirichlet density** $\frac{|S|}{|G|}$.

In the case when K is a cyclotomic field, Chebotarev's theorem becomes the classical theorem of Dirichlet about primes in arithmetical progressions (see Sect. 5.2.2).

A similar statement holds in the relative case. Let K be a number field. Euler's product (A.7) together with the residue formula (A.8) implies that

$$\sum_{\mathfrak{p}} \mathcal{N}(\mathfrak{p})^{-s} = \log \frac{1}{s-1} + O(1),$$

where the sum extends to all the prime ideals of K. We say that a set A of prime ideals is *regular* of *Dirichlet density* a if

$$\sum_{\mathfrak{p} \in A} \mathcal{N}(\mathfrak{p})^{-s} = a \log \frac{1}{s-1} + O(1)$$

for $s > 1$.

---

[1] Indeed, if $\mathfrak{p}$ is an ideal above p, then the full set of ideals above p is $\{\mathfrak{p}^{\sigma} : \sigma \in G\}$, and a straightforward verification shows that $\varphi_{\mathfrak{p}^{\sigma}} = \sigma^{-1} \varphi_{\mathfrak{p}} \sigma$.

One simple remark would be relevant here: the set of prime ideals of degree 1 (over $\mathbb{Q}$) is regular and of Dirichlet density 1. Indeed, since there is at most $n = [K : \mathbb{Q}]$ prime ideals over every rational prime, we have, for $s > 1$,

$$\sum_{f_{\mathfrak{p}} \geq 2} \mathcal{N}(\mathfrak{p})^{-s} = \sum_{p} \sum_{\substack{\mathfrak{p} \mid p \\ f_{\mathfrak{p}} \geq 2}} p^{-f_{\mathfrak{p}} s} \leq n \sum_{p} p^{-2s} = O(1).$$

Hence, the prime ideals of degree at least 2 form a regular set of Dirichlet density 0, and those of degree 1 form a regular set of density 1.

Now let L/K be a finite Galois extension with Galois group G, and let $\mathfrak{p}$ be a prime ideal of K unramified in L. Then the Artin symbol of $\mathfrak{p}$ (with respect to the extension L/K) is

$$\left[\frac{\mathfrak{p}}{L/K} \right] = \{\varphi_{\mathfrak{P}/\mathfrak{p}} : \mathfrak{P} \mid \mathfrak{p}\};$$

it is a full conjugacy class of G. The Chebotarev density theorem asserts that the set of prime ideals whose Artin symbol is a given conjugacy class S is regular and of Dirichlet density $\frac{|S|}{|G|}$. In particular, the Artin map

$$\{\text{prime ideals of } K \text{ unramified in } L\} \rightarrow \{\text{conjugacy classes of } G\}$$

$$\mathfrak{p} \mapsto \left[\frac{\mathfrak{p}}{L/K} \right] \tag{A.11}$$

is surjective.

Since the prime ideals of degree 1 have Dirichlet density 1, **every conjugacy class of G serves as the Artin symbol for infinitely many prime ideals of K of degree** 1.

A.11   Hilbert Class Field

According to the theorem of Minkowski, mentioned in Sect. A.6, for every nontrivial extension of $\mathbb{Q}$, there exists a prime number ramified in this extension. This theorem, however, does not extend to arbitrary number fields: a number field K may have, in general, nontrivial extensions where no prime ideal of K ramifies. For example, so is the extension $\mathbb{Q}(\sqrt{-5}, \sqrt{-1})$ of the field $\mathbb{Q}(\sqrt{-5})$. Such extensions are called *unramified at finite places*, because prime ideals correspond to the *finite places* of the field K (see Sect. A.8). We want to extend this to infinite places as well.

There are two types of infinite places: those corresponding to real embeddings of K (*real* infinite places) and those corresponding to pairs of complex conjugate embeddings (*complex* infinite places).

Let L/K be a finite extension of number fields. We say that an infinite place w of L is *ramified over* K if w is a complex place, but the underlying place v of K is real. If both v and w are real, or both are complex, we say that w is unramified over K. An infinite place v of K is *unramified in* L if all the places above v are unramified over K. This happens if either v is complex or v is real and so are all the places above.

We say that L is *unramified over* K *at infinity* if every infinite place of K is unramified in L. Equivalently, L is unramified over K at infinity if every real embedding of K extends only to real embeddings of L. In particular, if K is totally real then only totally real extensions K are unramified at infinity, and if K is totally imaginary, then any extension of K is unramified at infinity.

Finally, we say that L is *unramified over* K if it is unramified at all places, both finite and infinite. For example, the already mentioned field $\mathbb{Q}(\sqrt{-5}, \sqrt{-1})$ is an unramified extension of $\mathbb{Q}(\sqrt{-5})$.

Now let L be a finite *abelian* extension of a number field K. This means that L/K is a Galois extension and the Galois group $G = \mathrm{Gal}(K/\mathbb{Q})$ is abelian.

Let $\mathfrak{p}$ be a prime ideal of K unramified in L. In Sect. A.10 we associated to it the Artin symbol $\left[\frac{\mathfrak{p}}{L/K}\right]$, which is a conjugate class of G. Since G is abelian, this class consists of a single element. Thus, for the abelian extensions, the Artin map has values in G itself rather than in the set of conjugacy classes.

If, in addition, we assume that the abelian extension L/K is unramified at all finite places, then we have the map

$$\{\text{prime ideals of } K\} \to G.$$

Extending it by linearity, we obtain a group homomorphism $I \to G$, where I is the group of all ideals of the field K. It is called the *Artin homomorphism* of the abelian extension L/K. Chebotarev density theorem implies that the Artin homomorphism is surjective. Describing its kernel is the central problem of the *Class Field Theory*. The principal result (usually referred to as *Artin's reciprocity law*) states: if the abelian extension L/K is unramified (everywhere, at all finite and infinite places), then the kernel of the Artin map contains the group of principal ideals.[2] This defines a surjective homomorphism $\mathcal{H}_K \to G$ (also called the *Artin map* or Artin homomorphism), where $\mathcal{H}_K$ is the ideal class group of K.

Conversely, let K be a number field, and fix a surjective homomorphism $\mathcal{H}_K \to G$ of the class group of K onto a finite abelian group G. Then there exists a unique (in the given algebraic closure) unramified abelian extension L/K with Galois group G and such that the fixed homomorphism is the Artin homomorphism

---

[2]It is not sufficient to assume that only the finite places of K are unramified in L. Under this weaker assumption, one can merely show that the kernel contains the principal ideals, generated by the elements $\alpha \in K$ with the property $\sigma(\alpha) > 0$ for any real embedding σ.

of L/K. In particular, there exists a **unique unramified abelian extension of K with Galois group canonically isomorphic to** $\mathcal{H}_K$, the isomorphism being given by the Artin map. It is called the *Hilbert class field* of K.

Certain analogues of the results described above hold for ramified abelian extensions as well. A very good account (without proofs) of the Class Field Theory can be found in an appendix to Washington's book [136]. Among the complete (with all proofs) expositions of the Class Field Theory, the most systematic one is, probably, due to Neukirch [104, 105]. The recent book of Childress [24] is a good introductory text for an unsophisticated reader. One may also consult the famous books of Artin and Tate [2], Cassels and Frölich [19], Weil [137] , and Lang [61].

Appendix B
Heights

In this appendix we introduce the *height function* on the field of algebraic numbers $\bar{\mathbb{Q}}$. We start with an informal discussion, which should motivate the definition.

Intuitively, the height should measure the "size" of an algebraic number. We expect our height function to have the following properties:

- The height of an algebraic number is a nonnegative real number.
- The height should "behave well" with respect to algebraic operations. (That is, one should have easy and efficient upper estimates for the heights of $\alpha + \beta$ and $\alpha\beta$ in terms of the heights (and, perhaps, degrees) of α and β.)
- For a given $C > 0$, there should exist only finitely many algebraic numbers of degree and height bounded by C.

Of course, one can imagine many functions with these properties. Our purpose is to define the one which is the most convenient to use.

On integers, the usual absolute value is an adequate measure of size. Thus, we define the height of a nonzero $\alpha \in \mathbb{Z}$ by $\mathrm{H}(\alpha) = |\alpha|$ and $\mathrm{H}(0) = 1$, the latter exception being made just for compatibility with the more general definition below.

On rational numbers, the absolute value is no longer adequate: there exist infinitely many rational numbers of bounded absolute value. To obtain finiteness, one should bound both the numerator and the denominator. Thus, for $\alpha = a/b \in \mathbb{Q}$, where $a, b \in \mathbb{Z}$ and $\gcd(a, b) = 1$, we define $\mathrm{H}(\alpha) = \max\{|a|, |b|\}$.

Next, we wish to extend this definition to all algebraic numbers. One idea is to observe that $bX - a$ is the minimal polynomial of the rational number a/b over $\mathbb{Q}$. Hence, for $\alpha \in \bar{\mathbb{Q}}$, we may put $\mathrm{H}(\alpha) = \max\{|a_0|, \ldots, |a_n|\}$, where $a_n X^n + \cdots + a_0$ is the minimal polynomial of α over $\mathbb{Z}$. Indeed, one extensively used this definition of height in the past. However, it is rather inconvenient from many points of view, and it was eventually abandoned in favor of a different definition, due to A. Weil.

The definition of Weil's height is motivated by the following observation: the height of a rational number $\alpha = a/b$, originally defined as $\max\{|a|, |b|\}$, satisfies the identity

$$\mathrm{H}(\alpha) = \prod_{v \in M_{\mathbb{Q}}} \max\{1, |\alpha|_v\}. \tag{B.1}$$

(The proof is an easy exercise.) Weil's height is the logarithm[1] of the right-hand side of this identity, properly generalized to number fields.

Now we are ready to give a formal definition of the height. Let K be a number field. Define the *logarithmic K-height* (or, simply, the K-height) $\mathrm{h}_K : K \to \mathbb{R}$ by

$$\mathrm{h}_K(\alpha) = \sum_{v \in M_K} \log \max\{1, |\alpha|_v\}.$$

In the sequel, we use the notation[2]

$$\|\alpha\|_v = \max\{1, |\alpha|_v\},$$

so that

$$\mathrm{h}_K(\alpha) = \sum_{v \in M_K} \log \|\alpha\|_v.$$

Proposition B.1. *If K and L are two number fields, then for any $\alpha \in K \cap L$ we have*

$$\frac{\mathrm{h}_K(\alpha)}{[K:\mathbb{Q}]} = \frac{\mathrm{h}_L(\alpha)}{[L:\mathbb{Q}]}. \tag{B.2}$$

Proof. Assume first that $K \subset L$. As follows from (A.3), for any $\alpha \in K$ and $v \in M_K$, we have

$$[L:K] \log \|\alpha\|_v = \sum_{w|v} \log \|\alpha\|_w,$$

where the sum extends to all $w \in M_L$ lying above this v. Hence for any $\alpha \in K$ we have

$$\frac{\mathrm{h}_L(\alpha)}{[L:\mathbb{Q}]} = \frac{1}{[L:\mathbb{Q}]} \sum_{w \in M_L} \log \|\alpha\|_w = \frac{1}{[L:\mathbb{Q}]} \sum_{v \in M_K} \sum_{w|v} \log \|\alpha\|_w$$

$$= \frac{[L:K]}{[L:\mathbb{Q}]} \sum_{v \in M_K} \log \|\alpha\|_v = \frac{\mathrm{h}_K(\alpha)}{[K:\mathbb{Q}]},$$

---

[1]One prefers the logarithm due to many reasons; for instance, because it is easier to add than to multiply.

[2]This notation is used only in this appendix. In the rest of the book $\| \cdot \|$ has a different meaning.

which proves (B.2) in the case $K \subset L$. In the general case, let M be a number field containing both K and L. Then for any $\alpha \in K \cap L$

$$\frac{h_K(\alpha)}{[K:\mathbb{Q}]} = \frac{h_M(\alpha)}{[M:\mathbb{Q}]} = \frac{h_L(\alpha)}{[L:\mathbb{Q}]},$$

which proves (B.2) in the general case. $\square$

Now let α be an algebraic number. We define the *absolute logarithmic height* (or, simply, *height*) of α by

$$h(\alpha) = \frac{h_K(\alpha)}{[K:\mathbb{Q}]}, \tag{B.3}$$

where K is any number field containing α. By Proposition B.1, the right-hand side of (B.3) depends only on α and is independent of the particular choice of the field K.

We have defined a function $h : \bar{\mathbb{Q}} \to \mathbb{R}$. Its properties are summarized in the following proposition.

Proposition B.2.   *1. For any $\alpha \in \bar{\mathbb{Q}}$ we have $h(\alpha) \geq 0$.*
2. *(**Kronecker's first theorem**) We have $h(\alpha) = 0$ if and only if $\alpha = 0$ or α is a root of unity.*
3. *(**Kronecker's second theorem**) For every positive integer d, there exists $\varepsilon(d) > 0$ with the following property. Let α be an algebraic number of degree not exceeding d. Then either $h(\alpha) = 0$ (in which case α is zero or a root of unity) or $h(\alpha) \geq \varepsilon(d)$.*
4. *(**Galois action**) If α and β are conjugate over $\mathbb{Q}$ then $h(\alpha) = h(\beta)$.*
5. *(**Height of a rational number**) Assume that $\alpha = a/b \in \mathbb{Q}$, where $a, b \in \mathbb{Z}$ and $\gcd(a,b) = 1$. Then*

$$h(\alpha) = \log\max\{|a|, |b|\}.$$

In particular, if α is a nonzero rational integer then $h(\alpha) = \log|\alpha|$.
6. *(**Height of a quotient**) More generally, let K be a number field, $\alpha, \beta \in K$, and $\beta \neq 0$. Then*

$$h(\alpha/\beta) = \frac{1}{[K:\mathbb{Q}]} \sum_{v \in M_K} \log\max\{|\alpha|_v, |\beta|_v\}. \tag{B.4}$$

Also, if α and β are both algebraic integers, then

$$h(\alpha/\beta) \leq \frac{1}{[K:\mathbb{Q}]} \sum_{\sigma:K \to \mathbb{C}} \log\max\{|\sigma(\alpha)|, |\sigma(\beta)|\}, \tag{B.5}$$

the sum being extended to all embeddings $\sigma: K \to \mathbb{C}$.

7. **(Heights of sums and products)** *For any* $\alpha, \beta \in \bar{\mathbb{Q}}$ *we have*

$$h(\alpha\beta) \leq h(\alpha) + h(\beta), \quad h(\alpha + \beta) \leq h(\alpha) + h(\beta) + \log 2. \tag{B.6}$$

More generally, for $\alpha_1, \ldots, \alpha_n \in \bar{\mathbb{Q}}$, *we have*

$$h(\alpha_1 \cdots \alpha_n) \leq h(\alpha) + \cdots + h(\alpha_n), \tag{B.7}$$

$$h(\alpha_1 + \cdots + \alpha_n) \leq h(\alpha_1) + \cdots + h(\alpha_n) + \log n. \tag{B.8}$$

8. **(Height of a power)** *For any* $\alpha \in \bar{\mathbb{Q}}$ *and* $n \in \mathbb{Z}$ *(with* $\alpha \neq 0$ *if* $n < 0$*) we have*

$$h(\alpha^n) = |n| h(\alpha). \tag{B.9}$$

9. **(Height of a linear fraction)** *Let* $\begin{bmatrix} a & b \\ c & d \end{bmatrix}$ *be a nondegenerate matrix with algebraic entries. Then for any algebraic number* $x \neq -d/c$

$$h\left(\frac{ax + b}{cx + d}\right) = h(x) + O(1),$$

where the constant implied by $O(1)$ *may depend on* a, b, c, d *(but not on* x*).*

10. **(Northcott's finiteness theorem)** *For any* $C > 0$ *there exist only finitely many algebraic numbers* α *of degree and height bounded by* C.

Proof. Part 1 is obvious. To prove Part 4, put $K = \mathbb{Q}(\alpha)$, $L = \mathbb{Q}(\beta)$ and let $\tau : K \to L$ be the isomorphism defined by $\tau(\alpha) = \beta$. As we have seen in Sect. A.8, the isomorphism τ induces a bijection $\tau_* : M_K \to M_L$, and, by the definition of τ_*, we have $|\alpha|_v = |\beta|_{\tau_* v}$ for any $v \in M_K$. It follows that

$$h_L(\beta) = \sum_{v \in M_L} \log \|\beta\|_v = \sum_{v \in M_K} \log \|\beta\|_{\tau_* v} = \sum_{v \in M_K} \log \|\alpha\|_v = h_K(\alpha).$$

Since $[K:\mathbb{Q}] = [L:\mathbb{Q}]$, this implies $h(\alpha) = h(\beta)$.

For Part 6, observe that $\|\alpha/\beta\|_v = |\beta|_v^{-1} \max\{|\alpha|_v, |\beta|_v\}$. Hence

$$h_K(\alpha/\beta) = \sum_{v \in M_K} \log \max\{|\alpha|_v, |\beta|_v\} - \log \prod_{v \in M_K} |\beta|_v,$$

and the second term vanishes by the product formula. This proves (B.4).

If α and β are algebraic integers, then $\log \max\{|\alpha|_v, |\beta|_v\} \leq 0$ for any finite v. Hence, (B.4) implies that

$$h(\alpha/\beta) \leq \frac{1}{[K:\mathbb{Q}]} \sum_{\substack{v \in M_K \\ v|\infty}} \log \max\{|\alpha|_v, |\beta|_v\}, \tag{B.10}$$

where the sum extends to infinite valuations of K. It remains to notice that the right-hand side of (B.10) is equal to the right-hand side of (B.5).

Part 5 [which is a reformulation of identity (B.1)] is a particular case of (B.4).

To prove Part 7, fix a number field K containing $\alpha_1, \ldots, \alpha_n$ and observe that for any $v \in M_K$

$$\|\alpha_1 \cdots \alpha_n\|_v \leq \|\alpha_1\|_v \cdots \|\alpha_n\|_v \, ,$$

$$\|\alpha_1 + \cdots + \alpha_n\|_v \leq \|n\|_v \max\{\|\alpha_1\|_v, \ldots, \|\alpha_n\|_v\} \leq \|n\|_v \|\alpha_1\|_v \cdots \|\alpha_n\|_v \, .$$

Taking the logarithm and summing up over $v \in M_K$, we obtain

$$h_K(\alpha_1 \cdots \alpha_n) \leq h_K(\alpha) + \cdots + h_K(\alpha_n),$$

$$h_K(\alpha_1 + \cdots + \alpha_n) \leq h_K(\alpha_1) + \cdots + h_K(\alpha_n) + h_K(n).$$

It remains to divide by $[K:\mathbb{Q}]$ and to observe that $h(n) = \log n$.

Part 8 for $n > 0$ is obvious, because $\|\alpha^n\|_v = \|\alpha\|_v^n$. To extend (B.9) to negative n, it suffices to observe that $h(1/\alpha) = h(\alpha)$ by (B.4).

In Part 9 put

$$y = \frac{ax + b}{cx + d}.$$

If $c = 0$ then the statement is immediate from Part 7. If $c \neq 0$ then

$$y = \frac{a}{c} + \frac{bc - ad}{cx + d} \cdot \frac{1}{c},$$

and, using Parts 7 and 8, we find $h(y) \leq h(x) + O(1)$. Writing

$$x = \frac{-dy + b}{cy - a},$$

we prove similarly that $h(x) \leq h(y) + O(1)$, and Part 9 follows.

To prove Part 10, fix an algebraic number α of degree and height bounded by C. Let $x^n + a_{n-1}x^{n-1} + \cdots + a_0$ be the minimal polynomial of α over $\mathbb{Q}$. Using Parts 4 and 7, we may estimate the heights of the coefficients $a_0, \ldots, a_{n-1}$ in terms of C. By Part 5, the numerators and denominators of the rational numbers $a_0, \ldots, a_{n-1}$ are bounded in terms of C. Hence we have only finitely many possibilities for $a_0, \ldots, a_{n-1}$. This proves Northcott's theorem.

We are left with Parts 2 and 3 (Kronecker's theorems). Obviously, the height of 0 is 0, as well as the height of any root of unity. Conversely, let α be an algebraic number of height 0. Then its powers $1, \alpha, \alpha^2, \ldots$ are of height 0 as well. Northcott's theorem implies that among the numbers $1, \alpha, \alpha^2, \ldots$, only finitely many are distinct. Hence there exist distinct integers ℓ and m such that $\alpha^\ell = \alpha^m$. This implies that $\alpha = 0$ or α is a root of unity, proving Kronecker's first theorem.

Kronecker's second theorem is proved similarly. Let $N = N(d)$ be the number of algebraic numbers of degree bounded by d and height bounded by 1. If $h(\alpha) \neq 0$ then all the powers $1, \alpha, \alpha^2, \ldots$ are pairwise distinct, which implies that $h(\alpha^N) > 1$. Hence Kronecker's second theorem holds with $\varepsilon(d) = N(d)^{-1}$. $\square$

The following statement is almost trivial, but it plays an important role in Diophantine analysis and, in particular, in the solution of Catalan's problem.

Proposition B.3 (Liouville's inequality). *Let K be a number field and $\alpha \in K^\times$. Then for any $v \in M_K$ we have*

$$|\alpha|_v \geq e^{-[K:\mathbb{Q}]h(\alpha)}. \tag{B.11}$$

More generally, for any $S \subseteq M_K$, we have

$$\prod_{v \in S} |\alpha|_v \geq e^{-[K:\mathbb{Q}]h(\alpha)}. \tag{B.12}$$

In particular, assume that K is a subfield of $\mathbb{C}$. Then

$$|\alpha|^d \geq e^{-[K:\mathbb{Q}]h(\alpha)}, \tag{B.13}$$

where $d = 1$ if $K \subset \mathbb{R}$ and $d = 2$ otherwise.

Proof. We have

$$\left(\prod_{v \in S} |\alpha|_v \right)^{-1} = \prod_{v \in S} |\alpha^{-1}|_v \leq \prod_{v \in S} \|\alpha^{-1}\|_v \leq \prod_{v \in M_K} \|\alpha^{-1}\|_v = e^{[K:\mathbb{Q}]h(\alpha^{-1})}.$$

Since $h(\alpha^{-1}) = h(\alpha)$, this proves (B.12) and, a fortiori, (B.11). For (B.13), let $v \in M_K$ be the infinite valuation corresponding to the identical embedding $K \hookrightarrow \mathbb{C}$. Then $|\alpha|_v = |\alpha|$ if the embedding is real (that is, $K \subset \mathbb{R}$) and $|\alpha|_v = |\alpha|^2$ otherwise. Applying (B.11) with this v, we obtain (B.13). $\square$

Remark B.4. Liouville's inequality goes back to the celebrated work of Liouville [74] (published the same year as Catalan's *note extraite*). Liouville proved that algebraic numbers cannot be well approximated by rationals. Precisely, if α is a real algebraic number of degree $n > 1$ and p/q is a rational number, then

$$|\alpha - p/q| \geq c|q|^{-n}, \tag{B.14}$$

where c is a positive constant depending on α. One may deduce (B.14) from Proposition B.3 by applying (B.13) with $\alpha - p/q$ instead of α and estimating $h(\alpha - p/q)$ using Proposition B.2.

Appendix C
Commutative Rings, Modules, and Semi-simplicity

In this appendix, we recall several basic facts about rings and modules over them. In particular, we give a very brief treatise of semi-simplicity. The notion of semi-simplicity is central in the theory of noncommutative rings. In this outline we assume that the rings in question are commutative. This is irrelevant in some cases: for instance, the contents of Sect. C.3 extend without changes to the (left) modules over noncommutative base rings. On the other hand, the structure theory of semi-simple commutative rings (see Theorem C.11) is drastically simpler than its noncommutative counterpart.

All rings in this appendix are commutative and with 1. We assume that the reader is familiar with the definitions and basic facts about modules over a (commutative) ring, which can be found in any reasonable textbook of algebra.

One piece of notation: if S is a subset of an R-module, then the *annihilator* of S is, by definition, the set of all $a \in R$ such that $aS = 0$. It is an ideal of R, which will be denoted by $\mathrm{ann}_R(S)$ or simply by $\mathrm{ann}(S)$. An R-module M is called *exact* if $\mathrm{ann}(M) = 0$. Every R-module M admits a natural structure of an exact $R/\mathrm{ann}(M)$-module and, more generally, of $R/\mathfrak{a}$-module, where $\mathfrak{a}$ is any ideal annihilating M. If M' is another R-module annihilated by $\mathfrak{a}$ then M is R-isomorphic to M' if and only if M is $R/\mathfrak{a}$-isomorphic to M'.

Warning. In this appendix the abelian group law on modules is written additively, and the ring action is written multiplicatively. This is different from the rest of the book, where we normally use the multiplicative notation for the abelian group law and the exponential notation for the ring action.

C.1 Cyclic Modules

Let R be a (commutative) ring. An R-module M is called *cyclic* if it is generated over R by a single element. That is, there exists $g \in M$ such that $M = Rg$.

If $M = Rg$ is a cyclic module, then $\text{ann}(M) = \text{ann}(g)$, and M is isomorphic, as an R-module, to the quotient ring $R/\text{ann}(M)$. In particular, the annihilator of a cyclic module defines it up to an isomorphism.

Since the submodules of $R/\text{ann}(M)$ are the ideals of the ring $R/\text{ann}(M)$, we have a one-to-one correspondence between the submodules of M and the ideals of $R/\text{ann}(M)$, or the ideals of R containing $\text{ann}(M)$. Explicitly, the latter correspondence can be described as follows: if $\mathfrak{a}$ is an ideal of R containing $\text{ann}(M)$ then the corresponding submodule of M is $\mathfrak{a}g$; conversely, if N is submodule of M then the corresponding ideal $\mathfrak{a}$ consists of all $a \in R$ such that $ag \in N$.

If N is a submodule of a cyclic module $M = Rg$ then the quotient M/N is a cyclic module generated by the image of g. Also, if R is a principal ideal ring then so is $R/\text{ann}(M)$. We obtain the following statement.

Proposition C.1. *A quotient of a cyclic R-module is cyclic as well. If R is a principal ideal ring, then every submodule of a cyclic R-module is cyclic.*

Maximal ideals of $R/\text{ann}(M)$ correspond to maximal (proper) submodules of M. Since every ring has a maximal ideal, we obtain the following assertion, which will be used in Sect. C.3.

Proposition C.2. *Every cyclic module has a proper maximal submodule.*

This statement does not extend to arbitrary modules; for instance, the additive group $\mathbb{Q}$, viewed as a $\mathbb{Z}$-module, does not have a proper maximal submodule.

C.2   Finitely Generated Modules

Let R be a commutative ring. If M is a cyclic R-module and $\mathfrak{a}$ an ideal of R, then the following property is immediate: $a \in R$ satisfies $aM \subset \mathfrak{a}M$ if and only if $a \in \mathfrak{a} + \text{ann}(M)$. We want to extend this to finitely generated modules.

Theorem C.3. *Let M be a finitely generated R-module and let $\mathfrak{a}$ be an ideal of R. Further, let $a \in R$ satisfy $aM \subseteq \mathfrak{a}M$. Then $a^m \in \mathfrak{a} + \text{ann}(M)$ for some positive integer m.*

(It will follow from the proof that m does not exceed the minimal number of generators of M. In particular, for the cyclic modules, we recover the statement above.)

For the proof, we need a lemma.

Lemma C.4. *Let M be an R-module generated by its elements $g_1, \ldots, g_m$ and let $A = [\alpha_{ij}]_{1 \le i, j \le m}$ be an $m \times m$-matrix over R. Assume that*

$$\sum_{j=1}^{m} \alpha_{ij} g_j = 0 \qquad (i = 1, \ldots, m). \tag{C.1}$$

Then $\det A \in \text{ann}(M)$.

Proof. Rewrite (C.1) as $AG = 0$, where G is the column $[g_1, \ldots, g_m]$, and multiply this from the left by the adjoint matrix[1] of A. We obtain the equality $(\det A)G = 0$, that is, $(\det A)g_1 = \cdots = (\det A)g_m = 0$. Hence $\det A$ annihilates M.   □

Proof of Theorem C.3. Let $g_1, \ldots, g_m$ be a system of R-generators of M. Since every element of M is an R-linear combination of $g_1, \ldots, g_m$, every element of $\mathfrak{a}M$ is an $\mathfrak{a}$-linear combination of $g_1, \ldots, g_m$. In particular, if $\mathfrak{a}M \subseteq \mathfrak{a}M$ then there exists a matrix $A = [\alpha_{ij}]_{1 \le i, j \le m}$ with entries in $\mathfrak{a}$ such that

$$ag_i = \sum_{j=1}^{m} \alpha_{ij} g_j \qquad (i = 1, \ldots, m).$$

Applying Lemma C.4, we find that $\det(aI - A) \in \operatorname{ann}(M)$. On the other hand, since the entries of A belong to $\mathfrak{a}$, we have $\det(aI - A) \equiv a^m \bmod \mathfrak{a}$. This implies that $a^m \in \mathfrak{a} + \operatorname{ann}(M)$.   □

If M is an R-module, and $\mathfrak{a}$ is an ideal of R, then $\bar{M} = M/\mathfrak{a}M$ has the natural structure of a module over the quotient ring $\bar{R} = R/\mathfrak{a}$. Obviously, if $a \in R$ annihilates M, then its image in $\bar{R}$ annihilates $\bar{M}$. It is natural to ask whether the converse is true, that is, whether every $\bar{a} \in \bar{R}$, annihilating $\bar{M}$, is an image of some $a \in R$ annihilating M. In other words, we ask whether $\operatorname{ann}_{\bar{R}}(\bar{M})$ is the $\bar{R}$-image of $\operatorname{ann}_R(M)$. Of course, this is not true in general, but Theorem C.3 provides a sufficient condition for this.

Recall that $\mathfrak{a}$ is a *radical ideal* of R if for any $a \in R$ and for any positive integer m we have $a^m \in \mathfrak{a} \Rightarrow a \in \mathfrak{a}$. Equivalently, $\mathfrak{a}$ is a radical ideal if the quotient ring $R/\mathfrak{a}$ has no nilpotent elements.

Corollary C.5. *Let M be a finitely generated R-module, and let $\mathfrak{a}$ be an ideal of R. Put, as above, $\bar{R} = R/\mathfrak{a}$ and $\bar{M} = M/\mathfrak{a}M$. Assume that $\mathfrak{a} + \operatorname{ann}_R(M)$ is a radical ideal of R. Then $\operatorname{ann}_{\bar{R}}(\bar{M})$ is the $\bar{R}$-image of $\operatorname{ann}_R(M)$.*

Proof. As we have seen above, the image of $\operatorname{ann}_R(M)$ is contained in $\operatorname{ann}_{\bar{R}}(\bar{M})$. Conversely, let $\bar{a} \in \bar{R}$ annihilate $\bar{M}$, and let $a \in R$ be a pullback of $\bar{a}$. Then $aM \subset \mathfrak{a}M$, and Theorem C.3 implies that $a^m \in \mathfrak{a} + \operatorname{ann}_R(M)$ for some positive integer m. Since $\mathfrak{a} + \operatorname{ann}_R(M)$ is a radical ideal, we obtain $a \in \mathfrak{a} + \operatorname{ann}_R(M)$. Write $a = \alpha + \beta$ with $\alpha \in \mathfrak{a}$ and $\beta \in \operatorname{ann}_R(M)$. Then the image of β in $\bar{R}$ is $\bar{a}$.   □

---

[1]Recall that the adjoint matrix is $A' = [A_{ij}]_{1 \le i, j \le m}$, where A_{ij} is $(-1)^{i+j}$ times the $(m-1) \times (m-1)$-determinant, obtained from A by removing its jth line and ith column. We have $A'A = AA' = (\det A)I$, where I is the unity matrix.

C.3   Semi-simple Modules

In this section we closely follow Lang [58, Sect. 17.2].

Let R be a (commutative) ring. A nonzero R-module is called *simple* if it has no submodules except 0 and itself.

For example, if R is a field then the simple modules are exactly the vector spaces of dimension 1. Also, simple $\mathbb{Z}$-modules are cyclic groups of prime order.

Simple modules are rather "rigid" objects. For instance, a morphism of a simple module (into another module) is either injective or a zero map. (Indeed, its kernel is a submodule of our simple module; hence it is either 0 or the module itself.) This property is usually called "Schur's lemma."

Another evidence for this "rigidity" is the fact that a sum of simple modules can always be made direct.[2] More precisely, we have the following property.

Proposition C.6. *Let $\sum_{\lambda \in \Lambda} M_\lambda$ be a sum of simple submodules of a certain R-module. Then there is a subset $\Lambda' \subseteq \Lambda$ such that $\sum_{\lambda \in \Lambda} M_\lambda = \bigoplus_{\lambda' \in \Lambda'} M_{\lambda'}$.*

Proof. Let Λ' be a maximal subset of Λ such that the sum $\sum_{\lambda' \in \Lambda'} M_{\lambda'}$ is direct. We have to show that every M_λ is contained in $M := \bigoplus_{\lambda' \in \Lambda'} M_{\lambda'}$. We cannot have $M_\lambda \cap M = 0$ because otherwise we can add λ to the set Λ' and still have a direct sum, contradicting the maximal choice of Λ'. Thus, $M_\lambda \cap M \neq 0$, and, since M_λ is simple, we have $M_\lambda \cap M = M_\lambda$, that is, $M_\lambda \subset M$.   □

Now we are ready to define semi-simple modules.

Proposition C.7. *Let R be a commutative ring and M a module over R. Then the following three properties are equivalent:*

1. *The module M is a sum of its simple submodules.*
2. *The module M is a direct sum of its simple submodules.*
3. *Each submodule of M has a direct complement. (That is, if N is a submodule of M then there exists another submodule N' such that $M = N \oplus N'$.)*

An R-module with these properties is called *semi-simple.*

Proof. Implication (1)$\Rightarrow$(2) is Proposition C.6. To deduce implication (2)$\Rightarrow$(3), write $M = \bigoplus_{\lambda \in \Lambda} M_\lambda$, where each M_λ is simple, and let N be a submodule of M. Let Λ' be a maximal subset of Λ such that the sum $N + \bigoplus_{\lambda' \in \Lambda'} M_{\lambda'}$ is direct. Arguing as in the proof of Proposition C.6, we show that $M = N \oplus \bigoplus_{\lambda' \in \Lambda'} M_{\lambda'}$.

We are left with the implication (3)$\Rightarrow$(1). First of all, let us show that Property 3 implies the following: *every nonzero submodule of M has a simple submodule.*

Thus, let N be a nonzero submodule of M. Since every nonzero module contains a cyclic submodule, we may assume that N is cyclic. By Proposition C.2, it has a maximal submodule S. By Property 3, the module S has a direct complement

---

[2]Recall that the sum $\sum_{\lambda \in \Lambda} M_\lambda$ is called *direct* and is denoted by $\bigoplus_{\lambda \in \Lambda} M_\lambda$ if for any $\mu \in \Lambda$ we have $M_\mu \cap \sum_{\substack{\lambda \in \Lambda \\ \lambda \neq \mu}} M_\lambda = 0$.

in M; that is, $M = S \oplus S'$, where S' is yet another submodule of M. It follows that $N = S \oplus P$, where $P = N \cap S'$. The module P, which is isomorphic to the quotient N/S, is simple, because S is a maximal submodule of N.

Now we are ready to deduce the implication $(3) \Rightarrow (1)$. Let M' be the sum of all simple submodules of M. If $M' \neq M$ then, by Property 3, we have $M = M' \oplus M''$, where M'' is a nonzero submodule. But, as we have just seen, M'' has a simple submodule, which contradicts our definition of M' as the sum of all simple submodules. Thus, $M' = M$. □

Remark C.8. Property 3 of semi-simple modules can also be restated as follows: *if M is a semi-simple module and N is a submodule of M then $M \cong N \oplus M/N$.*

Semi-simplicity is inherited by sub- and quotient modules.

Proposition C.9. *Submodules and quotient modules of a semi-simple module are semi-simple.*

Proof. Let M be a semi-simple module and N its submodule. Further, let S be a submodule of N. Since M is semi-simple, S has a direct complement in M; that is, $M = S \oplus S'$. It follows that $N = S \oplus (S' \cap N)$. We have shown that every submodule of N has a direct complement in N. Hence N is semi-simple.

Further, write $M = N \oplus N'$. Then $M/N \cong N'$. Since N' is semi-simple (as a submodule of M), so is M/N. □

C.4 Semi-simple Rings

A ring is called *semi-simple* if it is semi-simple as a module over itself. Since a submodule of R is an ideal of R, and a simple submodule of R is a *minimal* (nonzero) ideal, the following conditions are equivalent by Proposition C.7:

- R is semi-simple;
- R is a sum of its minimal ideals;
- R is a direct sum of its minimal ideals;
- Every ideal $\mathfrak{a}$ of R has a direct complement (that is, there is an ideal $\mathfrak{a}'$ such that $R = \mathfrak{a} \oplus \mathfrak{a}'$).

For instance, *a field is a semi-simple ring.* Further, *a direct product of finitely many semi-simple rings is again a semi-simple ring.*

Indeed, it suffices to verify that a direct product of two semi-simple rings is semi-simple. Thus, let R_1 and R_2 be semi-simple rings, and let $\mathfrak{a}$ be an ideal of $R = R_1 \times R_2$. Denote by $\mathfrak{a}_1$ and $\mathfrak{a}_2$ the projections of $\mathfrak{a}$ on R_1 and R_2, respectively, and let $\mathfrak{a}_1'$, respectively, $\mathfrak{a}_2'$ be a direct complement of $\mathfrak{a}_1$ in R_1, and, respectively, of $\mathfrak{a}_2$ in R_2. Then a straightforward verification shows that $R = \mathfrak{a} \oplus \mathfrak{a}'$, where $\mathfrak{a}' = \mathfrak{a}_1' \times \mathfrak{a}_2'$. Hence R is semi-simple.

In particular, *a direct product of finitely many fields is a semi-simple ring.*

On the other hand, $\mathbb{Z}$ is *not* a semi-simple ring: the submodule $2\mathbb{Z}$ has no direct complement.

Proposition C.10. *Any module over a semi-simple ring is semi-simple.*

Proof. If R is a semi-simple ring then any *free* R-module is semi-simple, because it is a direct sum of modules isomorphic to R. Since any R-module is a quotient of a free R-module, it is semi-simple by Proposition C.9.   □

It is remarkable that semi-simple commutative rings admit a very explicit classification.

Theorem C.11. *A (commutative) ring is semi-simple if and only if it is isomorphic to a direct product of finitely many fields.*

Proof. The "if" part is already proved in the beginning of the section. Now assume that R is a semi-simple ring and prove that it is isomorphic to a direct product of fields.

Write $R = \bigoplus_{\lambda \in \Lambda} \mathfrak{a}_\lambda$, where every $\mathfrak{a}_\lambda$ is a minimal nonzero ideal of R. It follows that every $x \in R$ can be (uniquely) presented as $x = \sum_{\lambda \in \Lambda} x_\lambda$, where $x_\lambda \in \mathfrak{a}_\lambda$, and for all but finitely many λ we have $x_\lambda = 0$.

In particular, write $1 = \sum_{\lambda \in \Lambda} 1_\lambda$ and let Λ' be the *finite* subset of Λ consisting of λ-s such that $1_\lambda \neq 0$. Then $1 = \sum_{\lambda \in \Lambda'} 1_\lambda$, and, in particular, 1 belongs to the ideal $\bigoplus_{\lambda \in \Lambda'} \mathfrak{a}_\lambda$. However, 1 cannot belong to a proper ideal. Hence $\bigoplus_{\lambda \in \Lambda'} \mathfrak{a}_\lambda = R$, that is, $\Lambda' = \Lambda$. Thus, we have proved that the set Λ is finite.

Further, since the sum is direct, we have $\mathfrak{a}_\lambda \cap \mathfrak{a}_\mu = 0$ for any distinct $\lambda, \mu \in \Lambda$. Hence $\mathfrak{a}_\lambda \mathfrak{a}_\mu = 0$ for $\lambda \neq \mu$. In particular, if $x \in \mathfrak{a}_\lambda$ then $1_\mu x = 0$ for any $\mu \neq \lambda$. It follows that for $x \in \mathfrak{a}_\lambda$,

$$1_\lambda x = 1_\lambda x + \sum_{\substack{\mu \in \Lambda \\ \mu \neq \lambda}} 1_\mu x = 1x = x.$$

Thus, every $\mathfrak{a}_\lambda$ is a ring, with 1_λ serving as its unity, and R is isomorphic to the direct product of the rings $\mathfrak{a}_\lambda$.

Moreover, let $\mathfrak{b}$ be an ideal of the ring $\mathfrak{a}_\lambda$, so that $\mathfrak{a}_\lambda \mathfrak{b} \subseteq \mathfrak{b}$. Then

$$R\mathfrak{b} = \mathfrak{a}_\lambda \mathfrak{b} + \sum_{\substack{\mu \in \Lambda \\ \mu \neq \lambda}} \mathfrak{a}_\mu \mathfrak{b} = \mathfrak{a}_\lambda \mathfrak{b} \subseteq \mathfrak{b};$$

that is, $\mathfrak{b}$ is an ideal of R as well. Since $\mathfrak{a}_\lambda$ is a minimal ideal of R, we have $\mathfrak{b} = \mathfrak{a}_\lambda$ or $\mathfrak{b} = 0$.

Thus, $\mathfrak{a}_\lambda$ is a ring without nontrivial ideals. Hence $\mathfrak{a}_\lambda$ is a field. The proof is complete.   □

It is easy to describe the ideals of a semi-simple ring. Let R be a semi-simple ring, and write it as a direct product of finitely many fields: $R = K_1 \times \cdots \times K_s$. Put $\Lambda = \{1, \dots, s\}$. For $\lambda \in \Lambda$ we denote by 1_λ the element $(x_1, \dots, x_s)$ such that $x_\lambda = 1$ and $x_\mu = 0$ for $\mu \neq \lambda$ (so that

$$1_1 = (1, 0, \dots, 0), \qquad 1_2 = (0, 1, 0, \dots, 0), \qquad (C.2)$$

etc.).

We leave to the reader the proof of the following proposition.

Proposition C.12. *1. For $\Lambda' \subset \Lambda$ let $\mathfrak{a}_{\Lambda'}$ consist of $x = (x_1, \dots, x_s) \in R$ such that $x_\lambda = 0$ for all $\lambda \notin \Lambda'$. Then $\mathfrak{a}_{\Lambda'}$ is an ideal of R. Conversely, any ideal of R is equal to $\mathfrak{a}_{\Lambda'}$ for some $\Lambda' \subseteq \Lambda$.*
2. The ideal $\mathfrak{a}_{\Lambda'}$ principal; it is generated by the element $1_{\Lambda'} = \sum_{\lambda \in \Lambda'} 1_\lambda$. More generally, $\mathfrak{a}_{\Lambda'}$ is generated by any $x = (x_1, \dots, x_s)$ such that $x_\lambda \neq 0$ for all $\lambda \in \Lambda'$ (and $x_\lambda = 0$ for all $\lambda \notin \Lambda'$).
3. In addition to this, $\mathfrak{a}_{\Lambda'}$ is itself a (semi-simple) ring, with the unity $1_{\Lambda'}$.
4. If Λ' and Λ'' are subsets of Λ then

$$\mathfrak{a}_{\Lambda'} \mathfrak{a}_{\Lambda''} = \mathfrak{a}_{\Lambda' \cap \Lambda''}, \qquad \mathfrak{a}_{\Lambda'} + \mathfrak{a}_{\Lambda''} = \mathfrak{a}_{\Lambda' \cup \Lambda''}.$$

Corollary C.13. *A semi-simple ring is a principal ideal ring without nilpotent elements. It has only finitely many ideals. Every prime ideal of a semi-simple ring is maximal. Any quotient of a semi-simple ring is again a semi-simple ring. Any ideals $\mathfrak{a}, \mathfrak{b}$ of a semi-simple ring satisfy $\mathfrak{a}\mathfrak{b} = \mathfrak{a} \cap \mathfrak{b}$.*

Another consequence of Proposition C.12 is that the direct complement of an ideal is defined uniquely.

Proposition C.14. *1. Let R be a semi-simple ring and $\mathfrak{a}$ an ideal of R. Then there is a uniquely defined ideal $\mathfrak{a}^\perp$ such that $\mathfrak{a} \oplus \mathfrak{a}^\perp = R$.*
2. For any ideals $\mathfrak{a}, \mathfrak{b}$ of R, one has

$$(\mathfrak{a}\mathfrak{b})^\perp = \mathfrak{a}^\perp + \mathfrak{b}^\perp, \qquad (\mathfrak{a} + \mathfrak{b})^\perp = \mathfrak{a}^\perp \mathfrak{b}^\perp. \qquad (C.3)$$

Also, $\mathfrak{a}\mathfrak{b} = 0$ if and only if $\mathfrak{b} \subseteq \mathfrak{a}^\perp$.

The ideal $\mathfrak{a}^\perp$ will be called the *complementary ideal*, or, simply, the *complement* of $\mathfrak{a}$.

Proof. Write $\mathfrak{a}$ as $\mathfrak{a}_{\Lambda'}$ and define $\mathfrak{a}^\perp$ as $\mathfrak{a}_{\Lambda \smallsetminus \Lambda'}$. Then all the statements of this proposition are immediate consequences of Proposition C.12(4). $\qquad \square$

It is equally easy to characterize R-modules. Let $R = K_1 \times \cdots \times K_s$ be a semi-simple ring, and fix vector spaces V_1 over $K_1, \dots, V_s$ over K_s. Then the direct product $M = V_1 \times \cdots \times V_s$ is an R-module, the R-action being defined componentwise: if $a = (a_1, \ldots, a_s) \in R$ and $v = (v_1, \ldots, v_s) \in M$, then $av = (a_1 v_1, \ldots, a_s v_s)$. It is easy to show that any R-module is of this type: for free modules this is obvious, and an arbitrary module is quotient of a free module.

Below we state several simple properties of modules over semi-simple rings. The proofs are left to the reader.

Proposition C.15. *Let* $R = K_1 \times \cdots \times K_s$ *be a semi-simple ring and let* $M = V_1 \times \cdots \times V_s$ *be an R-module. We put* $\Lambda = \{1, \ldots, s\}$:

1. *The module M is finitely generated if and only if all the vector spaces $V_1, \ldots, V_s$ are finite dimensional.*
2. *The module M is cyclic if and only if* $\dim_{K_\lambda} V_\lambda \leq 1$ *for all $\lambda \in \Lambda$.*
3. *The annihilator of M is $\mathfrak{a}_{\Lambda'}$, where the set Λ' consists of all $\lambda \in \Lambda$ with* $\dim_{K_\lambda} V_\lambda = 0$.

We have the following consequence for *finite* semi-simple rings.

Corollary C.16. *Let R be a finite semi-simple ring and let M be a finitely generated R-module. Then $|M| \geq |R/\mathrm{ann}(M)|$, with equality if and only if M is cyclic.*

Proof. Write $R = K_1 \times \cdots \times K_s$ and $M = V_1 \times \cdots \times V_s$, and let d_λ be the dimension of the K_λ-vector space V_λ. Then

$$|M| = \sum_{\lambda \in \Lambda} d_\lambda |K_\lambda|, \qquad |R/\mathrm{ann}(M)| = \sum_{\substack{\lambda \in \Lambda \\ d_\lambda > 0}} |K_\lambda|.$$

It is clear that the second sum does not exceed the first sum, and the two are equal only if $d_\lambda \leq 1$ for all λ. $\square$

C.5 The "Dual" Module

Let R be a commutative ring, S its subring, M an R-module, and N an S-module. Denote by $\mathrm{Hom}_S(M, N)$ the set of S-morphisms $M \to N$. Obviously, $\mathrm{Hom}_S(M, N)$ has a natural S-module structure. But one can say more: $\mathrm{Hom}_S(M, N)$ has an R-module structure, defined as follows: given $f \in \mathrm{Hom}_S(M, N)$ and $a \in R$, the morphism $af \in \mathrm{Hom}_S(M, N)$ is defined by $(af)(x) = f(ax)$ for $x \in M$. (Precisely speaking, if M is a *left* R-module, then $\mathrm{Hom}_S(M, N)$ becomes a *right* R-module. However, since the ring R is commutative, we need not distinguish between left and right.)

Now assume that the ring R has a field K as a subring. Then every R-module M is a K-vector space, and the "dual space" $M^* = \mathrm{Hom}_K(M, K)$ has a natural R-module structure, obtained by applying the previous paragraph with $S = N = K$. If our ring R is of finite K-dimension, and M is a finitely generated R-module, then M and the "dual module" M^* are finite-dimensional K-vector spaces of the same dimension; in particular, they are K-isomorphic. We shall see that when R is semi-simple, they are R-isomorphic as well.

Theorem C.17. *Let R be a semi-simple (commutative) ring, containing a field K as a subring. Assume that R is of finite dimension over K, and let M be a finitely generated R-module. Then $M^* = \mathrm{Hom}_K(M, K)$ is R-isomorphic to M.*

Proof. Write R as a direct product of finitely many fields: $R = L_1 \times \cdots \times L_s$. Each field L_λ contains a subfield isomorphic to K, given by $K1_\lambda$ (see (C.2) for the definition of 1_λ). Since R is finite dimensional over K, each L_λ is a finite extension of K.

Now write $M = V_1 \times \cdots \times V_s$, where each V_λ is a finite-dimensional L_λ-vector space. Each $V_\lambda^* = \mathrm{Hom}_K(V_\lambda, K)$ has an L_λ-module structure and is L_λ-isomorphic to V_λ, as follows by counting dimensions:

$$\dim_{L_\lambda} V_\lambda^* = \frac{\dim_K V_\lambda^*}{[L_\lambda : K]} = \frac{\dim_K V_\lambda}{[L_\lambda : K]} = \dim_{L_\lambda} V_\lambda.$$

Hence $V_1^* \times \cdots \times V_s^*$ is R-isomorphic to $M = V_1 \times \cdots \times V_s$. Finally, the map

$$M^* \to V_1^* \times \cdots \times V_S^*$$
$$f \mapsto (f|_{V_1}, \ldots, f|_{V_s})$$

is an R-isomorphism. The theorem is proved.   □

Appendix D
Group Rings and Characters

In this appendix we collect basic facts about group rings and characters of finite commutative groups.

Let A be a commutative ring and G a finite group. The *group ring* $A[G]$ is, by definition, the set of formal linear combinations $\sum_{g \in G} a_g g$, where $a_g \in A$, with the operations defined in the obvious way:

$$\sum_{g \in G} a_g g + \sum_{g \in G} b_g g = \sum_{g \in G} (a_g + b_g)g,$$

$$\left(\sum_{g \in G} a_g g\right)\left(\sum_{g \in G} b_g g\right) = \sum_{g \in G} \left(\sum_{\substack{h,k \in G \\ hk=g}} a_h b_k\right) g.$$

The unity of the group ring $A[G]$ is e, the neutral element of the group G. This suggests the following convention: *in the group ring, we identify e and 1.* Precisely speaking, we may write a typical element of the group ring $A[G]$ in both ways $\sum_{g \in G} a_g g$ or $a_e + \sum_{\substack{g \in G \\ g \neq e}} a_g g$, and we choose the writing most fit for the circumstances. In particular, we assume that A is a subring of $A[G]$.

In this book the group G is usually abelian, and A is either the ring of integers $\mathbb{Z}$ or a field (except Appendix E, where $A = \mathbb{Z}/p^s\mathbb{Z}$ occurs).

D.1   The Weight Function and the Norm Element

In this section we recall the most basic notions about the group rings. We define the *weight function* $\mathrm{w} : A[G] \to A$ by

$$\mathrm{w}\left(\sum_{g \in G} a_g g\right) = \sum_{g \in G} a_g.$$

One immediately verifies that the weight function is additive and multiplicative:

Proposition D.1. *For any $x, y \in A[G]$*

$$w(x + y) = w(x) + w(x), \qquad w(xy) = w(x)w(y).$$

Thus, the weight function is a ring homomorphism. Its kernel, consisting of elements of weight 0, is called the *augmentation ideal* of the group ring $A[G]$.

The *norm element* of $A[G]$ is

$$\mathcal{N} = \sum_{g \in G} g.$$

It is obvious that $x\mathcal{N} = \mathcal{N}x = \mathcal{N}$ for any $x \in G$. Extending this by linearity, we obtain the following property.

Proposition D.2. *For any $x \in A[G]$ we have $x\mathcal{N} = \mathcal{N}x = w(x)\mathcal{N}$. In particular, $A[G]\mathcal{N} = \mathcal{N}A[G] = A\mathcal{N}$.*

The ideal $A\mathcal{N}$ is called the *norm ideal* of the group ring $A[G]$. If the cardinality $|G|$ is an invertible element of A (which is, in particular, the case if A is a field of characteristic not dividing $|G|$) then, writing each $x \in A[G]$ as $(x - w(x)|G|^{-1}\mathcal{N}) + w(x)|G|^{-1}\mathcal{N}$, we obtain the following.

Proposition D.3. *Assume that $|G|$ is an invertible element of A. Then $A[G]$ is the direct sum of its augmentation ideal and its norm ideal.*

The terms *norm element* and *norm ideal* are not common, but they are most suited for the purposes of this book.

D.2   Characters of a Finite Abelian Group

Let G be a (finite or infinite, commutative or not) group, and let K be a field. Denote by $\bar{K}$ the algebraic closure of K. A *K-character*[1] of G is a group homomorphism $\chi : G \to \bar{K}^{\times}$. The character χ is called *trivial* if $\chi(g) = 1$ for any $g \in G$. Characters form a multiplicative group, with the trivial character as the unity.

If G is finite, then the values of a nontrivial character sum to 0.

Proposition D.4. *Let χ be a nontrivial character of a finite group G. Then $\sum_{x \in G} \chi(x) = 0$.*

---

[1] For noncommutative groups the word "character" has a wider meaning, and what we define here is usually called "linear character." Since we deal mainly with abelian groups, we use the word "character."

Proof. Since χ is nontrivial, there exists $g \in G$ such that $\chi(g) \neq 1$. We obtain

$$0 = \sum_{x \in G} \chi(x) - \sum_{x \in G} \chi(gx) = (1 - \chi(g)) \sum_{x \in G} \chi(x).$$

Since $1 - \chi(g) \neq 0$ by the choice of g, the result follows.   □

Denote by $\bar{K}^G$ the $\bar{K}$-vector space of $\bar{K}$-valued functions on G. It is well known that the characters of G form a linearly independent subset of $\bar{K}^G$. This statement is usually attributed to Artin (see [58, Sect. 7.4]).

Proposition D.5 (Artin). *The K-characters of G are linearly independent over $\bar{K}$. That is, given pairwise distinct K-characters $\chi_1, \ldots, \chi_m$, and $\alpha_1, \ldots, \alpha_m \in \bar{K}$, not all zero, the linear combination $\alpha_1 \chi_1 + \cdots + \alpha_m \chi_m$ is not identically zero on G.*

Proof. Assume that some nonzero linear combination $\alpha_1 \chi_1 + \cdots + \alpha_m \chi_m$ identically vanishes on G:

$$\alpha_1 \chi_1(x) + \cdots + \alpha_m \chi_m(x) = 0 \tag{D.1}$$

for any $x \in G$. We may assume that m is minimal; in particular, $\alpha_1 \neq 0$. Also, $m \geq 2$, and, in particular, $\chi_1 \neq \chi_m$, because the characters are pairwise distinct. Thus, there exists $g \in G$ such that $\chi_1(g) \neq \chi_m(g)$. Rewriting (D.1) with gx instead of x, we obtain

$$\alpha_1 \chi_1(g)\chi_1(x) + \cdots + \alpha_m \chi_m(g)\chi_m(x) = 0. \tag{D.2}$$

Multiplying (D.1) by $\chi_m(g)$ and subtracting the resulting identity from (D.2), we obtain

$$\beta_1 \chi_1(x) + \cdots + \beta_{m-1} \chi_{m-1}(x) = 0 \tag{D.3}$$

for any $x \in G$. Here $\beta_k = \alpha_k(\chi_k(g) - \chi_m(g))$, and, in particular,

$$\beta_1 = \alpha_1(\chi_1(g) - \chi_m(g)) \neq 0.$$

Identity (D.3) contradicts the minimal choice of m. The proposition is proved.

□

In the sequel, unless the contrary is stated explicitly, the group G will be **abelian and finite**, and we shall assume that **the characteristic of K does not divide $|G|$**. Under these assumptions, the group of characters is usually called *the dual group* of G and is denoted by $\hat{G}$. The structure of this group is well known.

Theorem D.6. *Let G be a finite abelian group and K a field of characteristic not dividing $|G|$. Then the group of characters $\hat{G}$ is isomorphic to G. In particular, there are exactly $|G|$ distinct characters.*

Proof. We use induction on $|G|$. Assume first that $G = \langle g \rangle$ is a cyclic group of finite order m. Then the map $\chi \mapsto \chi(g)$ is an isomorphism of $\hat{G}$ and the group of mth roots of unity in $\bar{K}$. Since char K does not divide m, the latter group is again cyclic of order m. This shows that $\hat{G} \cong G$.

Now let G be not cyclic. Then it is a direct sum of two nontrivial subgroups: $G = G_1 \oplus G_2$. Consider the homomorphism $\hat{G} \to \hat{G}_1 \times \hat{G}_2$ defined by $\chi \mapsto (\chi|_{G_1}, \chi|_{G_2})$, and the homomorphism $\hat{G}_1 \times \hat{G}_2 \to \hat{G}$, which to each pair (χ_1, χ_2) associates the character $\chi \in \hat{G}$ defined by $\chi(x_1 x_2) = \chi_1(x_1)\chi_2(x_2)$. A routine verification shows that the two homomorphisms are inverse one to the other. Hence $\hat{G} \cong \hat{G}_1 \times \hat{G}_2$. Since, by induction, $\hat{G}_1 \cong G_1$ and $\hat{G}_2 \cong G_2$, we obtain $\hat{G} \cong G$.   □

Theorem D.6 implies that the characters form a basis of the space $\bar{K}^G$ of $\bar{K}$-valued functions on G: indeed, the characters are linearly independent, and their number is equal to the dimension $|G|$ of this space. We obtain the following statement.

Proposition D.7. *Let G be a finite abelian group and K a field of characteristic not dividing $|G|$. Then the K-characters of the group G form a $\bar{K}$-basis of the vector space $\bar{K}^G$. In particular, the $|G| \times |G|$ matrix $[\chi(g)]_{\substack{\chi \in \hat{G} \\ g \in G}}$ is nondegenerate.*

In the case $K \subseteq \mathbb{C}$ one can say more: the characters form an orthogonal basis of the space $\bar{K}^G$, with respect to the natural inner product

$$(f_1, f_2) = \frac{1}{|G|} \sum_{x \in G} f_1(x)\overline{f_2(x)}.$$

Proposition D.8. *Assume that $K \subseteq \mathbb{C}$. Then for any two characters χ_1 and χ_2 we have $(\chi_1, \chi_2) = 1$ if $\chi_1 = \chi_2$ and $(\chi_1, \chi_2) = 0$ if $\chi_1 \neq \chi_2$.*

Proof. Since the values of any character χ are roots of unity, we have $\bar{\chi} = \chi^{-1}$. Hence $\chi_1 \bar{\chi}_2$ is a trivial character if and only if $\chi_1 = \chi_2$. Applying Proposition D.4 to the character $\chi_1 \bar{\chi}_2$, we obtain the result.   □

D.3   Conjugate Characters

For a character $\chi \in \hat{G}$ we denote by K_χ the extension of K generated by the values of χ. It is the dth cyclotomic extension of K, where d is the order of χ. In particular, K_χ is a Galois extension of K. If $\sigma \in \text{Gal}(K_\chi/K)$, then $\sigma \circ \chi$ is also a K-character of G.

Let us say that two characters $\chi, \chi' \in \hat{G}$ are *conjugate* (over K) if $K_\chi = K_{\chi'}$ and there exists $\sigma \in \text{Gal}(K_\chi/K)$ such that $\chi' = \sigma \circ \chi$. The conjugacy relation is an equivalence on $\hat{G}$, the class of every $\chi \in \hat{G}$ containing exactly $[K_\chi : K]$ elements. Thus, if we pick a representative in every conjugacy class of characters, and denote by M the set of chosen representatives, then we obtain the equality

$$\sum_{\chi \in M} [K_\chi : K] = |G|. \tag{D.4}$$

Remark D.9. Let us mention several simple facts, which are relevant here, but are not used in the book:

1. If χ and χ' are conjugate characters, then they are of the same order; moreover, $\chi' = \chi^a$, where a is an integer coprime with the order.
2. The converse of the last statement is, in general, not true: for instance, if K is algebraically closed then the only character conjugate to χ is χ itself.
3. However, the converse is true if $K = \mathbb{Q}$; that is, if χ is a $\mathbb{Q}$-character of order m then for every $a \in \mathbb{Z}$, coprime with m, the character χ^a is conjugate to χ.
4. In particular, two $\mathbb{Q}$-characters of a finite cyclic group are conjugate if and only if they are of the same order.

The proofs are left to the reader.

D.4 Semi-simplicity of the Group Ring

In this section G is a finite abelian group and K is a field of characteristic not dividing $|G|$. The K-characters of G extend by linearity to the group ring $K[G]$: given $\chi \in \hat{G}$, we define the map $K[G] \to K_\chi$ by

$$\sum_{g \in G} a_g g \mapsto \sum_{g \in G} a_g \chi(g).$$

As one immediately verifies, this is a ring homomorphism. We denote it also by χ, and we call it a *character* of the group ring $K[G]$. The set of all characters of $K[G]$ will again be denoted by $\hat{G}$.

Non-degeneracy of the matrix $[\chi(g)]_{\substack{\chi \in \hat{G} \\ g \in G}}$ (Proposition D.7) implies that the common kernel of all characters is trivial.

Proposition D.10. *If $x \in K[G]$ satisfies $\chi(x) = 0$ for every character $\chi \in \hat{G}$, then $x = 0$.*

Using this property, we show that the group ring $K[G]$ is *semi-simple*, as defined in Appendix C.4.

Theorem D.11 (the "abelian Maschke theorem").   *Let G be a finite abelian group and K a field of characteristic not dividing $|G|$. Choose a system M of representatives of conjugacy classes of characters. Then the ring homomorphism*

$$K[G] \rightarrow \prod_{\chi \in M} K_\chi$$

$$\text{(D.5)}$$

$$x \mapsto (\chi(x))_{\chi \in M}.$$

is an isomorphism. In particular, the ring $K[G]$ is semi-simple.

Proof. Let x be in the kernel of the map (D.5), that is, $\chi(x) = 0$ for any character $\chi \in M$. Since conjugate characters vanish simultaneously at x, we obtain $\chi(x) = 0$ for any $\chi \in \hat{G}$. Proposition D.10 implies that $x = 0$. Hence (D.5) is a monomorphism and, by (D.4), the K-dimensions of both parts of (D.5) are equal. Therefore we have an isomorphism.   □

Remark D.12. In the case "G is a cyclic group of order m and $K = \mathbb{Q}$" Theorem D.11 implies the isomorphism $\mathbb{Q}[G] \cong \prod_{d \mid m} \mathbb{Q}(\zeta_d)$. We do not use this in the present book.

As Proposition C.12 suggests, the ideals of $K[G]$ can be characterized as common kernels of characters from the set M: for a subset S of M let $\mathcal{I}_S$ be the common kernel of the characters from the complement $M \setminus S$. Then $\mathcal{I}_S$ is an ideal of $K[G]$, and any ideal of $K[G]$ is equal to $\mathcal{I}_S$ for some $S \subseteq M$.

The ideal $\mathcal{I}_S$ is isomorphic, as a $K[G]$-module, to $\prod_{\chi \in S} K_\chi$. It follows that it is a K-vector space, and

$$\dim_K \mathcal{I}_S = \sum_{\chi \in S} [K_\chi : K].$$

If $\theta \in K[G]$ then the principal ideal (θ) is equal to $\mathcal{I}_S$, where S consists of characters $\chi \in M$ with $\chi(\theta) \neq 0$. Hence

$$\dim_K (\theta) = \sum_{\chi(\theta) \neq 0} [K_\chi : K].$$

$$\text{(D.6)}$$

Since conjugate characters vanish at θ simultaneously, and since for every χ there are exactly $[K_\chi : K]$ characters, conjugate to χ, the sum in (D.6) is equal to the number of $\chi \in \hat{G}$ with $\chi(\theta) \neq 0$. We have proved the following statement.

Proposition D.13. *Let θ be an element of $K[G]$. Then the K-dimension of the principal ideal (θ) is equal to the number of characters $\chi \in \hat{G}$ nonvanishing at θ.*

Finally, we state one more consequence of Theorem D.11, to be used in Appendix F.4. Let N be a module over the group ring $K[G]$. Then N is a K-vector space, and the "dual space" $N^* = \operatorname{Hom}_K(N, K)$ has a natural G-module structure (see Appendix C.5). The following result is a direct application of Theorems C.17 and D.11.

Corollary D.14. *Let G be a finite abelian group and K a field of characteristic not dividing $|G|$. Let N be a finitely generated $K[G]$-module. Then the "dual module" $N^* = \mathrm{Hom}_K(N, K)$ is $K[G]$-isomorphic to N.*

D.5 Idempotents

We retain the setup of the previous section. It is useful to have an explicit basis of the ideal $\mathcal{I}_S$ as a K-vector space and an explicit generator of $\mathcal{I}_S$ as a principal ideal. In this section we produce both, under the additional assumption

$$K \text{ contains the } |G|\text{th roots of unity,} \tag{D.7}$$

which is sufficient for our purposes. The reader is invited to examine the general case.

Assumption (D.7) implies that $K_\chi = K$ for all characters χ, or, equivalently, every character is conjugate only to itself. It follows that in Theorem D.11 we have $M = \hat{G}$, and the map (D.5) becomes the isomorphism

$$\psi : K[G] \to K^{\hat{G}},$$

where $K^{\hat{G}}$ is the ring of K-functions on $\hat{G}$.

For every character χ, we define the element $\varepsilon_\chi \in K[G]$ as follows:

$$\varepsilon_\chi = \frac{1}{|G|} \sum_{g \in G} \chi(g) g^{-1}.$$

These elements have many remarkable properties; here are some of them.

Proposition D.15. *1. For any characters χ and χ' we have*

$$\chi'(\varepsilon_\chi) = \begin{cases} 1 & \text{if } \chi = \chi', \\ 0 & \text{if } \chi \neq \chi'. \end{cases}$$

2. For any $x \in K[G]$ we have $x\varepsilon_\chi = \chi(x)\varepsilon_\chi$.
3. For any χ we have $\varepsilon_\chi^2 = \varepsilon_\chi$.

Proof. Applying Proposition D.4 to the character $\chi(\chi')^{-1}$, we find

$$\chi'(\varepsilon_\chi) = \frac{1}{|G|} \sum_{g \in G} \chi(g)\chi'(g)^{-1} = \begin{cases} 1 & \text{if } \chi = \chi', \\ 0 & \text{if } \chi \neq \chi'. \end{cases}$$

This proves Part 1. Part 2 is trivially true for any $x \in G$. By linearity, it extends to $x \in K[G]$. Part 3 is an immediate consequence of the previous parts.   □

The element ε_χ is called *the idempotent of* χ. Notice that the idempotent of the trivial character is $|G|^{-1}\mathcal{N}$.

Since $x\varepsilon_\chi \in K\varepsilon_\chi$ for any $x \in K[G]$, the set $K\varepsilon_\chi$ is an ideal of $K[G]$. We have decomposed $K[G]$ into a direct sum of one-dimensional ideals:

$$K[G] = \bigoplus_{\chi \in \hat{G}} K\varepsilon_\chi .$$

The ideal $\mathcal{I}_S$ decomposes as

$$\mathcal{I}_S = \bigoplus_{\chi \in S} K\varepsilon_\chi ,$$

which, in particular, implies that $\{\varepsilon_\chi : \chi \in S\}$ is a K-basis of $\mathcal{I}_S$.

Also, it is easy to see that

$$\mathcal{I}_S = K[G] \sum_{\chi \in S} \varepsilon_\chi ,$$

which gives an explicit generator of the principal ideal $\mathcal{I}_S$. Since we do not use this result, its proof is left to the reader.

Appendix E
Reduction and Torsion of Finite G-Modules

In this appendix we study the reduction of G-modules modulo a prime number. Like in Appendix C (and unlike in the rest of the book) here we write the abelian group law on modules additively and the ring action multiplicatively.

Let M be a finite $\mathbb{Z}$-module (written additively), and let p be a prime number. It is easy to see that M/pM is isomorphic to the p-torsion submodule $M[p] = \{a \in M : p\,a = 1\}$. In this appendix we show that this automorphism extends, under a mild assumption, to finite G-modules.

Theorem E.1. *Let G be a finite abelian group and let M be a finite G-module. Further, let p be a prime number not dividing $|G|$. Then M/pM and $M[p]$ are G-isomorphic.*

It is easy to see (see end of Sect. E.4) that M can be replaced by its p-Sylow submodule and the group ring $\mathbb{Z}[G]$ can be replaced by the finite ring $\mathbb{Z}/p^s\mathbb{Z}[G]$ with sufficiently large s. We introduce a very special class of rings (called here *telescopic* rings) and describe, in terms of these rings, the structure of the group ring $\mathbb{Z}/p^s\mathbb{Z}[G]$ and of the finitely generated $\mathbb{Z}/p^s\mathbb{Z}[G]$-modules. After this preparation Theorem E.1 becomes immediate.

When writing this appendix we profited from very useful discussions with Jean Fresnel. Many arguments below are due to him or based on his ideas.

E.1 Telescopic Rings

Call a (commutative) ring R *telescopic* if R has an ideal $\mathfrak{m}$ such that any other ideal is a power of $\mathfrak{m}$. A basic example of a telescopic ring is $\mathbb{Z}/p^s\mathbb{Z}$, where p is a prime number: in this ring every ideal is a power of the principal ideal (p).

Obviously, R is a local ring with the maximal ideal $\mathfrak{m}$. Also, since the zero ideal is a power of $\mathfrak{m}$, the latter is nilpotent: $\mathfrak{m}^s = 0$ for some s. If we choose the *minimal s*

with this property (called the *index of nilpotency* of $\mathfrak{m}$) then $\mathfrak{m}^k \neq \mathfrak{m}^{k+1}$ for $k < s$: indeed, if $\mathfrak{m}^k = \mathfrak{m}^{k+1}$ then, multiplying by $\mathfrak{m}^{s-k-1}$, we obtain $\mathfrak{m}^{s-1} = \mathfrak{m}^s = 0$, contradicting the minimal choice of s. It follows that

$$R = \mathfrak{m}^0 \supsetneq \mathfrak{m} \supsetneq \ldots \supsetneq \mathfrak{m}^s = 0 \tag{E.1}$$

is the complete list of ideals of the telescopic ring R. Thus, we may regard telescopic rings as a "nilpotent analogue" of discrete valuation rings.

It is important that the maximal ideal of a telescopic ring is principal (and hence so are all its ideals). Indeed, if $s = 1$ then $\mathfrak{m} = 0$ (in which case R is a field), and if $s > 1$ then $\mathfrak{m} \supsetneq \mathfrak{m}^2$, and we may choose an element $\pi \in \mathfrak{m}$ which does not belong to $\mathfrak{m}^2$. Then $\mathfrak{m} \supseteq \pi R \supsetneq \mathfrak{m}^2$, which shows that $\mathfrak{m} = \pi R$.

Conversely, telescopic rings can be characterized as rings having a principal and nilpotent maximal ideal.

Proposition E.2. *A commutative ring with a principal and nilpotent maximal ideal is telescopic.*

Proof. Let $\mathfrak{m}$ be a principal and nilpotent maximal ideal of a ring R. Since $\mathfrak{m}$ is nilpotent, it is contained in any other maximal ideal of R, which means that $\mathfrak{m}$ is the only maximal ideal of R. Thus, R is a local ring; in particular, every element outside $\mathfrak{m}$ is invertible.

It remains to show that every proper ideal of R is a power of $\mathfrak{m}$. Thus, let $\mathfrak{a}$ be a nonzero proper ideal of R, and let k be the maximal integer with the property $\mathfrak{a} \subseteq \mathfrak{m}^k$ (since $\mathfrak{m}$ is nilpotent, the set of integers with this property is finite). We are going to show that $\mathfrak{a} = \mathfrak{m}^k$.

By the maximal choice of k we have $\mathfrak{a} \not\subseteq \mathfrak{m}^{k+1}$. Let α be an element of $\mathfrak{a}$ not contained in $\mathfrak{m}^{k+1}$. Recall that $\mathfrak{m}$ is a principal ideal, and let π be its generator. Since α belongs to $\mathfrak{m}^k$, but not to $\mathfrak{m}^{k+1}$, we have $\alpha = \pi^k \varepsilon$, where $\varepsilon \notin \mathfrak{m}$. Since R is a local ring, the element ε is invertible. Thus, $\pi^k \in \mathfrak{a}$, which proves that $\mathfrak{a} = \mathfrak{m}^k$. $\square$

Remark E.3. It is not sufficient to assume only that the maximal ideal is nilpotent (without assuming it principal), as shows the following example, kindly communicated to us by Jean Fresnel. Let K be a field, and put $R = K[x, y]/(x^2, y^2, xy)$. Then $\mathfrak{m} = xR + yR$ is a maximal ideal of R satisfying $\mathfrak{m}^2 = 0$, but R is not a telescopic ring, because the ideal xR is not a power of $\mathfrak{m}$.

E.2   Products of Telescopic Rings

Let R be a direct product of finitely many telescopic rings: $R = R_1 \times \cdots \times R_n$, where every R_j is telescopic. Let $\mathfrak{m}_j = \pi_j R_j$ be the maximal ideal of R_j and $K_j = R_j/\mathfrak{m}_j$ be the residue field. Then $\pi = (\pi_1, \ldots, \pi_n)$ is a nilpotent element of R, and $R/\pi R$ is isomorphic to $K_1 \times \cdots \times K_n$. In particular, $R/\pi R$ is a semi-simple ring (see Appendix C.4).

It turns out that this property characterizes products of telescopic rings.

Theorem E.4. *Let R be a (commutative) ring, having a nilpotent element π such that the quotient ring $R/\pi R$ is semi-simple. Then R is a direct product of finitely many telescopic rings.*

As a consequence, we obtain the structure of the group ring $\mathbb{Z}/p^s\mathbb{Z}[G]$.

Corollary E.5. *Let R be a telescopic ring with residue field K, and let G be a finite abelian group. Assume that the characteristic of K does not divide G. Then the group ring $R[G]$ is a direct product of finitely many telescopic rings.*

In particular, if G is a finite abelian group and p is a prime number not dividing $|G|$, then for any positive integer s the ring $\mathbb{Z}/p^s\mathbb{Z}[G]$ is a direct product of telescopic rings.

Proof. Let π generate the maximal ideal of the telescopic ring R. Then $R[G]/\pi R[G]$ is isomorphic to $K[G]$, which is a semi-simple ring by Theorem D.11. Whence the result. □

For the proof of Theorem E.4 we need a lemma (a familiar reader will quickly recognize in it a version of "Hensel's lemma").

Lemma E.6. *Let π be a nilpotent element of a ring R and let α_0, β_0 be coprime elements of R satisfying $\alpha_0\beta_0 \equiv 0 \bmod \pi$. Then there exists coprime $\alpha, \beta \in R$ such that $\alpha \equiv \alpha_0 \bmod \pi$, $\beta \equiv \beta_0 \bmod \pi$, and $\alpha\beta = 0$.*

Proof. Write $\alpha_0\beta_0 = \pi\lambda$. Since α_0 and β_0 are coprime, there exist $u, v \in R$ such that $u\alpha_0 + v\beta_0 = 1$. Put $\alpha_1 = \alpha_0 - v\lambda\pi$ and $\beta_1 = \beta_0 - u\lambda\pi$. Then $\alpha_1\beta_1 \equiv 0 \bmod \pi^2$. Also, α_1 and β_1 are coprime. Indeed, $u\alpha_1 + v\beta_1 \equiv 1 \bmod \pi$, and an element congruent to 1 modulo π is invertible[1].

Iterating the process, we find, for every k, coprime elements $\alpha_k, \beta_k \in R$ such that $\alpha_k \equiv \alpha_0 \bmod \pi$, $\beta_k \equiv \beta_0 \bmod \pi$, and $\alpha_k\beta_k \equiv 0 \bmod \pi^{2^k}$. Since π is nilpotent, we shall eventually obtain $\alpha_k\beta_k = 0$. □

We shall also use the "Chinese remainder theorem": if $\mathfrak{a}$ and $\mathfrak{b}$ are coprime ideals of a ring R, then the natural homomorphism $R \to R/\mathfrak{a} \times R/\mathfrak{b}$ is surjective and defines an isomorphism $R/\mathfrak{ab} \cong R/\mathfrak{a} \times R/\mathfrak{b}$. See, for instance, [58, Sect. 2.2] or [3, Proposition 1.10].[2]

---

[1] Write it as $1 - \pi t$. Then its inverse is $1 + \pi t + \cdots + (\pi t)^{s-1}$, where s is the nilpotency index of π.

[2] In both these references $R/(\mathfrak{a} \cap \mathfrak{b})$ appears instead of $R/\mathfrak{ab}$. However, for coprime ideals $\mathfrak{a}$ and $\mathfrak{b}$, we have $\mathfrak{a} \cap \mathfrak{b} = \mathfrak{ab}$. Indeed, since $\mathfrak{a}$ and $\mathfrak{b}$ are coprime, there exist $\alpha \in \mathfrak{a}$ and $\beta \in \mathfrak{b}$ such that $\alpha + \beta = 1$. Now if $\gamma \in \mathfrak{a} \cap \mathfrak{b}$ then $\gamma\alpha, \gamma\beta \in \mathfrak{ab}$, and hence $\gamma = \gamma(\alpha + \beta) \in \mathfrak{ab}$, which proves that $\mathfrak{a} \cap \mathfrak{b} = \mathfrak{ab}$.

Proof of Theorem E.4. We write $R/\pi R = K_1 \times \cdots \times K_n$, where $K_1, \ldots, K_n$ are fields and, arguing by induction on n, we shall prove that R is a product of n telescopic rings. The case $n = 1$ is exactly Proposition E.2.

Now assume that $n > 1$. There exist coprime $\alpha, \beta \in R$ such that $\alpha\beta = 0$, the image of α in $K_1 \times \cdots \times K_n$ is $e_1 = (1, 0, \ldots, 0)$ and the image of β is $1 - e_1 = (0, 1, \ldots, 1)$. To find such α and β, fix an arbitrary $\alpha_0 \in R$ with image e_1, put $\beta_0 = 1 - \alpha_0$, and apply Lemma E.6.

By the Chinese remainder theorem, the ring R is isomorphic to $R' \times R''$, where $R' = R/\alpha R$ and $R'' = R/\beta R$. Let π', respectively π'', be the image of π in R', respectively R''. Then

$$R'/\pi' R' = R/(\alpha R + \pi R) = K_2 \times \cdots \times K_n,$$

and by induction we conclude that R' is a direct product of $n - 1$ telescopic rings. Similarly, $R''/\pi'' R'' = K_1$, whence R'' is a telescopic ring. The theorem is proved.
□

E.3   Elementary Divisors and Finitely Generated Modules

Like principal ideal domains, telescopic rings admit the "theory of elementary divisors," and a finitely generated module over a telescopic ring is a direct sum of its cyclic submodules.

Theorem E.7. *Let $n \geq 1$ be an integer, R a telescopic ring with the maximal ideal $\mathfrak{m}$, and H a submodule of the free module R^n. Then there exist $a_1, \ldots, a_n \in R^n$ such that*

$$R^n = Ra_1 \oplus \cdots \oplus Ra_n, \qquad H = \mathfrak{m}^{r_1} a_1 \oplus \cdots \oplus \mathfrak{m}^{r_n} a_n, \qquad (E.2)$$

where $r_1, \ldots, r_n$ are nonnegative integers.

Proof. Let $p_k : R^n \to R$ be the projection on the kth coordinate. Then $p_k(H) = \mathfrak{m}^{t_k}$, where $t_1, \ldots, t_k$ are nonnegative integers. We put

$$r_1 = \min\{t_1, \ldots, t_n\}$$

and we may assume that $r_1 = t_1$, so that $p_1(H) = \mathfrak{m}^{r_1}$ and $p_k(H) \subseteq \mathfrak{m}^{r_1}$ for $k = 2, \ldots, n$.

Let π be a generator of $\mathfrak{m}$. Then H has an element b such that $p_1(b) = \pi^{r_1}$. Write $b = (\pi^{r_1}, \pi^{r_1} x_2, \ldots, \pi^{r_1} x_n) = \pi^{r_1} a_1$, where $a_1 = (1, x_1, \ldots, x_n)$. Obviously, $R^n = Ra_1 \oplus \ker p_1$, and an easy verification shows that $H = Rb \oplus H'$, where $H' = \ker p_1 \cap H$. Since $Rb = \mathfrak{m}^{r_1} a_1$, we obtain

$$R^n = Ra_1 \oplus \ker p_1, \qquad H = \mathfrak{m}^{r_1} a_1 \oplus H'.$$

By induction, there exist $a_2, \dots, a_n \in \ker p_1$ such that

$$\ker p_1 = Ra_2 \oplus \cdots \oplus Ra_n, \qquad H' = \mathfrak{m}^{r_2} a_2 \oplus \cdots \oplus \mathfrak{m}^{r_n} a_n.$$

This proves the theorem. □

Corollary E.8. *A finitely generated module over a telescopic ring is a direct sum of its cyclic submodules.*

Proof. Let M be a finitely generated module over a telescopic ring R. Then for some n there is a surjective homomorphism $R^n \to M$. Let H be its kernel, so that $M \cong R^n / H$, and let $a_1, \dots, a_n \in R^n$ be such that (E.2) holds. Then $M = R\alpha_1 \oplus \cdots \oplus R\alpha_n$, where $\alpha_1, \dots, \alpha_n \in M$ are the images of $a_1, \dots, a_n$, respectively. □

E.4 Reduction and Torsion

Let R be a ring, let M be an R-module, and let $\mathfrak{a}$ be an ideal of R. Put $M[\mathfrak{a}] = \{x \in M : \mathfrak{a}x = 0\}$, and call it the $\mathfrak{a}$-*torsion submodule* of M.

Proposition E.9. *Let M be a finitely generated module over a telescopic ring R, and let $\mathfrak{a}$ be an ideal of R. Then $M/\mathfrak{a}M$ is R-isomorphic to $M[\mathfrak{a}]$.*

Proof. If $M = \bigoplus_{i=1}^n M_i$ is a direct sum of its submodules, then

$$M[\mathfrak{a}] \cong \bigoplus_{i=1}^n M_i[\mathfrak{a}], \qquad M/\mathfrak{a}M \cong \bigoplus_{i=1}^n M_i/\mathfrak{a}M_i.$$

By Corollary E.8 this reduces Proposition E.9 to the case when M is a cyclic module. Replacing R by the ring $R/\mathrm{ann}_R(M)$ (and $\mathfrak{a}$ by its image in this ring), we may assume that $\mathrm{ann}_R(M) = 0$ and thereby $M \cong R$.

Let $\mathfrak{m} = \pi R$ be the maximal ideal of R, and write $\mathfrak{a} = \mathfrak{m}^k = \pi^k R$. Then $M/\mathfrak{a}M = R/\pi^k R$ and $M[\mathfrak{a}] = \pi^{s-k} R$, where s is the nilpotency index of π. The map $R \to R$ defined by $x \mapsto \pi^{s-k} x$ has $\pi^{s-k} R$ as its image and $\pi^k R$ as its kernel. Hence $R/\pi^k R \cong \pi^{s-k} R$, as wanted. □

Proposition E.9 extends to direct products of telescopic rings.

Proposition E.10. *Let R be a direct product of finitely many telescopic rings, and let M be a finitely generated R-module. Then for any ideal $\mathfrak{a}$ of R we have $M/\mathfrak{a}M \cong M[\mathfrak{a}]$.*

Proof. Write $R = R_1 \times \cdots \times R_n$, where each R_k is a telescopic ring. Put

$$e_1 = (1, 0, \dots, 0), \qquad e_2 = (0, 1, 0, \dots, 0),$$

and so on, so that $1 = e_1 + \cdots + e_n$. Then an R-module M splits into a direct sum as $M = \bigoplus_{i=1}^{n} M_i$, where $M_i = e_i M$. As in the proof of Proposition E.9, it suffices to verify the required property for every M_i.

Since M_i is annihilated by $1 - e_i$ it can be viewed as a module over the ring $R/(1 - e_i)R$, which is isomorphic to the telescopic ring R_i. We reduced Propositions E.10–E.9, hereby completing the proof.   □

Now we are ready to prove Theorem E.1.

Proof of Theorem E.1. Let M_p be the p-Sylow subgroup of M. Then M_p is a G-submodule of M, and the embedding $M_p \hookrightarrow M$ induces G-isomorphisms $M[p] \cong M_p[p]$ and $M/pM \cong M_p/pM_p$. Hence we may assume that M is a p-group; in particular, M has a natural structure of a $\mathbb{Z}/p^s\mathbb{Z}[G]$-module for a suitable s.

By the assumption, p does not divide $|G|$. Hence $\mathbb{Z}/p^s\mathbb{Z}[G]$ is a direct product of finitely many telescopic rings (Corollary E.5), and our theorem becomes a direct consequence of Proposition E.10.   □

Appendix F
Radical Extensions

In this appendix we give an account of the theory of the *q-radical extensions*, that is, the extensions generated by qth roots of elements of the field. Everywhere in this appendix K is a field and q a prime number, distinct from the characteristic of K.

F.1   Field Generated by a Single Root

Fix $\beta \in K$ and a qth root $\beta^{1/q}$. Then all conjugates of $\beta^{1/q}$ over K are contained in the set

$$\left\{ \beta^{1/q}\xi : \xi \in \mu_q \right\}, \tag{F.1}$$

where $\mu_q = \left\{ 1, \zeta_q, \ldots, \zeta_q^{q-1} \right\}$ is the group of the qth roots of unity.

Proposition F.1. *If β is not a qth power in K, then $[K(\beta^{1/q}) : K] = q$ for any choice of the root $\beta^{1/q}$.*

In other words, the polynomial $x^q - \beta$ either has a root in K or is irreducible over K (in which case (F.1) is the full set of conjugates of $\beta^{1/q}$).

Proof. Assume that $[K(\beta^{1/q}) : K] = r < q$ for some choice of the root $\beta^{1/q}$. We have $\mathcal{N}_{K(\beta^{1/q})/K}(\beta^{1/q}) = \beta^{r/q}\xi$ with $\xi \in \mu_q$. Since q is prime, there exist integers a and b such that $ar + bq = 1$. Then $\left((\beta^{r/q}\xi)^a \beta^b \right)^q = \beta$; that is, β is a qth power in K. $\qquad\square$

Since $[K(\zeta_q) : K] \leq q - 1$, we obtain the following consequence, which will be systematically used in the sequel.

Corollary F.2. *If $\beta \in K$ is not a qth power in K, then it is not a qth power in $K(\zeta_q)$ either.*

227

F.2 Kummer's Theory

We wish to study the Galois group (over K) of the field $K(\zeta_q, \beta^{1/q})$ (the field, generated by all the qth roots of β). Actually, we consider a more general situation. We fix a subgroup B of the multiplicative group $K^\times$, and consider the field $L = K(\sqrt[q]{B})$, generated over K by the set

$$\sqrt[q]{B} = \{\rho \in \bar{K}^\times : \rho^q \in B\}.$$

It is a Galois extension of K (the composite of splitting fields of all polynomials $x^q - \beta$, where $\beta \in B$). The structure of the Galois group $\Gamma = \mathrm{Gal}(L/K)$ depends on whether or not ζ_q belongs to K.

In this section we assume that

$$\zeta_q \in K. \tag{F.2}$$

In this case the theory of q-radical extensions is called *Kummer's theory*. Fix $\beta \in B$ and a qth root $\beta^{1/q}$. Then all conjugates of $\beta^{1/q}$ over K are contained in the set $\{\beta^{1/q}\xi : \xi \in \mu_q\}$, where

$$\mu_q = \left\{1, \zeta_q, \dots, \zeta_q^{q-1}\right\}$$

is the group of the qth roots of unity. In other words, for any $\gamma \in \Gamma$, we have $(\beta^{1/q})^\gamma/\beta^{1/q} \in \mu_q$. The quotient $(\beta^{1/q})^\gamma/\beta^{1/q}$ depends only on β and γ, but not on the particular choice of the qth root $\beta^{1/q}$; indeed, if we replace $\beta^{1/q}$ by $\beta^{1/q}\xi$ with $\xi \in \mu_q$, the quotient is not changed because $\mu_q \subset K$ by the assumption (F.2).

To continue, recall the notion of pairing. Given groups V and W, and an abelian group A, an *A-pairing* of V and W is a map $f : V \times W \to A$ such that for any $y \in W$ the map $x \mapsto f(x, y)$ is a group homomorphism $V \to A$, and for any $x \in V$ the map $y \mapsto f(x, y)$ is a group homomorphism $W \to A$. The pairing induces group homomorphisms $V \to \mathrm{Hom}(W, A)$ and $W \to \mathrm{Hom}(V, A)$. The kernels of these homomorphisms are called *left* and *right kernels* of our pairing. The pairing is *left* (respectively, *right) faithful* if its left (right) kernel is 1 and *faithful* if it is both left and right faithful.

In our case we have *Kummer's pairing* $B \times \Gamma \to \mu_q$ defined by

$$(\beta, \gamma) \mapsto \frac{(\beta^{1/q})^\gamma}{\beta^{1/q}}.$$

It is obviously right faithful, which implies that Γ is isomorphic to a subgroup of $\mathrm{Hom}(B, \mu_q)$. It follows that Γ is a q-torsion abelian group; in particular, it has a natural structure of an $\mathbb{F}_q$-vector space.

However, Kummer's pairing is not left faithful in general, its right kernel being $B \cap (K^\times)^q$. Putting $\bar{B} = B/B \cap (K^\times)^q$, we obtain a faithful pairing $\bar{B} \times \Gamma \to \mu_q$, and the following holds.

Theorem F.3. *If the group B is finitely generated, then Γ is isomorphic to $\bar{B}$.*

This is a special case of a general statement in linear algebra. Let F be a field and V and W vector spaces over F. A pairing $V \times W \to F$ is *F-bilinear* if both the induced maps $V \to \operatorname{Hom}(W, F)$ and $W \to \operatorname{Hom}(V, F)$ are F-linear.

Proposition F.4. *Assume that one of the spaces V and W is of finite F-dimension and that there exists a faithful F-bilinear pairing $V \times W \to F$; then V and W are isomorphic vector spaces.*

Proof. Assume, for instance, that V is finite dimensional. Then the dual space $V^* = \operatorname{Hom}(V, F)$ is of the same dimension as V. Since our pairing is right faithful, we have a linear monomorphism $W \hookrightarrow V^*$. Hence, W is finite dimensional, and

$$\dim W \leq \dim V^* = \dim V.$$

Similarly, $\dim V \leq \dim W$. Hence $\dim V = \dim W$, as wanted.   □

Proof of Theorem F.3. Both groups $\bar{B}$ and Γ have a natural structure of an $\mathbb{F}_q$-vector space. Identifying μ_q with the additive group of $\mathbb{F}_q$, we obtain a faithful $\mathbb{F}_q$-bilinear pairing $\bar{B} \times \Gamma \to \mathbb{F}_q$. Since B is finitely generated, $\bar{B}$ is a finite-dimensional $\mathbb{F}_q$-space. Now apply Proposition F.4.   □

If B' is a subgroup of $B(K^\times)^q$ then the field $L' = K(\sqrt[q]{B'})$ is a subfield of L. In the case when B is finitely generated, any subfield of L, containing K, is of this form. More precisely, we have the following statement (which will not be used in the sequel).

Proposition F.5. *Assume that B is finitely generated. Then there is a one-to-one correspondence between the group towers $(K^\times)^q \leq B' \leq B(K^\times)^q$ and the field towers $K \subseteq L' \subseteq L$, given by $L' = K(\sqrt[q]{B'})$.*

Proof. If there is a faithful F-bilinear pairing $V \times W \xrightarrow{f} F$ of finite-dimensional vector spaces, then the subspaces V' of V and W' of W stay in the one-to-one correspondence given by[1] $W' = (V')^\perp$. In our case we obtain a one-to-one correspondence between the subgroups of $\bar{B}$ and of Γ. Since the former correspond to the group towers $(K^\times)^q \leq B' \leq B(K^\times)^q$ and the latter to the field towers $K \subseteq L' \subseteq L$, we obtain a one-to-one correspondence between the two types of towers. A straightforward inspection shows that $L' = K(\sqrt[q]{B'})$ for the corresponding towers.   □

---

[1] Where $(V')^\perp = \{w \in W : f(v, w) = 0 \text{ for all } v \in V'\}$.

Without assuming that B is finitely generated, the following statement still holds: there is a one-to-one correspondence between the group towers $(K^\times)^q \le B' \le B(K^\times)^q$ and the field towers $K \subseteq L' \subseteq L$ such that both the index $[B(K^\times)^q : B']$ and the degree $[L : L']$ are finite, and the correspondence is again given by $L' = K(\sqrt[q]{B'})$. We leave the proof to the reader.

F.3   General Radical Extensions

We no longer assume that $\zeta_q \in K$. Since $1 \in B$, we have $\zeta_q \in \sqrt[q]{B}$, which means that we have a tower of fields

$$K \subset K(\zeta_q) \subset L,$$

where ζ_q is a primitive qth root of unity and $L = K(\sqrt[q]{B})$.

Proposition F.6. *Assume that the group B is finitely generated. Then the group $\Gamma = \mathrm{Gal}(L/K(\zeta_q))$ is isomorphic to $\bar{B} = B/B \cap (K^\times)^q$.*

Proof. Corollary F.2 implies that $B \cap (K^\times)^q = B \cap (K(\zeta_q)^\times)^q$. Now apply Theorem F.3 with $K(\zeta_q)$ instead of K.   □

Putting $\Delta = \mathrm{Gal}(L/K)$ and $H = \mathrm{Gal}(K(\zeta_q)/K)$, we obtain the exact sequence

$$1 \to \Gamma \to \Delta \to H \to 1. \tag{F.3}$$

Both extensions $K(\zeta_q)/K$ and $L/K(\zeta_q)$ are abelian, but, as we shall see in a while, when $\zeta_q \notin K$, the extension L/K is badly non-abelian.

Theorem F.7. *Assume that $\zeta_q \notin K$. Let L_0 be an abelian extension of K contained in L. Then $L_0 \subset K(\zeta_q)$.*

The proof relies on a simple group-theoretic lemma. To state it, we need some preparation. Recall, first of all, the notion of *group extension*. Let

$$1 \to A \to E \to Q \to 1 \tag{F.4}$$

be an exact sequence of groups, the group A being abelian. The group E acts on its normal subgroup A by conjugation (we write this action exponentially, $a \mapsto a^e = e^{-1}ae$). Since A is abelian, it acts on itself trivially. This induces a natural right action of $Q = E/A$ on A (again written exponentially).

Now assume that we are given an abelian group A, a group Q, and a right Q-action on A. Then any exact sequence (F.4) inducing the given Q-action on A is called an *extension of Q by A* (with respect to the given action).

Now let F be a field and A an F-vector space. It will be convenient for us to write the group structure on A multiplicatively and the F-action on A exponentially (that is, instead of the familiar $\lambda x + \mu y$, we write $x^\lambda y^\mu$).

Lemma F.8. *Let Q be a subgroup of the multiplicative group $F^\times$, and let $1 \to A \to E \to Q \to 1$ be an extension of Q by A with respect to the standard action of Q on A. Assume that $Q \neq 1$. Then[2] $A = [E, E]$, the commutator subgroup of E.*

In other words, if E' is a normal subgroup of E with abelian quotient E/E', then $E' \geq A$.

The assumption $Q \neq 1$ is essential: if $Q = 1$ then $E = A$ is an abelian group and $[E, E] = 1$.

Proof. Since the quotient E/A is abelian, we have $[E, E] \leq A$, so it suffices to prove that $A \leq [E, E]$.

Recall that we write the group law on A multiplicatively and the F-action on A exponentially. Fix an element $x \neq 1$ of Q. Then, when a runs over A, the expression a^{x-1} runs over A as well. If for every $x \in Q$ we fix a lifting $\tilde{x} \in E$, then $a^x = \tilde{x}^{-1} a \tilde{x}$. It follows that

$$a^{x-1} = a^{-1+x} = a^{-1}\tilde{x}^{-1} a \tilde{x} = [a, \tilde{x}].$$

Thus, every element of A can be presented as a commutator $[a, \tilde{x}]$. Hence $A \leq [E, E]$. □

Proof of Theorem F.7. Consider the exact sequence (F.3). The group Γ is an $\mathbb{F}_q$-vector space (see Sect. F.2), and H can be viewed as a subgroup of $(\mathbb{Z}/q\mathbb{Z})^\times = \mathbb{F}_q^\times$, the action of H on Γ being exactly the (restricted to H) $\mathbb{F}_q$-vector space action. Since $\zeta_q \not\subset K$, we have $H \neq 1$ and Lemma F.8 applies. By the lemma, every subgroup of Δ with abelian quotient contains Γ. By the Galois theory, this means that every subfield of L, abelian over K, is contained in the subfield fixed by Γ, that is, in $K(\zeta_q)$. The theorem is proved. □

F.4 Equivariant Kummer's Theory

Now assume that K is a Galois extension of some field K_0, with Galois group $G = \mathrm{Gal}(K/K_0)$, and that B is a G-invariant subgroup of K. Then $L = K(\sqrt[q]{B})$ is a Galois extension of K_0 (the composite of splitting fields of polynomials

---

[2]We identify A with its image in E.

$\prod_{\sigma \in G}(x - \beta^\sigma)$, where $\beta \in B$). It follows that G acts naturally on the group $\Delta = \mathrm{Gal}(L/K)$, and the subgroup $\Gamma = \mathrm{Gal}(L/K(\zeta_q))$ is invariant under this action.[3]

Thus, both B and Γ are G-modules, and so is $\bar{B} = B/B \cap (K^\times)^q$. Recall that Γ and $\bar{B}$ are isomorphic as abelian groups when B is finitely generated. It is natural to ask whether they are isomorphic as G-modules. In general, this is not true. For instance, assume that $\zeta_q \notin K_0$, $K = K_0(\zeta_q)$ and $B \subset K_0$. Then B is a trivial G-module, but Γ is a nontrivial G-module.

Similar examples show that to have G-isomorphism we must assume that

$$K \cap K_0(\zeta_q) = K_0. \tag{F.5}$$

Adding to these some technical hypotheses, we indeed prove that $\bar{B}$ and Γ are G-isomorphic.

Theorem F.9. *In the above setup, assume that the group B is finitely generated and that (F.5) holds. Assume, in addition, that the group G is finite and abelian and that q does not divide $|G|$. Then $\bar{B}$ and Γ are G-isomorphic.*

First of all, we establish an equivariant version of Proposition F.4. Let V and W be G-modules, where G is a group, and let A be an abelian group. A pairing $V \times W \xrightarrow{f} A$ is called *G-equivariant* if for any $x \in V$, $y \in W$, and $\sigma \in G$, we have $f(x^\sigma, y^\sigma) = f(x, y)$.

Proposition F.10. *Let F be a field, G a finite abelian group, V, W two finitely generated $F[G]$-modules, and $V \times W \to F$ a G-equivariant faithful bilinear pairing. Assume that the characteristic of F does not divide $|G|$. Then V and W are isomorphic as $F[G]$-modules.*

Proof. According to Proposition F.4, the natural map $W \to V^*$ is an isomorphism of vector spaces. Moreover, G-equivariance of the pairing implies that this map is a G-morphism. Thus, W is $F[G]$-isomorphic to the dual space V^*. Finally, Corollary D.14 implies that V^* is G-isomorphic to V. $\square$

Proof of Theorem F.9. Assume first that $\zeta_q \in K_0$. Then we also have $\zeta_q \in K$, and Kummer's theory (Sect. F.2) applies. In particular, we have Kummer's pairing $B \times \Gamma \xrightarrow{f} \mu_q$, and it is easy to verify that Kummer's pairing is G-equivariant.

Indeed, for $\gamma \in \Gamma$ and $\sigma \in G$, we have $\gamma^\sigma = \tilde{\sigma}^{-1}\gamma\tilde{\sigma}$, where $\tilde{\sigma}$ is a lifting of σ to $\mathrm{Gal}(L/K_0)$. Further, if $\beta^{1/q}$ is a qth root of $\beta \in B$, then $(\beta^{1/q})^{\tilde{\sigma}}$ is a qth root of β^σ. It follows that

$$f(\beta^\sigma, \gamma^\sigma) = \frac{(\beta^{1/q})^{\tilde{\sigma}\gamma^\sigma}}{(\beta^{1/q})^{\tilde{\sigma}}} = \frac{(\beta^{1/q})^{\gamma\tilde{\sigma}}}{(\beta^{1/q})^{\tilde{\sigma}}} = f(\beta, \gamma)^{\tilde{\sigma}}.$$

---

[3]In fact, Γ is invariant under any automorphism of Δ, because Γ is the only q-Sylow subgroup of Δ.

But $f(\beta, \gamma) \in \mu_q \subset K_0$ (recall that $\zeta_q \in K_0$), whence $f(\beta, \gamma)^{\tilde{\sigma}} = f(\beta, \gamma)$. Thus, we have proved that $f(\beta^\sigma, \gamma^\sigma) = f(\beta, \gamma)$ for any $\beta \in B$, $\gamma \in \Gamma$, and $\sigma \in G$. Hence Kummer's pairing is G-equivariant.

It follows that the faithful pairing $\bar{B} \times \Gamma \to \mathbb{F}_q$, defined in the proof of Theorem F.3, is G-equivariant as well. Proposition F.10 implies that $\bar{B}$ is $\mathbb{F}_q[G]$-isomorphic to Γ. This proves the theorem in the special case $\zeta_q \in K_0$.

In the general case, assumption (F.5) implies that

$$\operatorname{Gal}\left(K(\zeta_q)/K_0(\zeta_q)\right) = G.$$

Hence we may replace K_0 by $K_0(\zeta_q)$ and K by $K(\zeta_q)$, reducing the general case to the case $\zeta_q \in K_0$, already proved.   □

Remark F.11.   The results of this section remain true without assuming the group G abelian. The proofs are, basically, the same, but one should use the full (non-abelian) version of the Maschke theorem (Theorem D.11).

References

1. Aaltonen, M., Inkeri, K.: Catalan's equation $x^p - y^q = 1$ and related congruences. Math. Comp. **56**, 359–370 (1991)
2. Artin, E., Tate, J.: Class Field Theory, 2nd edn. Addison-Wesley, Redwood City (1990)
3. Atiyah, M.F., Macdonald, I.G.: Introduction to Commutative Algebra. Addison-Wesley, Reading (1969)
4. Baker, A.: Linear forms in the logarithms of algebraic numbers I–IV. Mathematika **13**, 204–216 (1966); **14**, 102–107, 220–224, (1967); **15** 204–216 (1968)
5. Baker, A.: Contributions to the theory of Diophantine equations. I. On the representation of integers by binary forms. Philos. Trans. Roy. Soc. Lond. Ser. A **263**, 173–191 (1968)
6. Baker, A.: Bounds for solutions of hyperelliptic equations. Proc. Camb. Phil. Soc. **65**, 439–444 (1969)
7. Baker, A., Wüstholz, G.: Logarithmic forms and group varieties. J. Reine Angew. Math. **442**, 19–62 (1993)
8. Bennett M.A.: Review of "Catalan's equation with a quadratic exponent". Math. Rev. MR1816462 (2002a:11021)
9. Bilu, Yu.F.: Catalan's conjecture (after Mihăilescu). Séminaire Bourbaki, Exposé 909, 55ème année (2002–2003)
10. Bilu, Yu.F.: Catalan without logarithmic forms (after Bugeaud, Hanrot and Mihăilescu). J. Th. Nombres Bordeaux **17**, 69–85 (2005)
11. Bilu, Yu.F., Hanrot, G.: Solving superelliptic Diophantine equations by Baker's method. Comp. Math. **112**, 273–312 (1998)
12. Borevich, Z.I., Shafarevich, I.R.: Number theory. Academic Press, London, New York (1966)
13. Bugeaud, Y.: Bounds for the solutions of superelliptic equations. Comp. Math. **107**, 187–219 (1997)
14. Bugeaud, Y., Hanrot, G.: Un nouveau critère pour l'équation de Catalan. Mathematika **47**, 63–73 (2000)
15. Carmichael, R.D.: Diophantine Analysis. Wiley, Chapman Hall, New York (1915)
16. Cassels, J.W.S.: On the equation $a^x - b^y = 1$. Am. J. Math. **75**, 159–162 (1953)
17. Cassels, J.W.S.: An Introduction to Diophantine Approximation. Cambridge Tracts in Mathematics and Mathematical Physics, vol. 45. Cambridge University Press, New York (1957)
18. Cassels, J.W.S.: On the equation $a^x - b^y = 1$ II. Proc. Camb. Philos. Soc. **56**, 97–103 (1960); Corrigendum: Ibid. **57**, 187 (1961)
19. Cassels, J.W.S., Frölich, A. (eds.): Algebraic Number Theory. Academic, London; Thompson Book Co. Inc., Washington, D.C. (1967)
20. Catalan, E.: Problème 48. Nouv. Ann. Math. **1**, 520 (1842)

21. Catalan, E.: Note extraite d'une lettre adressée à l'éditeur. J. Reine Angew. Math. **27**, 192 (1844)

22. Chapman, R.: The Stickelberger Ideal, unpublished manuscript (1999)

23. Chein, E.Z.: A note on the equation $x^2 = y^q + 1$. Proc. Am. Math. Soc. **56**, 83–84 (1976)

24. Childress, N.: Class Field Theory. Universitext. Springer, Berlin (2009)

25. Cohen, H.: Démonstration de la conjecture de Catalan. In: Théorie Algorithmique des Nombres et Équations Diophantiennes, pp. 1–83. Éditions de l'École Polytechnique, Palaiseau (2005)

26. Cohen, H.: Number Theory II: Analytic and Modern Tools. Graduate Texts in Mathematics, vol. 240. Springer, New York (2007)

27. Delaunay, B.: Über die Darstellung der Zahlen durch die Binäre kubische Formen mit negativer Diskriminante. Math. Z. **31**, 1–26 (1930)

28. Dickson, L.E.: History of the Theory of Numbers II: Diophantine Analysis. G. E. Stechert & Co., New York (1934)

29. Ernvall, R., Metsänkylä, T.: On the p-divisibility of Fermat quotients. Math. Comp. **66**, 1353–1365 (1997)

30. Evertse, J.-H.: Upper Bounds for the Numbers of Solutions of Diophantine Equations. Mathematical Centre Tracts, vol. 168. Mathematisch Centrum, Amsterdam (1983)

31. Euler, L.: Theorematum quorundam arithmeticorum demonstrationes. Comm. Acad. Sci. Petrop. **10**, 125–146 (1738); Opera Omnia Ser. I, Vol. I, Commentationes Arithmeticae I, 38–58. Teubner, Basel (1915)

32. Euler, L.: Vollständige Anleitung zur Algebra (2 volumes). Royal Academy of Sciences, Sankt-Petersburg (1770); reprinted by Springer, New York (1984)

33. Gebel, J., Pethö, A., Zimmer, H.G.: On Mordell's equation. Comp. Math. **110**, 335–367 (1998)

34. Gelfond, A.O.: Sur le septième Problème de D. Hilbert. Comptes Rendus Acad. Sci. URSS Moscou **2**, 1–6 (1934)

35. Gelfond, A.O.: Sur le septième Problème de Hilbert. Bull. Acad. Sci. URSS Leningrad **7**, 623–634 (1934)

36. Gelfond, A.O.: Approximations of transcendental numbers by algebraic numbers (Russian). Dokl. Akad. Nauk SSSR **2**, 177–182 (1935)

37. Gelfond, A.O.: Transcendent and Algebraic Numbers (Russian). Moscow (1952); English translation: Dover, New York (1960)

38. Gelfond, A.O., Linnik, Yu.V.: Thue's method and the problem of effectiveness in quadratic fields (Russian). Dokl. Akad. Nauk SSSR (N.S.) **61**, 773–776 (1948)

39. Gerono, G.C.: Sur la résolution en nombres entiers et positifs de l'équation $x^m = y^n + 1$. Nouv. Ann. Math. **16**, 394–398 (1857)

40. Gerono, G.C.: Sur la résolution en nombres entiers et positifs de l'équation $x^m = y^n + 1$. Nouv. Ann. Math. (2) **9**, 469–471 (1870)

41. Gerono, G.C.: Sur la résolution en nombres entiers et positifs de l'équation $x^m = y^n + 1$. Nouv. Ann. Math. (2) **10**, 204–206 (1871)

42. Glass, A.M.W., Meronk, D.B., Okada, T., Steiner, R.P.: A small contribution to Catalan's equation. J. Number Theory **47**, 131–137 (1994)

43. Gloden, A.: Sur un problème de Catalan. Mathesis **61**, 302–303 (1952)

44. Gloden, A.: Histoire du 'problème de Catalan'. In: Actes du VIIe Congrès International d'histoire des Sciences (Jérusalem 4–12 Août 1953), pp. 316–319. Académie Internationale d'Histoire des Sciences, Paris (1953)

45. Hampel, R.: On the solution in natural numbers of the equation $x^m - y^n = 1$. Ann. Polon. Math. **3**, 1–4 (1956)

46. Hawkins, Th.: New light on Frobenius' creation of the theory of group characters. Arch. History Exact Sci. **12**, 217–243 (1974)

47. Hyyrö, S.: Catalan's problem (Finnish). Arkhimedes **1**, 53–54 (1963)

48. Hyyrö, S.: Über das Catalan'sche problem. Ann. Univ. Turku Ser. AI **79**, 3–10 (1964)

49. Hyyrö, S.: Über die Gleichung $ax^n - by^n = z$ und das Catalansche problem. Ann. Acad. Sci. Fenn. Ser. A I No. **355**, 50 pp (1964)

50. Inkeri, K: On Catalan's problem. Acta Arith. **9**, 285–290 (1964)

51. Inkeri, K: On Catalan's conjecture. J. Number Theory **34**, 142–152 (1990)

52. Inkeri, K., Hyyrö, S.: On the congruence $3^{p-1} \equiv 1$ (mod p^2) and the Diophantine equation $x^2 - 1 = y^p$. Ann. Univ. Turku. Ser. A I No. **50**, 1–4 (1961)

53. Iwasawa, K.: A class number formula for cyclotomic fields. Ann. Math. **76**, 171–179 (1962)

54. Keller, W., Richstein, J.: Solutions of the congruence $a^{p-1} \equiv 1$ (mod p^r). Math. Comp. **74**, 927–936 (2005)

55. Ko, C.: On the diophantine equation $x^2 = y^n + 1$ (Chinese). Acta Sci. Natur. Univ. Szechuan. **2**, 57–64 (1960)

56. Ko, C.: On the diophantine equation $x^2 = y^n + 1$, $xy \neq 0$. Sci. Sinica **14**, 457–460 (1965)

57. Kučera, R.: On a certain subideal of the Stickelberger ideal of a cyclotomic field. Arch. Math. (Brno) **22**, 7–19 (1986)

58. Lang, S.: Algebra. Addison-Wesley, Reading (1965)

59. Lang, S.: Fundamentals of Diophantine Geometry. Springer, New York (1983)

60. Lang, S.: Cyclotomic Fields, I and II (Combined Second Edition). Graduate Texts in Mathematics, vol. 121. Springer, New York (1990)

61. Lang, S.: Algebraic Number Theory, 2nd edn. Graduate Texts in Mathematics, vol. 110. Springer, New York (1994)

62. Langevin, M.: Quelques applications de nouveaux résultats de Van der Poorten. In: Séminaire Delange-Pisot-Poitou, 17e année (1975/1976), Théorie des nombres: Fasc. 2, Exp. No. G12, 11 pp. Secrétariat Math., Paris (1977)

63. Laurent, M.: Linear forms in two logarithms and interpolation determinants. Acta Arith. **66**, 181–199 (1994)

64. Laurent, M.: Linear forms in two logarithms and interpolation determinants. II. Acta Arith. **133**, 325–348 (2008)

65. Laurent, M., Mignotte, M., Nesterenko, Yu.: Formes linéaires en deux logarithmes et déterminants d'interpolation. J. Number Theory **55**, 285–321 (1995)

66. Lebesgue, V.A.: Théorèmes nouveaux sur l'équation indéterminée $x^5 + y^5 = az^5$. J. Math. Pures Appl. **8**, 49–70 (1843)

67. Lebesgue, V.A.: Sur l'impossibilité en nombres entiers de l'équation $x^m = y^2 + 1$. Nouv. Ann. Math. **9**, 178–181 (1850)

68. Legendre, A.M.: Essai sur la Théorie des Nombres. Duprat, Paris (1798)

69. Lemmermeyer, F.: Reciprocity Laws. Springer, Berlin (2000)

70. LeVeque, W.J.: On the equation $a^x - b^y = 1$. Am. J. Math. **74**, 325–331 (1952)

71. LeVeque, W.J.: Topics in Number Theory. Vol. II. Addison-Wesley Publishing Co. Inc., Reading (1956)

72. van der Linden, F.J.: Class number computations of real abelian number fields. Math. Comp. **39**, 693–707 (1982)

73. Lionnet, E.: Question 884. Nouv. Ann. Math. (2) **7**, 240 (1868)

74. Liouville, J.: Sur des classes très étendues de quantités dont la valeur n'est ni algébrique, ni même réductible à des irrationelles algébriques". C. R. Acad. Sci. Paris **18**, 883–885, 910–911 (1844)

75. Lubelski, S.: Studien über den grossen Fermatschen Satz. Prace mat.-fiz. **42**, 11–44 (1935)

76. Mąkowski, A.: Three consecutive integers cannot be powers. Colloq. Math. **9**, 297 (1962)

77. Masley, J.M.: Class numbers of real cyclic number fields with small conductor. Comp. Math. **37**, 297–319 (1978)

78. Masser, D., et al.: Diophantine approximation. In: Amoroso, F., Zannier, U. (eds.) Lectures from the C.I.M.E. Summer School Held in Cetraro, June 28–July 6, 2000. Lecture Notes in Mathematics, vol. 1819. Fondazione C.I.M.E.. [C.I.M.E. Foundation] Springer, Berlin; Centro Internazionale Matematico Estivo (C.I.M.E.), Florence (2003)

79. Matveev, E.M.: An explicit lower bound for a homogeneous rational linear form in logarithms of algebraic numbers I, II (Russian). Izv. Ross. Akad. Nauk Ser. Mat. **62**, 81–136 (1998); **64**, 125–180 (2000); English translation: Izv. Math. **62**, 723–772 (1998); **64**, 1217–1269 (2000)

80. McCallum, W.: Consecutive Perfect Powers, Honours Project, 54 pp. University NSW, Sydney (1977, unpublished)
81. Metsänkylä, T.: Catalan's equation with a quadratic exponent. C. R. Math. Acad. Sci. Soc. R. Can. **23**, 28–32 (2001)
82. Metsänkylä, T.: Catalan's conjecture: another old Diophantine problem solved. Bull. Am. Math. Soc. (N.S.) **41**, 43–57 (2004)
83. Meyl, A.J.F.: Question 1196. Nouv. Ann. Math. (2) **15**, 545–547 (1876)
84. Mignotte, M.: Sur l'équation de Catalan. C. R. Acad. Sci. Paris **314**, 165–168 (1992)
85. Mignotte, M.: Un critère élémentaire pour l'équation de Catalan. C. R. Math. Rep. Acad. Sci. Can. **15**, 199–200 (1993)
86. Mignotte, M.: A criterion on Catalan's equation. J. Number Theory **52**, 280–283 (1995)
87. Mignotte, M.: Catalan's equation just before 2000. In: Number Theory (Turku, 1999), pp. 247–254. de Gruyter, Berlin (2001)
88. Mignotte, M.: A new proof of Ko Chao's theorem (Russian). Mat. Zametki **76**, 384–395 (2004); English translation: Math. Notes **76**, 358–367 (2004)
89. Mignotte, M., Roy, Y.: Catalan's equation has no new solution with either exponent less than 10651. Exp. Math. **4**, 259–268 (1995)
90. Mignotte, M., Roy, Y.: Lower bounds for Catalan's equation. The Ramanujan J. **1**, 351–356 (1997)
91. Mignotte, M., Roy, Y.: Minorations pour l'équation de Catalan. C. R. Acad. Sci. Paris **324**, 377–380 (1997)
92. Mihăilescu, P.: A class number free criterion for Catalan's conjecture. J. Number Theory **99**, 225–231 (2003)
93. Mihăilescu, P.: Primary cyclotomic units and a proof of Catalan's conjecture. J. Reine Angew. Math. **572**, 167–195 (2004)
94. Mihăilescu, P.: On the class groups of cyclotomic extensions in the presence of a solution to Catalan's equation. J. Number Theory **118**, 123–144 (2006)
95. Mirimanoff, D.: Sur le dernier théorème de Fermat. J. Reine Angew. Math. **139**, 309–324 (1911)
96. Mordell, L.J.: Diophantine equations. In: Pure and Applied Mathematics, vol. 30. Academic, London (1969)
97. Moret-Blanc, M.: Question 1175. Nouv. Ann. Math. (2) **15**, 44–46 (1876)
98. Nagell, T.: Problem 39. Norsk Mat. Tidsskrift **1**, 164 (1919)
99. Nagell, T.: Des équations indéterminées $x^2 + x + 1 = y^n$ et $x^2 + x + 1 = 3y^n$. Norsk. Mat. Forenings Skrifter (1) **2**, 12–14 (1921)
100. Nagell, T.: Sur l'impossibilité de l'équation indéterminée $z^p + 1 = y^2$. Norsk Mat. Forenings Skrifter (1) **4**, 14 pp (1921)
101. Nagell, T.: Résultats nouveaux de l'analyse indéterminée. Norsk Mat. Forenings Skrifter (1) **8**, 19 pp (1922)
102. Nagell, T.: Über die rationalen Punkte auf einigen kubischen Kurven. Tôhoku Math. J. **24**, 48–53 (1924/1925)
103. Nagell, T.: Sur une équation diophantienne à deux indéterminées. Norsk. Vid. Selsk. Forenings Trondheim **7**, 137–139 (1934)
104. Neukirch, J.: Class Field Theory. Grundlehren der Mathematischen Wissenschaften, vol. 280. Springer, Berlin (1986)
105. Neukirch, J.: Algebraic Number Theory. Grundlehren der Mathematischen Wissenschaften, vol. 322. Springer, Berlin (1999)
106. Nesterenko, Yu., Zudilin, W.: Catalan's problem (Russian). http://wain.mi.ras.ru/cp/
107. Notari, Ch.: Une résolution élémentaire de l'équation diophantienne $x^3 = y^2 - 1$. Expo. Math. **21**, 279–283 (2003)
108. Obláth, R.: On the numbers $x^2 - 1$ (Hungarian). Mat. Fiz. Lapok **47**, 58–77 (1940)
109. Obláth, R.: Über die Zahl $x^2 - 1$. Math. Zutphen. B. **8**, 161–172 (1940)

110. Obláth, R.: On impossible Diophantine equations of the form $x^m + 1 = y^n$ (Spanish). Revista Mat. Hisp.-Am. **1**, 122–140 (1941)
111. Obláth, R.: Über die Gleichung $x^m + 1 = y^n$. Ann. Polon. Math. **1**, 73–76 (1954)
112. Odlyzko, A.M.: Lower bounds for discriminants of number fields. Acta Arith. **29**, 275–297 (1976)
113. Odlyzko, A.M.: Lower bounds for discriminants of number fields. II. Tôhoku Math. J. **29**, 209–216 (1977)
114. Pillai, S.S.: On the inequality $0 < a^x - b^y < n$. J. Indian Math. Soc. **19**, 1–11 (1931)
115. Pillai, S.S.: On $a^x - b^y = c$. J. Indian Math. Soc. (N.S.) **2**, 119–122, 215 (1936)
116. Poirier, H.: Il a démontré la conjecture de Catalan. Science et Vie **1020**, 70–73 (2002)
117. Ribenboim, P.: Catalan's Conjecture. Academic, Boston (1994)
118. Ribet, K.A.: On the equation $a^p + 2^\alpha b^p + c^p = 0$. Acta Arith. **79**, 7–16 (1997)
119. Roth, K.F.: Rational approximations to algebraic numbers. Mathematika **2**, 1–20; corrigendum, 168 (1955)
120. Rotkiewicz, A.: Sur l'équation $x^z - y^t = a^t$, où $|x - y| = a$. Ann. Polon. Math. **2**, 7–8 (1956)
121. Schinzel, A.: Sur l'équation $x^z - y^t = 1$, où $|x - y| = 1$. Ann. Polon. Math. **3**, 5–6 (1956)
122. Schinzel, A., Tijdeman, R.: On the equation $y^m = P(x)$. Acta Arith. **31**, 199–204 (1976)
123. Schneider, T.: Transzendenzuntersuchungen periodischer Funktionen. I, II. J. Reine Angew. Math. **172**, 65–74 (1934)
124. Schoof, R.: Catalan's Conjecture. Universitext. Springer, London (2008)
125. Schwarz, W.: A note on Catalan's equation. Acta Arith. **72**, 277–279 (1995)
126. Selberg, S.: Solution of problem 39 (Norvegian). Norsk Mat. Tidsskrift **14**, 79–80 (1932)
127. Siegel, C.L. (under the pseudonym X): The integer solutions of the equation $y^2 = ax^n + bx^{n-1} + \cdots + k$ (Extract from a letter to Prof. L. J. Mordell). J. Lond. Math. Soc. **1**, 66–68 (1926)
128. Sierpiński, W.: On some unsolved problems of arithmetics. Scripta Math. **25**, 125–136 (1960)
129. Sinnott, W.: On the Stickelberger ideal and the circular units of a cyclotomic field. Ann. Math. (2) **108**, 107–134 (1978)
130. Skolem, T.: Diophantische Gleichungen. Springer, New York (1938)
131. Stickelberger, L.: Über eine Verallgemeinerung der Kreisteilung. Math. Ann. **37**, 321–367 (1890)
132. Thaine, F.: On the ideal class groups of real abelian number fields. Ann. Math. **128**, 1–18 (1988)
133. Thue, A.: Über Annäherungswerte Algebraischer Zahlen. J. Reine Angew. Math. **135**, 284–305 (1909)
134. Tijdeman, R.: On the equation of Catalan. Acta Arith. **29**, 197–209 (1976)
135. Waldschmidt, M.: Minorations de combinaisons linéaires de logarithmes de nombres algébriques. Can. J. Math. **45**, 176–224 (1993)
136. Washington, L.: Introduction to Cyclotomic Fields, 2nd edn. Springer, New York (1997)
137. Weil, A.: Basic Number Theory, 3rd edn. Grundlehren der Mathematischen Wissenschaften, vol. 144. Springer, New York, Berlin (1974)
138. Wieferich, A.: Zum letzten Fermat'schen theorem. J. Reine Angew. Math. **136**, 293–302 (1909)
139. Wüstholz, G. (ed.): A Panorama of Number Theory or the View from Baker's Garden. Cambridge University Press, Cambridge (2002)

Author Index

A
Artin, E., 195

b
Baker, A., 6, 159, 162, 168, 169
Bessy, B. F. de, 3
Bilu, Yu., 8
Bugeaud, Y., 7, 67, 109

C
Carmichael, R. D., 2
Cassels, J. W. S., 3, 5, 27, 28, 195
Catalan, E. C., v, 1, 4
Chapman, R., 95
Chebotarev, N. G., 192
Chein, E. Z., 15
Childress, N., 195
Cohen, H., v
Crelle, A. L., 1

D
Delaunay, B. N., 22
Dickson, L. E., 2, 4
Diophantus, 165
Dirichlet, J. P. G. Lejeune, 54, 139, 163

E
Euler, L., 3, 4, 15, 21, 27, 190
Evertse, J. H., 6

F
Fermat, P. de, 3, 21

Fresnel, J., 221
Frölich, A., 195

G
Gelfond, A., 159, 161, 162
Gerono, G. C., 1, 4
Glass, A. M. W., 8
Gloden, A, 3
Grantham, 8

H
Hampel, R., 4
Hanrot, G., 7, 8, 67, 109
Hyyrö, S., 4–6, 19, 27, 29, 32, 34

I
Inkeri, K., 4, 7, 66, 67, 69
Iwasawa, K., 93

K
Ko Chao, 4, 15, 18, 27
Kučera, R., 95
Kummer, E. E., 19, 62, 86

L
Lang, S., 37, 147, 195
Langevin, M., 6
Laurent, M., 162
Lebesgue, V. A., 3, 11, 27
Legendre, A. M., 22
Lemmermeyer, F., 95
Lenstra Jr, H. W. L., 17

© Springer International Publishing Switzerland 2014
Y.F. Bilu et al., *The Problem of Catalan*, DOI 10.1007/978-3-319-10094-4

Subject Index

© Springer International Publishing Switzerland 2014
Y.F. Bilu et al., *The Problem of Catalan*, DOI 10.1007/978-3-319-10094-4

Printed in the United States
By Bookmasters